Android 炫酷应用300例

提升篇

罗帅 罗斌 编著

清华大学出版社
北京

内 容 简 介

本书以"问题描述＋解决方案"的模式,以 Android 5.0 为核心列举了 300 个实用性极强的移动端应用开发技术实例,旨在帮助广大读者快速解决实际开发过程中面临的诸多问题,从而不断提高开发效率、拓展应用领域。全书根据实例功能将内容分为常用控件、通知栏、菜单、图形和图像、动画、文件和数据、系统和设备、Intent、第三方 SDK 开发等 9 章,以所见即所得、所学即所用的速成思维展示了个性化控件、定制通知栏、沉浸式状态栏、折叠式工具栏、分享菜单、抽屉菜单、底部导航菜单、悬挂式窗口、雷达扫描图、波纹扩散图、点九图、PorterDuff 特效、ColorMatrix 特效、内外阴影、图像裁剪、矢量动画、过渡动画、转场动画、网格动画、异步下载、图形验证码、数据解析和传递、Intent、使用方向传感器自制指南针、使用加速度传感器实现微信的摇一摇功能、使用 Camera 捕捉前置和后置摄像头画面等精彩实例的实现过程和代码,以及使用第三方 SDK 进行应用开发,如使用腾讯 SDK 将图像、视频等分享至 QQ 好友、QQ 空间及微信朋友圈等,使用百度地图 SDK 规划出行线路、骑行导航、自定义热力图、查询指定地点的卫星图、查询指定城市的兴趣点和街景,在百度地图上添加自定义动画、自定义颜色突出显示行政区域,根据经纬度值计算面积等,使用新浪微博 SDK 将信息内容发布到微博、执行第三方登录等。

为了突出实用性和简洁性,本书在演示或描述这些实例时,力求有针对性地解决问题,并且所有实例均配有插图。本书适合作为 Android 应用开发人员的案头参考书,无论是编程初学者,还是编程高手,本书都极具参考价值。

本书封面贴有清华大学出版社防伪标签,无标签者不得销售。
版权所有,侵权必究。侵权举报电话: 010-62782989 13701121933

图书在版编目(CIP)数据

Android 炫酷应用 300 例. 提升篇/罗帅,罗斌编著. —北京:清华大学出版社,2020.1
ISBN 978-7-302-53391-7

Ⅰ. ①A… Ⅱ. ①罗… ②罗… Ⅲ. ①移动终端—应用程序—程序设计 Ⅳ. ①TN929.53

中国版本图书馆 CIP 数据核字(2019)第 178700 号

责任编辑:黄　芝
封面设计:刘　键
责任校对:时翠兰
责任印制:刘海龙

出版发行:清华大学出版社
　　网　　址: http://www.tup.com.cn, http://www.wqbook.com
　　地　　址: 北京清华大学学研大厦 A 座
　　社 总 机: 010-62770175
　　投稿与读者服务: 010-62776969, c-service@tup.tsinghua.edu.cn
　　质量反馈: 010-62772015, zhiliang@tup.tsinghua.edu.cn
　　课件下载: http://www.tup.com.cn, 010-83470236
邮　编: 100084
邮　购: 010-62786544

印 装 者: 三河市铭诚印务有限公司
经　　销: 全国新华书店
开　　本: 210mm×285mm　　印　张: 29.75　　字　数: 858 千字
版　　次: 2020 年 1 月第 1 版　　印　次: 2020 年 1 月第 1 次印刷
印　　数: 1～2500
定　　价: 99.80 元

产品编号: 082528-01

Android 是一个以 Linux 为基础的开源操作系统，主要用于管理智能手机、智能手表、智能眼镜、智能电视等设备。Android 操作系统最初由 Andy Rubin 开发，2005 年 8 月由 Google 收购。2007 年 11 月，Google 与 84 家硬件制造商、软件开发商及电信营运商组建开放手机联盟，共同研发改良 Android 操作系统。第一部 Android 智能手机发布于 2008 年 10 月，在随后的十几年中，Android 开始了迅猛发展的历程，并很快成为全球最主要的移动端操作系统。

随着支持 Android 操作系统的智能设备的不断普及和推广，运行在 Android 操作系统上的智能应用也如雨后春笋般地涌现。从 IT 发展潮流来看，越来越丰富的移动应用是大势所趋，手机支付、手机拍照、手机游戏、手机导航、物联网等不断改变着人们的生活方式和工作方式。一个优秀的 Android 应用即可造就一家 IT 公司、打造一个产业链，甚至诞生数十个富豪，这已不再是神话。如何将最新的技术、理念和创意融入应用开发中，是每个 Android 程序员需要不断思考的问题，也是本书创作的初衷。

本书以"问题描述＋解决方案"的模式，以 Android 5.0 为核心列举了 300 个实用性极强的移动端应用开发技术案例，旨在帮助广大读者快速解决实际开发过程中面临的诸多问题，从而不断提高开发效率，拓展应用领域。全书根据实例功能将内容分为常用控件、通知栏、菜单、图形和图像、动画、文件和数据、系统和设备、Intent、第三方 SDK 开发等 9 章，以所见即所得、所学即所用的速成思维展示了个性化控件、定制通知栏、沉浸式状态栏、折叠式工具栏、分享菜单、抽屉菜单、底部导航菜单、悬挂式窗口、雷达扫描图、波纹扩散图、点九图、PorterDuff 特效、ColorMatrix 特效、内外阴影、图像裁剪、矢量动画、过渡动画、转场动画、网格动画、异步下载、图形验证码、数据解析和传递、Intent、使用方向传感器自制指南针、使用加速度传感器实现微信的摇一摇功能、使用 Camera 捕捉前置和后置摄像头画面等精彩实例的实现过程和代码，以及使用第三方 SDK 进行应用开发（例如，使用腾讯 SDK 将图像、视频等分享至 QQ 好友、QQ 空间、微信朋友圈等；使用百度地图 SDK 规划出行线路、骑行导航、自定义热力图、查询指定地点的卫星图、指定城市的兴趣点和街景，在百度地图上添加自定义动画、自定义颜色突出显示行政区域，根据经纬度值计算面积等；使用新浪微博 SDK 将信息内容发布到微博、执行第三方登录等）。

本书所有实例基于 Android 5.0，在 Android Studio 2.2 集成开发环境中使用 Java 和 XML 编写。因此，测试手机或模拟器的 Android 版本不得低于 5.0，部分实例在模拟器上无法测试，建议在学习时使用屏幕分辨率为 1920×1080 像素、操作系统为 Android 5.0 及以上版本的手机作为主要测试工具。

全书所有内容和思想并非一人之力所能及，而是凝聚了众多热心人士的智慧并经过充分的提炼和总结而成，在此对他们表示崇高的敬意和衷心的感谢！参与本书案例收集、整理、编写等工作的人员包括罗帅、罗斌、汪明云、曹勇、陈宁、邓承惠、邓小渝、范刚强、何守碧、洪亮、洪沛林、江素芳、蓝洋、雷国忠、雷惠、雷玲、雷平、雷治英、刘恭德、刘兴红、罗聃、唐静、唐兴忠、童缙嘉、汪兰、王彬、王伯芳、王年素、王正建、吴多、吴诗华、杨开平、杨琴、易伶、张志红、郑少文等，终稿由罗斌统筹完成。

由于作者水平有限和时间仓促，少量内容可能存在偏颇以及疏漏之处，敬请读者批评指正。

本书提供所有案例源代码，读者可将购书凭证发送至邮箱 huangzh@tup.tsinghua.edu.cn，索取源代码。

编　者

2019 年于重庆渝北

目录 Contents

第1章 常用控件 ·· 1

- 001 以折叠方式实现隐藏或显示 TextView ·· 1
- 002 使用可拉伸 9patch 图设置 TextView 背景 ·· 3
- 003 使用 TextSwitcher 平滑切换多个标题 ·· 4
- 004 在 EditText 中弹出输入电话号码的键盘 ·· 5
- 005 动态创建多个自定义风格的 CheckBox ·· 6
- 006 使用 RadioButton 创建单选按钮 ·· 8
- 007 使用 CheckedTextView 创建单选按钮 ·· 10
- 008 在 ListView 列表项中自定义单选按钮 ·· 11
- 009 使用资源创建自定义背景的椭圆按钮 ·· 14
- 010 使用 ShapeDrawable 创建渐变圆角按钮 ·· 15
- 011 在 ListView 列表项文本的右端添加按钮 ·· 16
- 012 将 FloatingActionButton 锚定到指定控件 ·· 18
- 013 使用 ZoomControls 实现放大和缩小图像 ·· 19
- 014 使用 StackView 实现堆叠显示多幅图像 ·· 20
- 015 使用 ScrollView 实现上下滑动切换图像 ·· 22
- 016 使用 HorizontalScrollView 水平切换图像 ·· 24
- 017 在 RecyclerView 中按照文件属性排序 ·· 25
- 018 使用 RecyclerView 实现简单的 ListView ·· 27
- 019 使用 ExpandableListView 分类显示文件 ·· 29
- 020 使用 NestedScrollView 实现嵌套滚动 ·· 30
- 021 在 ViewPager 轮播时实现立体旋转切换 ·· 31
- 022 自定义 selector 以透明前景切换控件 ·· 33
- 023 使用 ViewOutlineProvider 裁剪控件外形 ·· 35
- 024 使用 SwipeRefreshLayout 扫动刷新 UI ·· 36
- 025 使用自定义 Behavior 实现滑动遮盖效果 ·· 38
- 026 创建百分比数字跟随进度改变的进度条 ·· 39
- 027 使用 BottomNavigationBar 实现底部导航 ·· 42
- 028 使用两个 Spinner 实现省市两级联动 ·· 44
- 029 使用 BackgroundColorSpan 定制背景 ·· 46

第2章 通知栏 ·· 48

- 030 动态设置当前应用的标题栏文本 ·· 48

031	动态设置当前应用的标题栏背景	49
032	动态隐藏或显示当前应用标题栏	50
033	自定义 TextView 创建渐变标题栏	51
034	使用自定义布局创建个性化标题栏	52
035	在标题栏左侧添加默认的后退按钮	54
036	在标题栏右侧添加分享按钮分享文本	55
037	使用 SearchView 在标题栏添加搜索框	57
038	使用 SearchManager 实现标题栏搜索	58
039	使用 ActionBar 通过布局定制标题栏	60
040	使用 UI 标志动态隐藏或显示通知栏	62
041	使用 UI 标志动态隐藏或显示通知栏图标	63
042	使用窗口标志实现半透明显示通知栏	64
043	使用指定颜色动态设置通知栏背景色	64
044	将应用的背景图像扩展至通知栏	65
045	在通知栏上添加通知并实现跳转功能	66
046	使用 Notification 在通知栏上添加图标	68
047	使用 RemoteViews 自定义通知栏视图	69
048	以悬挂式窗口显示新增的通知栏任务	70
049	允许直接在通知栏上显示消息内容	72
050	禁止在通知栏上以右滑方式移除通知	73
051	在向通知栏发送消息时同时振动手机	75
052	在滚动文本时自动隐藏或显示工具栏	76
053	使用 Toolbar 在工具栏上添加查找按钮	77
054	使用 Toolbar 为导航图标添加关闭功能	79
055	在拖动改变控件大小时实现工具栏跟随	81
056	创建 CollapsingToolbarLayout 工具栏	82
057	使用 Snackbar 在底部创建浮出信息栏	84
058	自定义 Snackbar 文本颜色和字体大小	85
059	在 Snackbar 上新增自定义风格布局	87

第 3 章 菜单 ... 89

060	在 ActionBar 上以按钮风格显示菜单	89
061	使用 ActionBar 在标题栏添加下拉菜单	90
062	在 ActionBar 上使用 XML 文件创建菜单	91
063	使用 Toolbar 在工具栏上添加下拉菜单	93
064	使用 ActionProvider 创建二级菜单	94
065	在右上角二级菜单中实现单选按钮风格	96
066	在右上角二级菜单中实现多选框风格	99
067	使用 DrawerLayout 创建抽屉式侧滑菜单	101
068	使用手机菜单键控制侧滑菜单是否显示	103
069	在侧滑菜单中使用 NavigationView 导航	106
070	使用 TabLayout 高仿微信底部导航菜单	108

| | 071 | 在弹出底部菜单时主窗口立即变暗 | 112 |
| | 072 | 在长时间按住控件时弹出上下文菜单 | 114 |

第4章　图形和图像 …… 116

	073	通过像素操作在图像上添加马赛克特效	116
	074	通过像素操作实现为图像添加冰冻效果	118
	075	通过像素操作将彩色图像改变为怀旧风格	120
	076	使用 PorterDuffXfermode 裁剪六边形	122
	077	使用 PorterDuffXfermode 抠取异形图像	123
	078	使用 ColorMatrix 增强图像颜色对比度	124
	079	使用 ColorMatrix 为图像添加加亮效果	126
	080	使用 ColorMatrix 调整图像的红色色调	127
	081	使用 ColorMatrix 旋转图像的颜色色相	128
	082	自定义 ColorMatrix 改变图像对比度	130
	083	使用 Matrix 实现按照指定角度旋转图像	131
	084	通过改变图像透明度重叠显示两幅图像	132
	085	根据指定颜色过滤 ImageView 的图像	134
	086	使用高斯矩阵模板实现图像的柔化特效	135
	087	使用正弦函数创建波浪起伏风格的图像	136
	088	使用 BitmapFactory 控制图像采样比例	138
	089	使用 SweepGradient 创建多色扫描图	140
	090	使用 RadialGradient 绘制电波扩散图	141
	091	使用 BlurMaskFilter 为图像添加轮廓阴影	143
	092	使用 ComposeShader 实现内阴影图像	144
	093	使用 EmbossMaskFilter 强化图像轮廓	146
	094	使用 GradientDrawable 创建渐变色边框	147
	095	使用 VectorDrawable 调整矢量图形亮度	148
	096	使用 ClipDrawable 裁剪图像实现拉幕效果	150
	097	使用 ShapeDrawable 裁剪五角星图像	153
	098	使用 NinePatchDrawable 设置背景	154
	099	使用 DashPathEffect 创建虚线边框	155
	100	使用 ComposePathEffect 组合路径特效	157
	101	使用 ImageView 显示 XML 路径矢量图形	158
	102	使用 Region 的 INTERSECT 裁剪扇形图像	159
	103	使用裁剪路径将图像从矩形裁剪成椭圆	161
	104	在自定义 View 中使用扇形裁剪图像	162
	105	根据行列数量将图像切割成碎片并拼图	163
	106	使用 BitmapRegionDecoder 加载大图	165

第5章　动画 …… 166

| | 107 | 使用 ObjectAnimator 创建坐标平移动画 | 166 |
| | 108 | 使用 ObjectAnimator 创建波纹扩散动画 | 167 |

109	使用 ValueAnimator 动态绘制桃心图形	168
110	使用 AnimationSet 组合多个不同的动画	170
111	自定义 TypeEvaluator 合成多方向的位移	171
112	使用 PropertyValuesHolder 实现弹簧动画	173
113	自定义 selector 实现以动画形式改变透明度	174
114	使用 StateListAnimator 实现状态切换动画	176
115	自定义 TypeEvaluator 以加速动画显示字母	178
116	使用 BounceInterpolator 实现弹跳动画	179
117	使用矢量(Vector)动画模拟闹钟耳朵的摆动	180
118	控制 trimPathEnd 动态生成非连续矢量图	181
119	改变矢量数据实现不同图形数字的平滑过渡	183
120	自定义 TimeInterpolator 控制转圈进度动画	185
121	使用 animated-selector 实现轮播多幅图像	186
122	使用 animation-list 实现两幅图像的切换	189
123	使用 AnimationDrawable 逐帧播放图像	190
124	使用 AnimatedVectorDrawable 旋转图形	192
125	以旋转淡出的动画效果切换两个 Activity	194
126	在切换 Activity 的转场动画中共享不同元素	195
127	在过渡 Activity 时禁止部分控件产生动画	198
128	使用指定的裁剪区域动态切换两个 Activity	200
129	在关闭应用(Activity)时显示退场动画	201
130	使用转场动画 Slide 切换两个 Activity	202
131	使用 TransitionSet 组合 Explode 和 Fade 动画	204
132	使用 TransitionManager 实现缩放过渡动画	206
133	使用 TransitionManager 实现绕 Y 轴旋转动画	208
134	使用多个 TranslateAnimation 实现抖动窗口	210
135	使用 LayoutTransition 实现布局改变动画	212
136	使用 TransitionDrawable 动态改变图像颜色	214
137	在 GridView 的各个网格中实现 Explode 动画	215
138	使用 layoutAnimation 平移 RecyclerView 网格	217
139	在 ListView 列表项上实现抽屉式滑动动画	219
140	在 ViewPager 中实现渐变淡入的转场动画	221
141	使用 FragmentTransaction 实现转场动画	223
142	使用 PatternPathMotion 实现路径过渡动画	224
143	使用 RippleDrawable 创建波纹扩散动画	226
144	自定义 GLSurfaceView 实现波浪起伏的动画	228
145	自定义 Animation 实现硬币正反面绕 Y 轴旋转	229

第 6 章 文件和数据 232

146	采用 DOM 方式解析 XML 文件的内容	232
147	采用 Pull 方式解析 XML 文件的内容	234
148	使用 JSONArray 解析 JSON 串的多个对象	237

149	使用JSONArray解析JSON串的多个键值	238
150	使用JSONTokener获取JSON的不同对象	239
151	使用JSONTokener解析JSON非对象文本	241
152	使用Gson解析JSON字符串的单个对象	242
153	使用Intent在Activity之间传递基本数据	243
154	使用Intent在Activity之间传递数组数据	244
155	使用Intent在Activity之间传递图像数据	246
156	使用Intent在Activity之间传递多幅图像	247
157	在Intent传递数据时使用Bundle携带数据	249
158	使用Bundle从Activity向Fragment传递数据	251
159	根据指定网址下载应用安装包到手机SD卡	252
160	仅在WiFi时执行DownloadManager下载	254
161	使用AsyncTask实现异步访问网络图像	255
162	在进度条上显示AsyncTask的下载进度	257
163	以数据流形式加载并显示指定网址的图像	258
164	使用正则表达式校验在输入框的输入内容	260
165	使用随机数生成验证码图像并提交验证	262
166	将涂鸦内容在存储卡上保存为图像文件	264
167	使用BitmapFactory读取SD卡图像文件	266
168	在选择照片窗口中选择图像文件并显示	267
169	使用CookieManager读取和保存数据	268
170	使用PreferenceScreen跳转到显示设置	270
171	使用PreferenceFragment实现页面切换	271
172	使用EditTextPreference实现文本读写	274
173	使用SwitchPreference读写开关状态值	275
174	使用CheckBoxPreference实现多选功能	277
175	使用MultiSelectListPreference实现多选	279

第7章 系统和设备 ... 282

176	使用ContentResolver获取手机短信信息	282
177	使用ContentResolver获取所有联系人信息	283
178	使用ContentResolver查询联系人电话号码	285
179	使用ContentResolver动态新增联系人信息	287
180	使用ContentResolver动态修改联系人信息	289
181	使用ContentResolver动态删除联系人信息	291
182	使用PhoneStateListener监听来电号码	292
183	使用BroadcastReceiver监听拨出号码	293
184	动态注册BroadcastReceiver监听网络状态	295
185	使用BroadcastReceiver实现开机自启动	296
186	使用BroadcastReceiver获取电量百分比	298
187	使用ConnectivityManager检测数据连接	299
188	使用WifiManager动态打开或关闭WiFi	301

189	使用 LocationManager 判断 GPS 是否开启	302
190	使用 TelephonyManager 获取运营商等信息	303
191	使用 TelephonyManager 检测卡槽类型	305
192	使用 PackageManager 获取包名版本等信息	306
193	使用 WallpaperManager 随机更换壁纸	307
194	使用 RingtoneManager 自定义来电铃声	309
195	通过重力传感器控制飞行器的轨迹和速度	311
196	使用加速度传感器实现微信的摇一摇功能	313
197	使用传感器监测手机周围光线亮度变化	316
198	使用方向传感器实现自制指南针	317
199	使用 DisplayMetrics 获取屏幕分辨率	319
200	使用 StatFs 获取存储卡的空间大小信息	320
201	使用 Camera 实现打开或关闭手电筒	321
202	使用 Camera 捕捉前置和后置摄像头画面	322
203	使用 TextureView 实现照相机的预览功能	324
204	通过处理按键实现双击后退键退出应用	327
205	使用 GestureDetector 实现横向滑动切换	328
206	使用锁屏标志实现在锁屏时是否显示窗口	331
207	在当前应用中实现关机和重启功能	332

第 8 章 Intent … 334

208	使用 Intent 启动百度地图进行骑行导航	334
209	使用 Intent 启动百度地图查询公交线路	335
210	使用 Intent 启动百度地图查询步行线路	337
211	使用 Intent 启动百度地图查询兴趣点	338
212	使用 Intent 启动百度地图根据地名定位	340
213	使用 Intent 启动百度地图助手搜索地点	341
214	使用 Intent 在百度地图中展示详情页	342
215	使用 Intent 启动百度地图查询实时公交	344
216	使用 Intent 启动百度地图查询实时路况	345
217	使用 Intent 启动百度地图显示实时汇率	346
218	使用 Intent 直接跳转到百度地图 App 界面	347
219	使用 Intent 启动腾讯地图查询驾车线路	348
220	使用 Intent 启动腾讯地图搜索感兴趣内容	350
221	使用 Intent 启动腾讯地图显示指定位置	351
222	使用 Intent 启动 QQ 浏览器显示腾讯地图	352
223	使用 Intent 将文本内容仅分享到微信	354
224	使用 Intent 将本地图像发送到微信朋友圈	355
225	使用 Intent 将图像发送到微信我的收藏	357
226	使用 Intent 将视频发送到微信我的收藏	358
227	使用 Intent 将本地视频分享给微信好友	360
228	使用 Intent 直接调启微信的扫一扫功能	361

229 使用Intent直接跳转到微信主操作界面 ……………………………………… 362
230 使用Intent根据号码启动QQ聊天界面 ……………………………………… 363
231 使用Intent直接跳转到QQ主操作界面 ……………………………………… 364
232 使用Intent根据组件名称启动QQ ………………………………………… 365
233 使用Intent直接跳转到QQ的我的电脑 ……………………………………… 366
234 使用Intent将本地图像发送到QQ的我的电脑 ……………………………… 367
235 使用Intent将多首歌曲发送到QQ的我的电脑 ……………………………… 369
236 使用Intent将音乐文件分享到QQ好友 ……………………………………… 371
237 使用Intent将多幅图像发送到QQ好友 ……………………………………… 372
238 使用Intent实现截取屏幕部分区域 ………………………………………… 374
239 使用Intent调用照相机拍照并裁剪头像 …………………………………… 376
240 使用Intent实现允许或禁止按键截屏 ……………………………………… 378
241 使用Intent在应用市场中查找包名详情 …………………………………… 379
242 使用Intent根据包名卸载手机应用 ………………………………………… 380
243 使用Intent根据内容跳转到搜索工具 ……………………………………… 381
244 使用Intent指定应用打开PDF文件 ………………………………………… 381
245 使用Intent启动应用打开文本文件 ………………………………………… 384
246 使用Intent启动应用打开Excel文件 ………………………………………… 385
247 使用Intent在文件窗口中筛选安装文件 …………………………………… 386
248 使用Intent在文件窗口中选择图像文件 …………………………………… 388
249 使用Intent查询支持多个图像分享包名 …………………………………… 389
250 使用Intent启用默认网络文件下载器 ……………………………………… 390
251 使用Intent发送带附件的邮件 ……………………………………………… 391
252 使用Intent跳转到系统无障碍设置界面 …………………………………… 392

第9章 第三方SDK开发 …………………………………………………………… 394

253 使用腾讯SDK将指定图像分享给QQ好友 ………………………………… 394
254 使用腾讯SDK将指定链接分享到QQ空间 ………………………………… 396
255 使用腾讯SDK将本地视频发布到QQ空间 ………………………………… 398
256 使用微信SDK将本地图像分享到朋友圈 ………………………………… 399
257 使用微信SDK将本地图像分享至微信好友 ……………………………… 401
258 使用微信SDK将音乐链接分享至微信好友 ……………………………… 403
259 使用微信SDK将视频链接分享到朋友圈 ………………………………… 404
260 使用新浪SDK将文本分享到当前微博 …………………………………… 405
261 使用新浪SDK实现获取最新发布的微博 ………………………………… 407
262 使用新浪SDK实现第三方登录微博账号 ………………………………… 408
263 使用新浪SDK实现分享链接地址至微博 ………………………………… 409
264 使用新浪SDK实现跳转到微博账户简介 ………………………………… 410
265 使用百度SDK获取当前手机的经纬度值 ………………………………… 411
266 使用百度SDK在地图中定位指定的地名 ………………………………… 414
267 使用百度SDK查询指定地点的卫星图 …………………………………… 417
268 使用百度SDK在地图上自定义热力图 …………………………………… 418

269	使用百度SDK实现计算指定范围的面积	419
270	使用百度SDK在地图上叠加圆点覆盖物	420
271	使用百度SDK在地图上添加半透明椭圆	421
272	使用百度SDK在地图的指定位置添加标记	423
273	使用百度SDK实现在地图上添加图像按钮	424
274	使用百度SDK在地图的城市之间绘制虚线	425
275	使用百度SDK实现在地图上绘制多边形	426
276	使用百度SDK在地图的三点位置绘制弧线	427
277	使用百度SDK在地图上添加生长型动画	428
278	使用百度SDK在地图上添加降落型动画	429
279	使用百度SDK在地图上添加淡入放大动画	430
280	使用百度SDK在地图上添加水平展开动画	431
281	使用百度SDK在地图上查询省市行政中心	432
282	使用百度SDK判断某地是否在指定区域内	434
283	使用百度SDK在地图上自定义行政区颜色	435
284	使用百度SDK查询城市兴趣点并显示街景	437
285	使用百度SDK查询指定位置附近的兴趣点	438
286	使用百度SDK查询在指定区域内的兴趣点	440
287	使用百度SDK根据起止地点规划出行线路	442
288	使用百度SDK在地图中搜索指定公交线路	443
289	使用百度SDK查询百度地图的公交线规划	445
290	使用百度SDK调用百度地图的步行导航	446
291	使用百度SDK调用百度地图的骑行导航	447
292	使用百度SDK调用百度地图的Web导航	448
293	使用百度SDK实现POI检索并分享相关地址	449
294	使用百度SDK实现将公交线路分享给好友	451
295	使用百度SDK实现将骑行线路分享给好友	453
296	使用百度SDK将当前地图分享给QQ好友	456
297	使用百度SDK实现在输入框滑出建议列表	457
298	使用百度SDK实现隐藏或显示地图比例尺	459
299	使用百度SDK实现隐藏或显示地图缩放按钮	460
300	使用百度SDK实现自定义地图缩放按钮的位置	461

第1章

常用控件

001 以折叠方式实现隐藏或显示 TextView

此实例主要通过在 Animation 的 applyTransformation（float interpolatedTime，Transformation t）方法中根据 interpolatedTime 参数的变化设置 TextView 的 height 属性值，从而实现以动态展开或折叠 TextView 的风格显示或隐藏 TextView 控件的动画效果。当实例运行之后，单击向下箭头，TextView 控件将在 2s 内展开所有内容，同时向下箭头旋转 180°变成向上箭头，如图 001-1 的右图所示；单击向上箭头，则 TextView 控件将在 2s 内折叠所有内容，同时向上箭头旋转 180°变成向下箭头，如图 001-1 的左图所示。

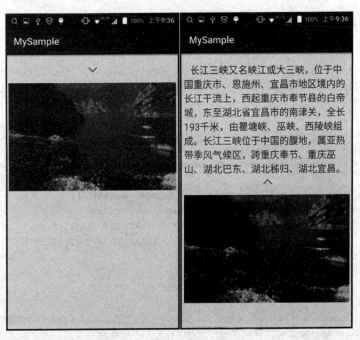

图 001-1

主要代码如下：

```
public class MainActivity extends Activity {
    private TextView myTextView;
```

```java
private ImageView myImageView;
int maxLine = 0;
boolean isExpand;
String myContent = "\t\t长江三峡又名峡江或大三峡,位于中国重庆市、恩施州、宜昌市地区境内的长江干流上,西起重庆市奉节县的白帝城,东至湖北省宜昌市的南津关,全长 193 千米,由瞿塘峡、巫峡、西陵峡组成.长江三峡位于中国的腹地,属亚热带季风气候区,跨重庆奉节、重庆巫山、湖北巴东、湖北秭归、湖北宜昌.";
@Override
protected void onCreate(Bundle savedInstanceState) {
    super.onCreate(savedInstanceState);
    setContentView(R.layout.activity_main);
    myTextView = (TextView) findViewById(R.id.myTextView);
    myImageView = (ImageView) findViewById(R.id.myImageView);
    myTextView.setText(myContent);
    myTextView.setHeight(myTextView.getLineHeight() * maxLine);
    myTextView.post(new Runnable() {
        @Override
        public void run() {
            myImageView.setVisibility(myTextView.getLineCount() >
                    maxLine ? View.VISIBLE : View.GONE);
        }
    });
    myImageView.setOnClickListener(new View.OnClickListener() {
        @Override
        public void onClick(View v) {
            isExpand = !isExpand;
            myTextView.clearAnimation();                    //清除动画
            final int myHeight;
            final int startHight = myTextView.getHeight();
            int myDurationTime = 2000;
            //在折叠或展开控件时让(向上或向下箭头)图标执行180°的正向或反向旋转动画
            if (isExpand) {
                myHeight = myTextView.getLineHeight() *
                             myTextView.getLineCount() - startHight;
                RotateAnimation myRotateAnimation = new RotateAnimation(0, 180,
                            Animation.RELATIVE_TO_SELF, 0.5f,
                            Animation.RELATIVE_TO_SELF, 0.5f);
                myRotateAnimation.setDuration(myDurationTime);
                myRotateAnimation.setFillAfter(true);
                myImageView.startAnimation(myRotateAnimation);
            } else {
                myHeight = myTextView.getLineHeight() * maxLine - startHight;
                RotateAnimation myRotateAnimation = new RotateAnimation(180, 0,
                            Animation.RELATIVE_TO_SELF, 0.5f,
                            Animation.RELATIVE_TO_SELF, 0.5f);
                myRotateAnimation.setDuration(myDurationTime);
                myRotateAnimation.setFillAfter(true);
                myImageView.startAnimation(myRotateAnimation);
            }
            Animation myAnimation = new Animation() {
                //根据图标旋转动画完成的百分比来显示 TextView 控件的高度
                protected void applyTransformation(float interpolatedTime, Transformation t){
                    myTextView.setHeight((int) (startHight + myHeight * interpolatedTime));
                }
            };
            myAnimation.setDuration(myDurationTime);
            myTextView.startAnimation(myAnimation);
        }});}}
```

上面这段代码在 MyCode\MySample691\app\src\main\java\com\bin\luo\mysample\MainActivity.java 文件中。在这段代码中，applyTransformation(float interpolatedTime, Transformation t)方法在绘制动画的过程中会反复被调用，每次调用时其参数 interpolatedTime 都会发生变化，该参数从 0 渐变为 1，当该参数为 1 时表明动画结束。此实例的完整项目存放在 MyCode\MySample691 文件夹中。

002　使用可拉伸 9patch 图设置 TextView 背景

此实例主要通过使用可在指定范围进行拉伸的 9patch 图作为 TextView 控件的背景（相框），从而使 TextView 控件能够实现不因内容的多少而改变背景边框大小的特殊效果。当实例运行之后，单击"显示原始背景图"按钮，显示 TextView 控件的原始背景图；单击"显示拉伸背景图"按钮，则使用 9patch 图设置的背景（相框图）将自动根据文本内容的大小进行纵向拉伸，效果分别如图 002-1 的左图和右图所示。

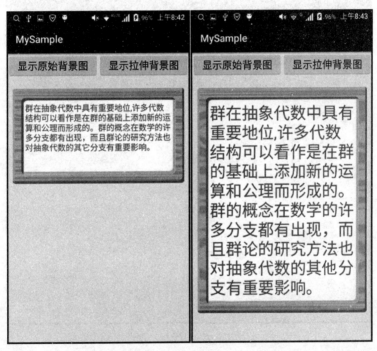

图　002-1

主要代码如下：

```xml
<TextView android:id = "@+id/myTextView"
        android:layout_width = "match_parent"
        android:layout_height = "wrap_content"
        android:layout_marginTop = "10dp"
        android:background = "@drawable/my9patch"
        android:text = "群在抽象代数中具有重要地位,许多代数结构可以看作是在群的基础上添加新的运算和公理而形成的.群的概念在数学的许多分支都有出现,而且群论的研究方法也对抽象代数的其他分支有重要影响."
        android:textSize = "24dp"/>
```

上面这段代码存放在 MyCode\MySample052\app\src\main\res\layout\activity_main.xml 文件中。在这段代码中，android：background=" @drawable/my9patch" 表示使用 drawable 下的 my9patch 资源作为 TextView 控件的背景。my9patch 资源是一个 9patch 图，这种格式的图像能按照设定来拉伸特定区域，而不是整体放大或缩小，从而保证了图像在各个分辨率的屏幕上都可以完美展示。此实例的完整项目在 MyCode\MySample052 文件夹中。

003 使用 TextSwitcher 平滑切换多个标题

此实例主要通过使用 TextSwitcher 控件，实现以动画的形式轮流显示多个文本标题。当实例运行之后，单击"启动轮播"按钮，将轮流显示多个电影片名；在轮流显示时，当前电影片名从右端滑出，新电影片名从左端滑入，效果分别如图 003-1 的左图和右图所示。单击"停止轮播"按钮，则将停止轮流显示电影片名。

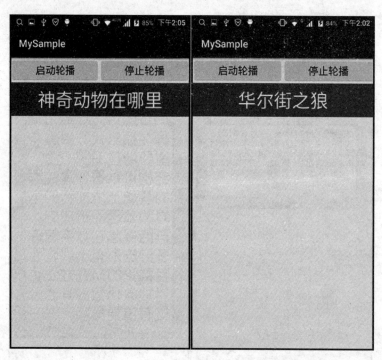

图 003-1

主要代码如下：

```
public class MainActivity extends Activity
                implements ViewSwitcher.ViewFactory{
    private TextSwitcher myTextSwitcher;
    private int myID = 0;
    String[] myItems = {"神奇动物在哪里","爱乐之城","血战钢锯岭",
                "海边的曼彻斯特","月光男孩","达拉斯买家俱乐部",
                "华尔街之狼","少年派奇幻漂流","雨果的秘密",
                "艺术家","被解放的姜戈"};
    private Timer myTimer;
    @Override
    protected void onCreate(Bundle savedInstanceState) {
        super.onCreate(savedInstanceState);
```

```java
    setContentView(R.layout.activity_main);
    myTextSwitcher = (TextSwitcher) findViewById(R.id.myTextSwitcher);
    myTextSwitcher.setFactory(this);
    Animation mySwitchIn = AnimationUtils.loadAnimation(this,
            android.R.anim.slide_in_left);              //文本滑入动画
    Animation mySwitchOut = AnimationUtils.loadAnimation(this,
            android.R.anim.slide_out_right);            //文本滑出动画
    myTextSwitcher.setInAnimation(mySwitchIn);
    myTextSwitcher.setOutAnimation(mySwitchOut);
    updateCounter();
}
public void onClickmyBtn1(View v) {                     //响应单击"启动轮播"按钮
    myID = 0;
    myTimer = new Timer();
    myTimer.schedule(new TimerTask() {
     @Override
     public void run() {
      runOnUiThread(new Runnable() {
       @Override
       public void run() {
           updateCounter();
       } });
} }, 0, 1000); }
public void onClickmyBtn2(View v) {                     //响应单击"停止轮播"按钮
    myTimer.cancel();
}
private void updateCounter() {
    myTextSwitcher.setText(myItems[myID]);
    if (myID < 10) { myID++; }
else { myID = 0;}
}
public View makeView() {
TextView myTextView = new TextView(this);
    myTextView.setGravity(Gravity.TOP | Gravity.CENTER_HORIZONTAL);
    myTextView.setTextSize(36);
    myTextView.setTextColor(Color.YELLOW);
    return myTextView;
} }
```

上面这段代码在 MyCode\MySample387\app\src\main\java\com\bin\luo\mysample\MainActivity.java 文件中。在这段代码中，TextSwitcher 包含 TextView，它被用来使屏幕上的标题文字产生动画效果，每当 TextSwitcher 的 setText() 方法被调用时，TextSwitcher 就采用动画方式使当前的文字内容消失并显示新的文字内容。此实例的完整项目在 MyCode\MySample387 文件夹中。

004 在 EditText 中弹出输入电话号码的键盘

此实例主要通过设置 EditText 控件的 inputType 属性为 phone，实现在单击 EditText 控件时，弹出输入电话号码的专用数字键盘。当实例运行之后，单击"联系电话："输入框，将在下面弹出数字键盘，单击键盘上的数字按键，这些数字将添加到输入框中，如图 004-1 的左图所示。单击数字键盘右上角的倒三角形按钮，将关闭数字键盘，如图 004-1 的右图所示。

图　004-1

主要代码如下：

```
< EditText android:layout_width = "match_parent"
           android:layout_height = "wrap_content"
           android:layout_toRightOf = "@id/myTextView"
           android:inputType = "phone"/>
```

上面这段代码在 MyCode\MySample024\app\src\main\res\layout\activity_main.xml 文件中。在这段代码中，android:inputType="phone"表示在 EditText 控件中输入内容（获得焦点）时，弹出专用于输入电话号码的数字键盘。在 Android 中，inputType 属性有点类似于 HTML 的<input>标签的 type 属性，它支持大量的属性值，不同的属性值用于指定特定的输入类型。此实例的完整项目在 MyCode\MySample024 文件夹中。

005　动态创建多个自定义风格的 CheckBox

此实例主要实现了根据选项的数量，通过循环的方式动态创建多个 CheckBox，并且在 CheckBox 控件的选中状态监听事件中调用 setButtonDrawable()方法设置当前勾选标记图像，从而实现定制 CheckBox 的选中样式。当实例运行之后，对任意 CheckBox 选中或不选，将会显示对应的符号标记（图像），效果分别如图 005-1 的左图和右图所示。

主要代码如下：

```
public class MainActivity extends Activity {
    String[] myBooks = new String[]{"深入理解 Android 自动化测试",
            "Android 开发秘籍","实践者的研究方法","互联网算法面试宝典"};
```

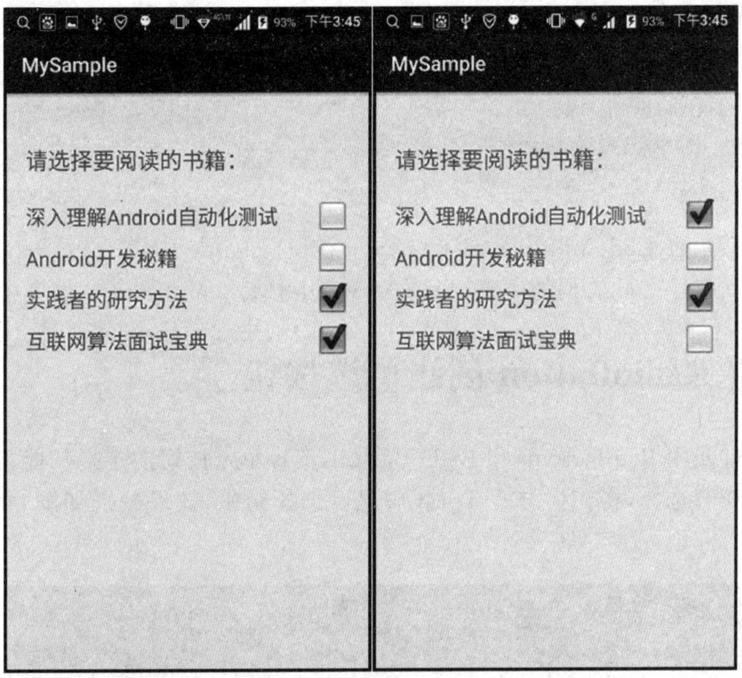

图 005-1

```java
@Override
protected void onCreate(Bundle savedInstanceState) {
    super.onCreate(savedInstanceState);
    setContentView(R.layout.activity_main);
    LinearLayout myLinearLayout = (LinearLayout) findViewById(R.id.myLayout);
    InitBookList(myLinearLayout);
}
public void InitBookList(LinearLayout myLinearLayout) {
    for (int i = 0; i < myBooks.length; i++) {              //动态添加多个 CheckBox
        RelativeLayout myContainer = new RelativeLayout(this);
        myContainer.setLayoutParams(new RelativeLayout.LayoutParams(
                        RelativeLayout.LayoutParams.MATCH_PARENT, 120));
        TextView myTextView = new TextView(this);
        myTextView.setLayoutParams(new ViewGroup.LayoutParams(
            ViewGroup.LayoutParams.MATCH_PARENT, ViewGroup.LayoutParams.MATCH_PARENT));
        myTextView.setText(myBooks[i]);
        myTextView.setTextSize(20);
        myTextView.setPadding(0,20,0,0);
        CheckBox myCheckBox = new CheckBox(this);
        myCheckBox.setGravity(Gravity.CENTER_VERTICAL);
        RelativeLayout.LayoutParams myParams = new RelativeLayout.LayoutParams(
            ViewGroup.LayoutParams.WRAP_CONTENT, ViewGroup.LayoutParams.MATCH_PARENT);
        myParams.addRule(RelativeLayout.ALIGN_PARENT_RIGHT);   //设置图像在右边
        myCheckBox.setLayoutParams(myParams);
        myCheckBox.setOnCheckedChangeListener(                 //选中状态监听
                new CompoundButton.OnCheckedChangeListener() {
            @Override
            public void onCheckedChanged(CompoundButton buttonView, boolean isChecked) {
                //根据选中状态设置显示 CheckBox 标记图像,以达到定制 CheckBox 的效果
                buttonView.setButtonDrawable(
                        isChecked ? R.mipmap.mychecked : R.mipmap.myunchecked);
```

```
    }});
    myCheckBox.setChecked(true);
    myContainer.addView(myTextView);
    myContainer.addView(myCheckBox);
    myLinearLayout.addView(myContainer);
}}}
```

上面这段代码在 MyCode\MySample802\app\src\main\java\com\bin\luo\mysample\MainActivity.java 文件中。此实例的完整项目在 MyCode\MySample802 文件夹中。

006 使用 RadioButton 创建单选按钮

此实例主要通过使用 RadioGroup 和 RadioButton，实现单选按钮的选择功能。当实例运行之后，在 4 个单选按钮中任意选择，将在弹出的 Toast 中显示选择结果，效果分别如图 006-1 的左图和右图所示。

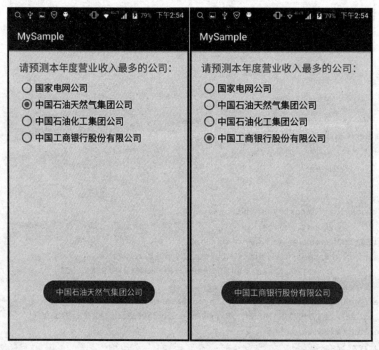

图 006-1

主要代码如下：

```xml
<RadioGroup android:id = "@ + id/myRadioGroup"
            android:layout_width = "wrap_content"
            android:layout_height = "wrap_content"
            android:layout_x = "3px"
            android:orientation = "vertical">
    <RadioButton android:textSize = "18dp"
            android:id = "@ + id/myRadioButton1"
            android:layout_width = "wrap_content"
            android:layout_height = "wrap_content"
            android:text = "国家电网公司" />
```

```xml
< RadioButton android:textSize = "18dp"
            android:id = "@ + id/myRadioButton2"
            android:layout_width = "wrap_content"
            android:layout_height = "wrap_content"
            android:text = "中国石油天然气集团公司"/>
< RadioButton android:textSize = "18dp"
            android:id = "@ + id/myRadioButton3"
            android:layout_width = "wrap_content"
            android:layout_height = "wrap_content"
            android:text = "中国石油化工集团公司"/>
< RadioButton android:textSize = "18dp"
            android:id = "@ + id/myRadioButton4"
            android:layout_width = "wrap_content"
            android:layout_height = "wrap_content"
            android:text = "中国工商银行股份有限公司"/>
</RadioGroup>
```

上面这段代码在 MyCode\MySample050\app\src\main\res\layout\activity_main.xml 文件中。在这段代码中，android:orientation = "vertical"表示在 RadioGroup 中的多个 RadioButton 以垂直方式排列，如果是 android:orientation = "horizontal"，则表示在 RadioGroup 中的多个 RadioButton 以水平方式排列。获取用户选择的 RadioButton，则以 Java 代码实现。主要代码如下：

```java
public class MainActivity extends Activity {
 RadioGroup myRadioGroup;
 @Override
 protected void onCreate(Bundle savedInstanceState) {
  super.onCreate(savedInstanceState);
  setContentView(R.layout.activity_main);
  myRadioGroup = (RadioGroup) findViewById(R.id.myRadioGroup);
  //监听用户单击单选按钮 RadioButton
  myRadioGroup.setOnCheckedChangeListener(
                       new RadioGroup.OnCheckedChangeListener() {
   @Override
   public void onCheckedChanged(RadioGroup group, int checkedId) {
    //checkedId 即是用户单击单选按钮 RadioButton 的 ID
    RadioButton myRadioButton = (RadioButton)findViewById(checkedId);
    Toast myToast = Toast.makeText(getApplicationContext(),
     myRadioButton.getText(), Toast.LENGTH_LONG);   //在 Toast 中显示选择结果
    myToast.setGravity(Gravity.TOP, 0, 1500);
    myToast.show();
   } });
} }
```

上面这段代码在 MyCode\MySample050\app\src\main\java\com\bin\luo\mysample\MainActivity.java 文件中。在这段代码中，myRadioGroup.setOnCheckedChangeListener 用于监听在 RadioGroup 中的任意 RadioButton 是否被单击，如果某个 RadioButton 被单击，则在 onCheckedChanged(RadioGroup group, int checkedId)事件响应方法的 checkedId 参数中返回该 RadioButton 的 ID，即哪个按钮被单击了。myRadioButton.getText()用于获取 RadioButton 的标题文本。此实例的完整项目在 MyCode\MySample050 文件夹中。

007 使用 CheckedTextView 创建单选按钮

此实例主要通过使用 CheckedTextView，实现支持单选按钮的选择功能。当实例运行之后，在 4 个单选按钮中任意选择，然后单击"提交信息"按钮，将在弹出的 Toast 中显示选择结果，效果分别如图 007-1 的左图和右图所示。

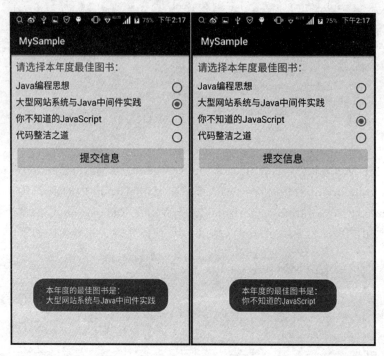

图 007-1

主要代码如下：

```
<CheckedTextView
    android:id = "@+id/checkedTextView1"
    android:layout_width = "fill_parent"
    android:layout_height = "wrap_content"
    android:checkMark = "?android:attr/listChoiceIndicatorSingle"
    android:textSize = "18dp"
    android:text = "Java 编程思想"/>
```

上面这段代码在 MyCode\MySample043\app\src\main\res\layout\activity_main.xml 文件中。在这段代码中，单选按钮的标记图像使用 android:checkMark＝"?android:attr/ listChoiceIndicatorSingle" 实现。检测单选按钮的标记状态变化则使用 Java 代码实现，主要代码如下：

```java
public class MainActivity extends Activity {
    private CheckedTextView myRadio1;
    private CheckedTextView myRadio2;
    private CheckedTextView myRadio3;
    private CheckedTextView myRadio4;
    private ArrayList<Integer> myIDArray = new ArrayList();
```

```java
@Override
protected void onCreate(Bundle savedInstanceState) {
    super.onCreate(savedInstanceState);
    setContentView(R.layout.activity_main);
    myRadio1 = (CheckedTextView)this.findViewById(R.id.checkedTextView1);
    myRadio2 = (CheckedTextView)this.findViewById(R.id.checkedTextView2);
    myRadio3 = (CheckedTextView)this.findViewById(R.id.checkedTextView3);
    myRadio4 = (CheckedTextView)this.findViewById(R.id.checkedTextView4);
    myRadio3.setChecked(true);                       //设置myRadio3处于选中状态
    myIDArray.add(myRadio1.getId());
    myIDArray.add(myRadio2.getId());
    myIDArray.add(myRadio3.getId());
    myIDArray.add(myRadio4.getId());
    //响应单击每个单选按钮
    View.OnClickListener myRef = new View.OnClickListener() {
     public void onClick(View arg0) {
      for (int i = 0; i < myIDArray.size(); i++) {
       if (myIDArray.get(i).intValue() !=
                         ((CheckedTextView) arg0).getId()) {
        ((CheckedTextView) MainActivity.this
              .findViewById(myIDArray.get(i))).setChecked(false);
       } else {
        ((CheckedTextView) MainActivity.this
              .findViewById(myIDArray.get(i))).setChecked(true);
} } } };
    myRadio1.setOnClickListener(myRef);
    myRadio2.setOnClickListener(myRef);
    myRadio3.setOnClickListener(myRef);
    myRadio4.setOnClickListener(myRef);
}
public void OnClickmyBtnSubmit(View v) {            //响应单击"提交信息"按钮
    String myText = "本年度的最佳图书是：";
    for (int i = 0; i < myIDArray.size(); i++) {
     CheckedTextView myCheckedTextView =
          ((CheckedTextView) MainActivity.this.findViewById(myIDArray.get(i)));
     if (myCheckedTextView.isChecked() == true) {
      myText += "\n" + myCheckedTextView.getText().toString();
     }
    }
    Toast.makeText(this, myText, Toast.LENGTH_LONG).show();
}
```

上面这段代码在 MyCode\MySample043\app\src\main\java\com\bin\luo\mysample\MainActivity.java 文件中。在这段代码中，myRadio3.setChecked(true)用于设置 myRadio3 处于选中状态，如果是 myRadio3.setChecked(false)，则表示 myRadio3 处于未选状态。myCheckedTextView.isChecked()用于判断单选按钮是否被选中，如果返回值为 true，则表示该单选按钮被选中；如果返回值为 false，则表示该单选按钮没有被选中。myCheckedTextView.getText()用于获取单选按钮的标题文本。此实例的完整项目在 MyCode\MySample043 文件夹中。

008　在 ListView 列表项中自定义单选按钮

此实例主要通过在 setChoiceMode()方法中设置 ListView.CHOICE_MODE_SINGLE 参数，从而在 ListView 多个列表项中实现单选功能。当实例运行之后，ListView 的每个列表项右边均有一个单选按钮，选中任意单选按钮，其他按钮立即取消选中，即 ListView 的列表项同一时刻只有一个被选中,效果分别如图008-1的左图和右图所示。

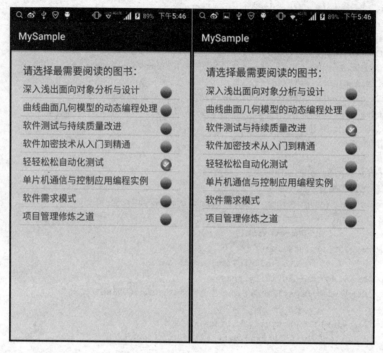

图 008-1

主要代码如下:

```
public class MainActivity extends Activity {
 ListView myListView;
 @Override
 protected void onCreate(Bundle savedInstanceState) {
  super.onCreate(savedInstanceState);
  setContentView(R.layout.activity_main);
  myListView = (ListView) findViewById(R.id.myListView);
  myListView.setChoiceMode(ListView.CHOICE_MODE_SINGLE);
  myListView.setAdapter(new MyAdapter(this));
} }
```

上面这段代码在 MyCode\MySample804\app\src\main\java\com\bin\luo\mysample\MainActivity.java 文件中。在这段代码中,myListView.setChoiceMode(ListView.CHOICE_MODE_SINGLE)表示 ListView 的列表项支持单选功能。myListView.setAdapter(new MyAdapter(this))表示使用创建的自定义适配器 MyAdapter 的对象作为 myListView 的数据源。自定义适配器 MyAdapter 类的主要代码如下:

```
public class MyAdapter extends BaseAdapter{
 Context myContext;
 ViewHolder myViewHolder;
 List<BookInfo> myBookList;
 public MyAdapter(Context context){
  myContext = context;
  myBookList = InitBookInfoList();
 }
```

```java
@Override
public int getCount() { return myBookList.size(); }
@Override
public Object getItem(int position) { return myBookList.get(position); }
@Override
public long getItemId(int position) { return position; }
@Override
public View getView(final int position, View convertView, ViewGroup parent) {
 if (convertView == null){
  convertView = LayoutInflater.from(myContext).inflate(R.layout.myitem,null);
  myViewHolder = new ViewHolder();
  myViewHolder.myTextView =
           (TextView)convertView.findViewById(R.id.myTextView);
  myViewHolder.myTextView.setText(myBookList.get(position).getBookName());
  myViewHolder.myCheckBox =
           (CheckBox)convertView.findViewById(R.id.myCheckBox);
  myViewHolder.myCheckBox.setChecked(myBookList.get(position).getChecked());
  myViewHolder.myCheckBox.setClickable(false);
  //初次显示时,设置所有列表项Item未选中
  myViewHolder.myCheckBox.setButtonDrawable(R.mipmap.myunchecked);
  //监听选中状态
  myViewHolder.myCheckBox.setOnCheckedChangeListener(
           new CompoundButton.OnCheckedChangeListener() {
   @Override
   public void onCheckedChanged(CompoundButton buttonView, boolean isChecked){
    //根据选中状态设置显示单选按钮图像,以达到定制效果
    buttonView.setButtonDrawable(
             isChecked ? R.mipmap.mychecked : R.mipmap.myunchecked);
   } });
  convertView.setTag(myViewHolder);
 } else {
  myViewHolder = (ViewHolder)convertView.getTag();
 }
 return convertView;
}
class ViewHolder {
 TextView myTextView;
 CheckBox myCheckBox;
}
static String[] myBooks = new String[]{"深入浅出面向对象分析与设计",
        "曲线曲面几何模型的动态编程处理","软件测试与持续质量改进",
        "软件加密技术从入门到精通","轻轻松松自动化测试",
        "单片机通信与控制应用编程实例","软件需求模式","项目管理修炼之道"};
public static List<BookInfo> InitBookInfoList(){
 ArrayList<BookInfo> myBookInfoList = new ArrayList<BookInfo>();
 for(int i = 0; i< myBooks.length; i++){
  BookInfo myBookInfo = new BookInfo();
  myBookInfo.setChecked(false);
  myBookInfo.setBookName(myBooks[i]);
  myBookInfoList.add(myBookInfo);
 }
 return myBookInfoList;
} }
```

上面这段代码在 MyCode\MySample804\app\src\main\java\com\bin\luo\mysample\MyAdapter.java 文件中。此实例的完整项目在 MyCode\MySample804 文件夹中。

009 使用资源创建自定义背景的椭圆按钮

此实例主要通过创建椭圆形状的资源文件设置按钮的背景,使按钮的外观呈现椭圆形状。当实例运行之后,单击椭圆按钮,将弹出 Toast 显示提示信息,效果分别如图 009-1 的左图和右图所示。

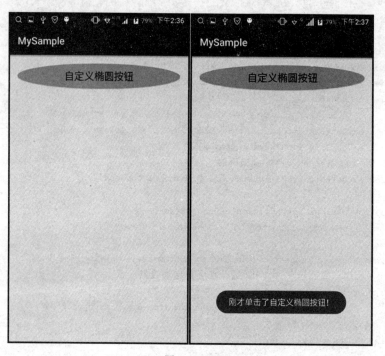

图 009-1

主要代码如下:

```
<Button android:layout_width = "match_parent"
        android:layout_height = "wrap_content"
        android:background = "@drawable/myoval"
        android:onClick = "onClickmyBtn"
        android:padding = "6dp"
        android:text = "自定义椭圆按钮"
        android:textSize = "20dp"/>
```

上面这段代码在 MyCode\MySample047\app\src\main\res\layout\activity_main.xml 文件中。在这段代码中,android:background = "@drawable/myoval" 用于获取椭圆背景资源 myoval。android:onClick = "onClickmyBtn" 用于将按钮单击事件与 onClickmyBtn 方法关联起来。onClickmyBtn() 方法的主要代码如下:

```
//响应单击自定义椭圆按钮
public void onClickmyBtn(View v) {
    Toast.makeText(getApplicationContext(),
        "刚才单击了自定义椭圆按钮!", Toast.LENGTH_SHORT).show();
}
```

上面这段代码在 MyCode\MySample047\app\src\main\java\com\bin\luo\mysample\MainActivity.java 文件中。在这段代码中，Toast.makeText(getApplicationContext()，"刚才单击了自定义椭圆按钮！"，Toast.LENGTH_SHORT).show()用于在单击椭圆按钮后显示提示信息"刚才单击了自定义椭圆按钮！"。椭圆背景资源 myoval 的主要内容如下：

```xml
<?xml version="1.0" encoding="utf-8"?>
<shape xmlns:android="http://schemas.android.com/apk/res/android"
    android:layout_width="match_parent"
    android:layout_height="match_parent"
    android:shape="oval" >
 <solid android:color="#00cc33" />
</shape>
```

上面这段代码在 MyCode\MySample047\app\src\main\res\drawable\myoval.xml 文件中。在这段代码中，android:shape="oval"用于设置背景的形状为椭圆，android:shape 属性可用的值有 rectangle|oval|line|ring。<solid android:color="#00cc33"/>用于设置背景的填充颜色。此实例的完整项目在 MyCode\MySample047 文件夹中。

010 使用 ShapeDrawable 创建渐变圆角按钮

此实例主要通过使用 ShapeDrawable 和 LinearGradient，实现将普通的直角按钮改变为渐变色的圆角按钮。当实例运行之后，将显示三个直角按钮，单击任意一个按钮，则三个直角按钮将改变为三个渐变色的圆角按钮，效果分别如图 010-1 的左图和右图所示。

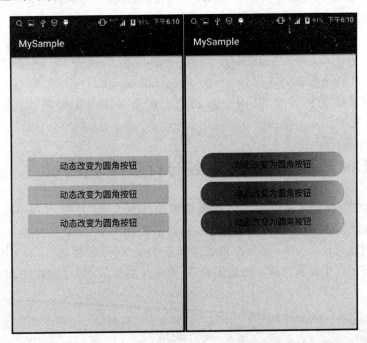

图 010-1

主要代码如下：

```java
public class MainActivity extends Activity {
```

```
@Override
protected void onCreate(Bundle savedInstanceState) {
    super.onCreate(savedInstanceState);
    setContentView(R.layout.activity_main);
}
public void onClickmyButton(View v) {       //响应单击"动态改变为圆角按钮"按钮
    Button myButton1 = (Button) findViewById(R.id.myButton1);
    Button myButton2 = (Button) findViewById(R.id.myButton2);
    Button myButton3 = (Button) findViewById(R.id.myButton3);
    int myWidth = myButton1.getWidth();
    //使用4对值设置按钮的左上、右上、右下、左下四个角的圆角半径都为100
    ShapeDrawable myShapeDrawable = new ShapeDrawable(
                    new RoundRectShape(new float[]{100,100,
                    100,100,100,100,100,100},null,null));
    //设置按钮背景为线性渐变颜色
    LinearGradient myGradient = new LinearGradient(0,0,myWidth,
                    0,Color.RED,Color.GREEN, Shader.TileMode.MIRROR);
    myShapeDrawable.getPaint().setShader(myGradient);
    myButton1.setBackgroundDrawable(myShapeDrawable);
    myButton2.setBackgroundDrawable(myShapeDrawable);
    myButton3.setBackgroundDrawable(myShapeDrawable);
}}
```

上面这段代码在 MyCode\MySample805\app\src\main\java\com\bin\luo\mysample\MainActivity.java 文件中。在这段代码中，ShapeDrawable myShapeDrawable＝new ShapeDrawable(new RoundRectShape(new float[]{100,100,100,100,100,100,100,100}, null,null))用于创建圆角半径为100 的圆角矩形。LinearGradient myGradient＝new LinearGradient(0,0,myWidth,0,Color.RED,Color.GREEN, Shader.TileMode.MIRROR)用于根据指定的参数创建线性渐变。myShapeDrawable.getPaint().setShader(myGradient)表示使用渐变色的着色器设置 myShapeDrawable 的 Paint。myButton1.setBackgroundDrawable(myShapeDrawable)表示使用渐变色的圆角矩形设置按钮的背景，从而使按钮产生圆角效果。此实例的完整项目在 MyCode\MySample805 文件夹中。

011　在 ListView 列表项文本的右端添加按钮

此实例主要通过在布局文件中为每个列表项添加 TextView 控件和 Button 控件，并在自定义 MyArrayAdapter 适配器中处理按钮单击事件，从而实现在 ListView 中单击列表项的按钮时弹出 Toast。当实例运行之后，ListView 的每个列表项都有 TextView 控件和 Button 控件，单击 Button 控件，则将在弹出的 Toast 中显示当前单击的列表项，效果分别如图 011-1 的左图和右图所示。

主要代码如下：

```
public class MainActivity extends ListActivity {
    private String[] myNames = {"神秘宇宙与微观世界珍贵图集",
    "写给所有人的极简统计学","很杂很杂的杂学知识","英国皇家园艺学会植物学指南"};
    @Override
    protected void onCreate(Bundle savedInstanceState) {
        MyArrayAdapter myArrayAdapter =
                    new MyArrayAdapter(this, R.layout.activity_main, myNames);
```

第1章 常用控件

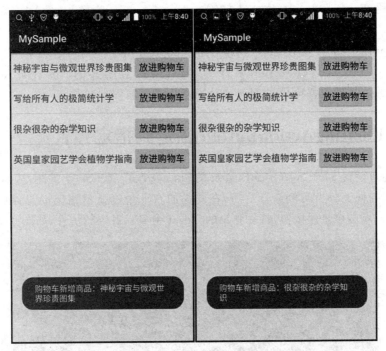

图 011-1

```
    ListView myListView = getListView();
    myListView.setAdapter(myArrayAdapter);
    super.onCreate(savedInstanceState);
}
private class MyArrayAdapter extends ArrayAdapter<String>{
    private int myID;
    public MyArrayAdapter(Context context, int myID, String[] myNames){
        super(context, myID, myNames);
        this.myID = myID;
    }
    @Override
    public View getView(int position, View convertView, ViewGroup parent){
        final String myItem = getItem(position);
        LinearLayout myLinearLayout = new LinearLayout(getContext());
        LayoutInflater myLayoutInflater = (LayoutInflater)getContext().
                        getSystemService(Context.LAYOUT_INFLATER_SERVICE);
        myLayoutInflater.inflate(myID, myLinearLayout, true);
        final TextView myTextView =
                (TextView)myLinearLayout.findViewById(R.id.myTextView);
        myTextView.setText(myItem);
        Button myButton = (Button)myLinearLayout.findViewById(R.id.myButton);
        myButton.setOnClickListener(new View.OnClickListener() {
            @Override
            public void onClick(View v){
                Toast.makeText(MainActivity.this,
                        "购物车新增商品:" + myItem,Toast.LENGTH_SHORT).show();
            } });
        return myLinearLayout;
} } }
```

上面这段代码在 MyCode\MySample636\app\src\main\java\com\bin\luo\mysample\MainActivity.java 文件中。在这段代码中,myListView = getListView()用于获取当前环境的 ListView 对象。myListView.setAdapter(myArrayAdapter)用于使用自定义的 myArrayAdapter 处理在 ListView 控件中加载数据及单击事件等动作,因此自定义的 myArrayAdapter 必须重写 getView()方法。此实例的完整项目在 MyCode\MySample636 文件夹中。

012　将 FloatingActionButton 锚定到指定控件

此实例主要通过设置 FloatingActionButton 的 app:layout_anchor 等属性,从而实现将悬浮按钮锚定到指定的布局控件上。当实例运行之后,在图像的右下角将显示锚定的悬浮(播放)按钮,如图 012-1 的左图所示。单击该悬浮按钮,将在弹出的 Toast 中显示对应的信息,如图 012-1 的右图所示。

图　012-1

主要代码如下:

```
<android.support.design.widget.FloatingActionButton
    android:layout_width="wrap_content"
    android:layout_height="wrap_content"
    android:src="@mipmap/myimage2"
    android:layout_marginBottom="10dp"
    android:layout_marginRight="7dp"
    android:onClick="onClickmyBtn1"
    app:layout_anchor="@+id/myFrameLayout"
    app:layout_anchorGravity="bottom|right"
    app:elevation="15dp"/>
```

上面这段代码在 MyCode\MySample495\app\src\main\res\layout\activity_main.xml 文件中。在这段代码中,app:layout_anchor="@+id/myFrameLayout" 表示将悬浮按钮锚定到

myFrameLayout 布局控件上。app:layout_anchorGravity="bottom|right"表示悬浮按钮位于布局控件的右下角。android:onClick="onClickmyBtn1"表示在单击悬浮按钮时响应 onClickmyBtn1()方法。onClickmyBtn1()方法的主要代码如下：

```java
public void onClickmyBtn1(View v) {            //响应单击图像右下角的悬浮按钮
    Toast.makeText(MainActivity.this,
            "刚才单击了悬浮按钮!",Toast.LENGTH_LONG).show();
}
```

上面这段代码在 MyCode\MySample495\app\src\main\java\com\bin\luo\mysample\MainActivity.java 文件中。此外，使用此实例的相关控件需要在 gradle 中引入 compile 'com.android.support:design:25.0.1'依赖项。此实例的完整项目在 MyCode\MySample495 文件夹中。

013　使用 ZoomControls 实现放大和缩小图像

此实例主要通过使用 ZoomControls 控件左右两边的"－"和"＋"按钮功能，从而实现单击"－"按钮缩小图像，单击"＋"按钮放大图像的效果。当实例运行之后，单击 ZoomControls 控件右边的"＋"按钮，则放大图像；单击 ZoomControls 控件左边的"－"按钮，则缩小图像，效果分别如图 013-1 的左图和右图所示。

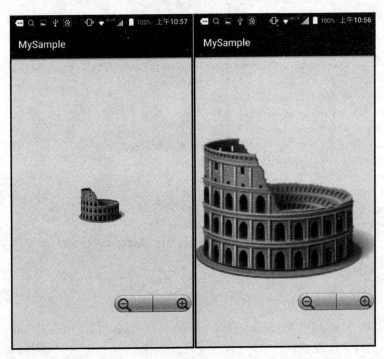

图　013-1

主要代码如下：

```java
public class MainActivity extends Activity {
    private ImageView myImageView;
    private ZoomControls myZoomControls;
    private float myScaleWidth = 1;
    private float myScaleHeight = 1;
```

```java
private Bitmap myBitmap;
@Override
protected void onCreate(Bundle savedInstanceState) {
    super.onCreate(savedInstanceState);
    setContentView(R.layout.activity_main);
    myBitmap = BitmapFactory.decodeResource(getResources(), R.mipmap.myimage);
    myImageView = (ImageView) findViewById(R.id.myImageView);
    myZoomControls = (ZoomControls) findViewById(R.id.myZoomControls);
    myZoomControls.setIsZoomInEnabled(true);
    myZoomControls.setIsZoomOutEnabled(true);
    //监听单击放大(+)按钮
    myZoomControls.setOnZoomInClickListener(new ZoomControls.OnClickListener(){
      public void onClick(View v) {
        myZoomControls.setIsZoomOutEnabled(true);
        int myBmpWidth = myBitmap.getWidth();
        int myBmpHeight = myBitmap.getHeight();
        double myScale = 1.5;                    //设置图像的放大比例
        myScaleWidth = (float) (myScaleWidth * myScale);
        myScaleHeight = (float) (myScaleHeight * myScale);
        Matrix myMatrix = new Matrix();
        myMatrix.postScale(myScaleWidth, myScaleHeight);
        Bitmap myBmpResize = Bitmap.createBitmap(myBitmap,
                    0, 0, myBmpWidth, myBmpHeight, myMatrix, false);
        myImageView.setImageBitmap(myBmpResize);
      } });
    //监听单击缩小(-)按钮
    myZoomControls.setOnZoomOutClickListener(new ZoomControls.OnClickListener(){
      public void onClick(View v) {
        myZoomControls.setIsZoomInEnabled(true);
        int myBmpWidth = myBitmap.getWidth();
        int myBmpHeight = myBitmap.getHeight();
        double myScale = 0.5;                    //设置图像的缩小比例
        myScaleWidth = (float) (myScaleWidth * myScale);
        myScaleHeight = (float) (myScaleHeight * myScale);
        Matrix myMatrix = new Matrix();
        myMatrix.postScale(myScaleWidth, myScaleHeight);
        Bitmap myBmpResize = Bitmap.createBitmap(myBitmap,
                    0, 0, myBmpWidth, myBmpHeight, myMatrix, true);
        myImageView.setImageBitmap(myBmpResize);
      } });
} }
```

上面这段代码在 MyCode\MySample383\app\src\main\java\com\bin\luo\mysample\MainActivity.java 文件中。在这段代码中,ZoomControls 是一组可缩放的控件,它包含放大按钮和缩小按钮;该控件通常用于图像浏览和地图定位的情形。此实例的完整项目在 MyCode\MySample383 文件夹中。

014 使用 StackView 实现堆叠显示多幅图像

此实例主要通过使用 StackView 控件,实现自动轮播堆叠在一起的多幅图像。当实例运行之后,单击"向下轮播"按钮,图像将从右上角滑入,再顺次从左下角滑出;单击"向上轮播"按钮,图像将从左下角滑入,再顺次从右上角滑出;单击左下角完全显示(最顶层)的图像,将在左上角提示图像的当前位置,如"第 3 部电影",效果分别如图 014-1 的左图和右图所示。

图 014-1

主要代码如下:

```java
public class MainActivity extends Activity {
 private StackView myStackView;
 private int[] myIDs = {R.mipmap.myimage1, R.mipmap.myimage2,
         R.mipmap.myimage3, R.mipmap.myimage4, R.mipmap.myimage5,
         R.mipmap.myimage6, R.mipmap.myimage7 };
 private List<Integer> myImages = new ArrayList<>();
 private ImageAdapter myImageAdapter;
 private TextView myTextView;
 private Timer myTimerDown;
 private Timer myTimerUp;
 @Override
 protected void onCreate(Bundle savedInstanceState) {
  super.onCreate(savedInstanceState);
  setContentView(R.layout.activity_main);
  myStackView = (StackView) findViewById(R.id.myStackView);
  myTextView = (TextView) findViewById(R.id.myTextView);
  for (int i = 0; i < myIDs.length; i++) { myImages.add(myIDs[i]); }
  myImageAdapter = new ImageAdapter(myImages, this);
  myStackView.setAdapter(myImageAdapter);
  myStackView.setOnItemClickListener(new AdapterView.OnItemClickListener(){
   @Override
   public void onItemClick(AdapterView<?> parent,
                      View view, int position, long id){
    myTextView.setText("第" + (position + 1) + "部电影");
   } });
```

```
}
public void onClickmyBtn1(View v){                //响应单击"向下轮播"按钮
 if(myTimerUp != null){ myTimerUp.cancel(); }
 myTimerDown = new Timer();
 myTimerDown.schedule(new TimerTask() {
  @Override
  public void run() {
   runOnUiThread(new Runnable() {
    @Override
    public void run() { myStackView.showNext(); }
   }); } },0,1000);
}
public void onClickmyBtn2(View v){                //响应单击"向上轮播"按钮
 if(myTimerDown != null){ myTimerDown.cancel(); }
 myTimerUp = new Timer();
 myTimerUp.schedule(new TimerTask() {
  @Override
  public void run() {
   runOnUiThread(new Runnable() {
    @Override
    public void run() { myStackView.showPrevious(); }
}); } },0,1000); } }
```

上面这段代码在 MyCode\MySample385\app\src\main\java\com\bin\luo\mysample\MainActivity.java 文件中。在这段代码中，StackView 是 AdapterViewAnimator 的子类，它以堆叠 (Stack) 的风格来显示多个 View 控件。StackView 主要提供下列两种方式来控制 View 控件的显示。

（1）如果拖走最顶层的 View 控件，则下一层的 View 控件自动成为最顶层的 View 控件。

（2）直接调用 showNext()方法显示下一个 View 控件，调用 showPrevious()方法显示上一个 View 控件。

此实例的完整项目在 MyCode\MySample385 文件夹中。

015　使用 ScrollView 实现上下滑动切换图像

此实例主要通过使用 GestureDetector 动态监听 ScrollView 的滑动操作，并重写 onFling()方法监听滑动事件，从而实现在 ScrollView 中上下滑动时切换其中的图像（ImageView 控件）。当实例运行之后，向上滑动手指，则滑入下一幅图像，向下滑动手指，则滑入上一幅图像，效果分别如图 015-1 的左图和右图所示。

主要代码如下：

```
public class MainActivity extends Activity {
 int[] myImages = {R.mipmap.myimage1, R.mipmap.myimage2,
                   R.mipmap.myimage3, R.mipmap.myimage4, R.mipmap.myimage5};
 int myIndex;
 @Override
 protected void onCreate(Bundle savedInstanceState) {
  super.onCreate(savedInstanceState);
  setContentView(R.layout.activity_main);
  final ScrollView myScrollView = (ScrollView) findViewById(R.id.myScrollView);
  final LinearLayout myLinearLayout =
                 (LinearLayout) findViewById(R.id.myLinearLayout);
```

第1章 常用控件

图 015-1

```
final int myWidth = getWindowManager().getDefaultDisplay().getWidth();
final int myHeight = getWindowManager().getDefaultDisplay().getHeight();
for (int i = 0; i < myImages.length; i++) {
 ImageView myImageView = new ImageView(MainActivity.this);
 myImageView.setScaleType(ImageView.ScaleType.FIT_XY);
 myImageView.setLayoutParams(
            new RelativeLayout.LayoutParams(myWidth, myHeight));
 myImageView.setImageResource(myImages[i]);
 myLinearLayout.addView(myImageView);
}
final GestureDetector myDetector =
      new GestureDetector(new GestureDetector.SimpleOnGestureListener(){
 @Override
 public boolean onFling(MotionEvent e1, MotionEvent e2,
                        float velocityX, float velocityY){
  if (e2.getY() > e1.getY()) {           //显示上一幅图像
   myScrollView.smoothScrollTo(0,myHeight * (myIndex < 0 ? 0 : --myIndex));
  } else if (e2.getY() < e1.getY()) {    //显示下一幅图像
   myScrollView.smoothScrollTo(0,
      myHeight * (myIndex >= myLinearLayout.getChildCount() ?
      myLinearLayout.getChildCount() - 1 : ++myIndex));
  }
  return true;
 } });
myScrollView.setOnTouchListener(new View.OnTouchListener() {
 @Override
 //传递事件至手势识别器对象
 public boolean onTouch(View v, MotionEvent event) {
  return myDetector.onTouchEvent(event);
 } }); } }
```

上面这段代码在 MyCode\MySample837\app\src\main\java\com\bin\luo\mysample\MainActivity.java 文件中。在这段代码中，onFling()方法用于监听手指在垂直方向上滑动时是向上滑动还是向下滑动，当起点 Y 坐标(e1.getY())小于终点 Y 坐标(e2.getY())时，表示向下滑动；当起点 Y 坐标(e1.getY())大于终点 Y 坐标(e2.getY())时，表示向上滑动。此实例的完整项目在 MyCode\MySample837 文件夹中。

016　使用 HorizontalScrollView 水平切换图像

此实例主要通过使用 GestureDetector 动态监听 HorizontalScrollView 的滑动操作，并重写 onFling()的方法监听滑动事件，从而实现在 HorizontalScrollView 中左右滑动时切换其中的图像（ImageView 控件）。当实例运行之后，向左滑动手指，则滑入下一幅图像，向右滑动手指，则滑入上一幅图像，效果分别如图 016-1 的左图和右图所示。

图　016-1

主要代码如下：

```java
public class MainActivity extends Activity {
    int[] myImages = {R.mipmap.myimage1, R.mipmap.myimage2,
              R.mipmap.myimage3, R.mipmap.myimage4, R.mipmap.myimage5};
    int myIndex;
    @Override
    protected void onCreate(Bundle savedInstanceState) {
      super.onCreate(savedInstanceState);
      setContentView(R.layout.activity_main);
      final HorizontalScrollView myHorizontalScrollView =
          (HorizontalScrollView) findViewById(R.id.myHorizontalScrollView);
```

```
    final LinearLayout myLinearLayout =
            (LinearLayout) findViewById(R.id.myLinearLayout);
    final int myWidth = getWindowManager().getDefaultDisplay().getWidth();
    int myHeight = getWindowManager().getDefaultDisplay().getHeight();
    for (int i = 0; i < myImages.length; i++) {
      ImageView myImageView = new ImageView(MainActivity.this);
      myImageView.setScaleType(ImageView.ScaleType.FIT_XY);
      myImageView.setLayoutParams(
                  new RelativeLayout.LayoutParams(myWidth, myHeight));
      myImageView.setImageResource(myImages[i]);
      myLinearLayout.addView(myImageView);
    }
    final GestureDetector myDetector =
          new GestureDetector(new GestureDetector.SimpleOnGestureListener(){
      @Override
      public boolean onFling(MotionEvent e1, MotionEvent e2,
                              float velocityX, float velocityY) {
        if (e2.getX() > e1.getX()) {         //显示上一幅图像
          myHorizontalScrollView.smoothScrollTo(myWidth * (
                              myIndex < 0 ? 0 : -- myIndex), 0);
        } else if (e2.getX() < e1.getX()) {   //显示下一幅图像
          myHorizontalScrollView.smoothScrollTo(
                  myWidth * (myIndex >= myLinearLayout.getChildCount() ?
                  myLinearLayout.getChildCount() - 1 : ++myIndex), 0);
        }
        return true;
      } });
    myHorizontalScrollView.setOnTouchListener(new View.OnTouchListener() {
      @Override
      //传递事件至手势识别器对象
      public boolean onTouch(View v, MotionEvent event) {
        return myDetector.onTouchEvent(event);
      }
});}}
```

上面这段代码在 MyCode\MySample836\app\src\main\java\com\bin\luo\mysample\MainActivity.java 文件中。在这段代码中，onFling()方法用于监听手指在水平方向上滑动时是向左滑动还是向右滑动，当起点 X 坐标(e1.getX())小于终点 X 坐标(e2.getX())时，表示向右滑动；当起点 X 坐标(e1.getX())大于终点 X 坐标(e2.getX())时，表示向左滑动。此实例的完整项目在 MyCode\MySample836 文件夹中。

017 在 RecyclerView 中按照文件属性排序

此实例主要通过使用 Collections 的 sort()方法，从而实现对 RecyclerView 的文件列表项按照文件名称和类型进行排序。当实例运行后，将在 RecyclerView 中显示存储卡根目录下的所有文件，单击"按照文件类型排序"按钮，则按照文件类型排序（即默认排序），排序之后的文件列表如图 017-1 的左图所示；单击"按照文件名称排序"按钮，则按照文件名称排序，排序之后的文件列表如图 017-1 的右图所示。

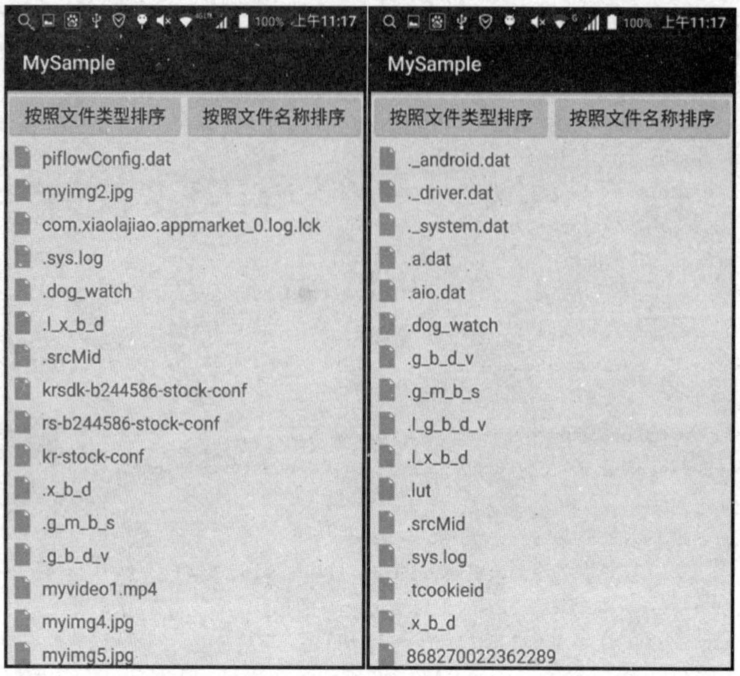

图 017-1

主要代码如下：

```java
public class MainActivity extends Activity {
 private MyAdapter myAdapter;
 @Override
 protected void onCreate(Bundle savedInstanceState) {
   super.onCreate(savedInstanceState);
   setContentView(R.layout.activity_main);
RecyclerView myRecyclerView = (RecyclerView) findViewById(R.id.myRecyclerView);
  List<File> myFileList = getFileList();
  myAdapter = new MyAdapter(this, myFileList);
  myRecyclerView.setAdapter(myAdapter);
  myRecyclerView.setLayoutManager(new LinearLayoutManager(this));
}
 public void onClickButtonByType(View view) {      //响应单击"按照文件类型排序"按钮
  List<File> myFileList = getFileList();
  myAdapter.setMyFileList(myFileList);
}
public void onClickButtonByName(View view) {      //响应单击"按照文件名称排序"按钮
  List<File> myFileList = getFileList();
  Collections.sort(myFileList, new Comparator<File>() {
   @Override
   public int compare(File file, File t1) {
    //文件名称排序,即按照符号、大写字母、小写字母、中文的顺序来进行排序
    return file.getName().compareTo(t1.getName());
  } });
  myAdapter.setMyFileList(myFileList);           //更新数据源
}
```

```
public static List<File> getFileList() {        //获取存储卡根目录中的文件
  List<File> myFileList = new ArrayList<File>();
  File myPath = Environment.getExternalStorageDirectory();
  File[] myFiles = myPath.listFiles();
  for (int i = 0; i < myFiles.length; i++) {
    if (myFiles[i].isFile())
      myFileList.add(myFiles[i]);
  }
  return myFileList;
} }
```

上面这段代码在 MyCode\MySample737\app\src\main\java\com\bin\luo\mysample\MainActivity.java 文件中。在这段代码中，Collections.sort(myFileList, new Comparator<File>() {...}) 表示按照第二个参数制定的排序规则对第一个参数代表的数据列表进行排序。myRecyclerView.setAdapter(myAdapter) 表示使用自定义适配器 MyAdapter 的实例管理 myRecyclerView 的列表项，关于 MyAdapter 的详细内容请参考 MyCode\MySample737\app\src\main\java\com\bin\luo\mysample\MyAdapter.java 文件。此外，在存储卡上访问文件需要在 AndroidManifest.xml 文件中添加< uses-permission android:name="android.permission.READ_EXTERNAL_STORAGE"/>权限。需要说明的是，使用此实例的相关类需要在 gradle 中引入 compile 'com.android.support:recyclerview-v7:25.3.1'依赖项。此实例的完整项目在 MyCode\MySample737 文件夹中。

018　使用 RecyclerView 实现简单的 ListView

此实例主要演示了使用 RecyclerView 实现简单的 ListView 功能。当实例运行之后，使用手指上下滑动屏幕，RecyclerView 的列表项也随之移动，效果分别如图 018-1 的左图和右图所示。

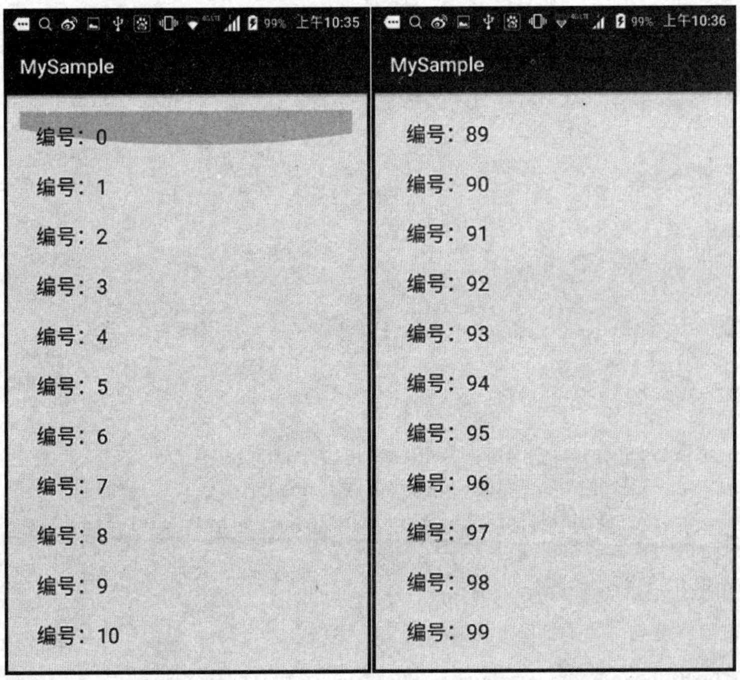

图　018-1

主要代码如下:

```xml
<android.support.v7.widget.RecyclerView
    android:id="@+id/myRecyclerView"
    android:layout_width="match_parent"
    android:layout_height="match_parent"/>
```

上面这段代码在 MyCode\MySample266\app\src\main\res\layout\activity_main.xml 文件中。当 RecyclerView 控件的布局完成之后,就应该向其添加列表项(Item),主要代码如下:

```java
public class MainActivity extends Activity {
    private RecyclerView myRecyclerView;
    private ItemAdapter myItemAdapter;
    @Override
    protected void onCreate(Bundle savedInstanceState) {
        super.onCreate(savedInstanceState);
        setContentView(R.layout.activity_main);
        myRecyclerView = (RecyclerView) findViewById(R.id.myRecyclerView);
        List<Item> myList = new ArrayList<>(); //存储列表项数据
        for (int i = 0; i < 100; i++) {
            Item myItem = new Item();
            myItem.setId(i);
            myList.add(myItem);
        }
        myItemAdapter = new ItemAdapter(myList);
        myRecyclerView.setAdapter(myItemAdapter);
        myRecyclerView.setLayoutManager(new LinearLayoutManager(this));
    }
    private class ItemHolder extends RecyclerView.ViewHolder {
        private TextView myTextView;
        public ItemHolder(View itemView) {
            super(itemView);
            myTextView = (TextView) itemView;
        }
    }
    public class Item {
        private int myID;
        public int getId() { return myID; }
        public void setId(int id) { this.myID = id; }
    }
    private class ItemAdapter extends RecyclerView.Adapter<ItemHolder> {
        private List<Item> myItems;
        public ItemAdapter(List<Item> items) { myItems = items; }
        @Override
        public ItemHolder onCreateViewHolder(ViewGroup viewGroup, int i) {
            LayoutInflater myLayoutInflater = LayoutInflater.from(MainActivity.this);
            View myView = myLayoutInflater.inflate(android.R.layout.simple_list_item_1,
                    viewGroup, false);
            return new ItemHolder(myView);
        }
        @Override
        public void onBindViewHolder(ItemHolder viewHolder, int i) {
            Item myItem = myItems.get(i);
            viewHolder.myTextView.setTextSize(20);
```

```
        viewHolder.myTextView.setText("编号:" + String.valueOf(myItem.getId()));
    }
    @Override
    public int getItemCount() { return myItems.size(); }
} }
```

上面这段代码在 MyCode\MySample266\app\src\main\java\com\bin\luo\mysample\MainActivity.java 文件中。在这段代码中，Item 类用于设置 RecyclerView 的列表项内容，ItemHolder 用于承载每个列表项，ItemAdapter 用于管理 RecyclerView 的列表项。此实例的完整项目在 MyCode\MySample266 文件夹中。

019　使用 ExpandableListView 分类显示文件

此实例主要通过使用 ExpandableListView 和自定义 BaseExpandableListAdapter，实现按照存储卡上的文件后缀名进行分类。当实例运行之后，将显示三组文件类别名称，如图 019-1 的左图所示。单击"音乐"组名（父列表项），将在子列表中显示存储卡上的所有音乐文件，如图 019-1 的右图所示。单击其他组名将实现类似的功能。

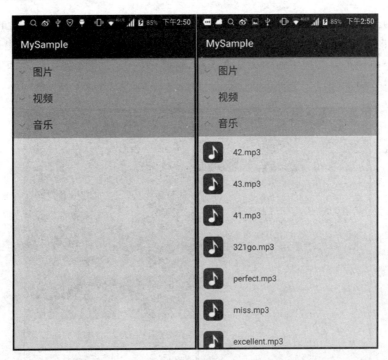

图　019-1

主要代码如下：

```
public class MainActivity extends Activity {
    @Override
    protected void onCreate(Bundle savedInstanceState) {
        super.onCreate(savedInstanceState);
        setContentView(R.layout.activity_main);
        HashMap<String,List<File>> myFiles = new HashMap();
```

```
        MyUtils.getData(this,myFiles);
        MyAdapter myAdapter = new MyAdapter(this,myFiles);
        ExpandableListView myExpandableListView =
                (ExpandableListView) findViewById(R.id.myExpandableListView);
        myExpandableListView.setAdapter(myAdapter);
    }
}
```

上面这段代码在 MyCode\MySample756\app\src\main\java\com\bin\luo\mysample\MainActivity.java 文件中。在这段代码中，myExpandableListView.setAdapter(myAdapter)表示使用自定义适配器 MyAdapter 的实例在 ExpandableListView 上加载文件组名列表及其分类文件子列表，关于 MyAdapter 的详细内容请参考 MyCode\MySample756\app\src\main\java\com\bin\luo\mysample\MyAdapter.java 文件。此外，访问外部存储卡需要在 AndroidManifest.xml 文件中添加 <uses-permission android:name="android.permission.READ_EXTERNAL_STORAGE"/>权限。此实例的完整项目在 MyCode\MySample756 文件夹中。

020 使用 NestedScrollView 实现嵌套滚动

此实例主要通过使用 NestedScrollView 实现工具栏和其他控件之间支持相同的嵌套滚动效果。当实例运行之后，向下拖动工具栏或下面的 NestedScrollView 大文本区域，将拖出图像，如图 020-1 的右图所示。向上拖动工具栏或下面的大文本区域，将隐藏图像，如图 020-1 的左图所示。当然，图像和文本可以在垂直方向的任意位置停留。

图　020-1

主要代码如下：

```
public class MainActivity extends AppCompatActivity {
    @Override
```

```
protected void onCreate(@Nullable Bundle savedInstanceState) {
    super.onCreate(savedInstanceState);
    setContentView(R.layout.activity_main);
    initView();
}
private void initView() {
    final Toolbar myToolbar = (Toolbar) findViewById(R.id.appbar_layout_toolbar);
    myToolbar.inflateMenu(R.menu.mymenu);
    myToolbar.setOnMenuItemClickListener(new Toolbar.OnMenuItemClickListener() {
        @Override
        public boolean onMenuItemClick(MenuItem item) {    //响应单击搜索按钮(菜单)
            if (item.getItemId() == R.id.mysearchmenu) {
                //获取 ToolBar 上的输入框内容
                EditText myEditCity = (EditText) myToolbar.findViewById(R.id.myEditCity);
                String myCity = myEditCity.getText().toString();
                Toast.makeText(getApplicationContext(), "搜索条件是: " + myCity,
                        Toast.LENGTH_SHORT).show();
            }
            return false;
        } });
} }
```

上面这段代码在 MyCode\MySample485\app\src\main\java\com\bin\luo\mysample\MainActivity.java 文件中。在这段代码中，setContentView(R.layout.activity_main)用于加载布局文件 activity_main.xml，NestedScrollView 就在该布局文件中，该布局文件的主要内容如下：

```xml
<android.support.v4.widget.NestedScrollView
    android:layout_width="match_parent"
    android:layout_height="match_parent"
    app:layout_behavior="@string/appbar_scrolling_view_behavior">
    <TextView
        android:layout_width="wrap_content"
        android:layout_height="wrap_content"
        android:layout_margin="18dp"
        android:text="重庆是国家历史文化名城.1189 年,宋光宗赵惇先封恭王再即帝位,自诩双重喜庆,重庆由此得名.重庆既是红岩精神起源地,又是巴渝文化发祥地,"火锅""吊脚楼"等影响深远；在 3000 余年历史中,曾三为国都,四次筑城,史称"巴渝"；抗战时期为国民政府陪都."
        android:textSize="20dp"/>
</android.support.v4.widget.NestedScrollView>
```

上面这段代码在 MyCode\MySample485\app\src\main\res\layout\activity_main.xml 文件中。在这段代码中，app:layout_scrollFlags="scroll|exitUntilCollapsed"表示支持折叠风格的滚动效果。此外，使用此实例的相关控件需要在 gradle 中引入 compile 'com.android.support:design:25.0.1'依赖项。此实例的完整项目在 MyCode\MySample485 文件夹中。

021 在 ViewPager 轮播时实现立体旋转切换

此实例主要通过在 setPageTransformer()方法中使用 setPivotX()方法设置锚点和使用 setRotationY()方法设置绕 Y 轴的旋转角度，从而实现 ViewPager 在轮播图像时，前后两幅图像呈现立体式的切换效果。当实例运行之后，7 个红色的圆点表示总共有 7 幅图像参与轮播，淡红色的圆点表示当前显示的图像；使用手指向左（或右）滑动，则前后两幅图像将根据锚点和绕 Y 轴的旋转角度呈现一个动态变化的立体，直到完全显示（或消失），效果分别如图 021-1 的左图和右图所示。

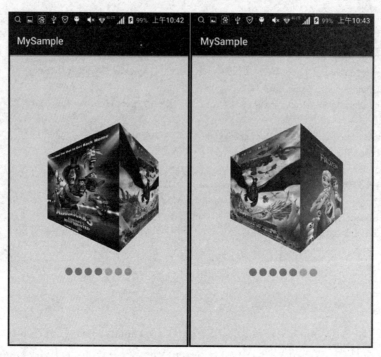

图 021-1

主要代码如下：

```java
public class MainActivity extends Activity {
 public List<ImageView> myImageViews;
 public LinearLayout myPointer;
 public MyAdapter myAdapter;
 @Override
 protected void onCreate(Bundle savedInstanceState) {
  super.onCreate(savedInstanceState);
  setContentView(R.layout.activity_main);
  myPointer = (LinearLayout) findViewById(R.id.myPointer);
  InitData();
  myAdapter = new MyAdapter(myImageViews, myPointer);
  ViewPager myViewPager = (ViewPager)findViewById(R.id.myViewPager);
  myViewPager.setAdapter(myAdapter);
  myViewPager.setCurrentItem((myImageViews.size()) * 100);
  myViewPager.setOnPageChangeListener(myAdapter);
  myViewPager.setPageTransformer(true, new ViewPager.PageTransformer(){
   @Override
   public void transformPage(View page, float position) {
    int myWidth = page.getWidth();
    int myPivotX = 0;
    if (position <= 1 && position > 0) {            //向右滚动
     myPivotX = 0;
    } else if (position < 0 && position >= -1) {     //向左滚动
     myPivotX = myWidth;
    }
```

```
        page.setPivotX(myPivotX);              //设置X轴的锚点
        page.setRotationY(90f * position);     //设置绕Y轴旋转的角度
    }});
}
private void InitData() {
    Integer[] MyImages = {R.mipmap.myimage1, R.mipmap.myimage2,
            R.mipmap.myimage3, R.mipmap.myimage4, R.mipmap.myimage5,
            R.mipmap.myimage6, R.mipmap.myimage7};
    myImageViews = new ArrayList<ImageView>();
    LinearLayout.LayoutParams myLayoutParams = new LinearLayout.LayoutParams(
                    LinearLayout.LayoutParams.WRAP_CONTENT,
                    LinearLayout.LayoutParams.WRAP_CONTENT);
    myLayoutParams.leftMargin = 15;
    myLayoutParams.topMargin = 2;
    for (int i = 0; i < MyImages.length; i++) {
        ImageView myImageView = new ImageView(this);
        myImageView.setBackgroundResource(MyImages[i]);
        myImageViews.add(myImageView);
        ImageView myPointView = new ImageView(this);
        myPointView.setImageResource(R.drawable.mypointer_inactive);
        myPointView.setLayoutParams(myLayoutParams);
        myPointer.addView(myPointView);
    }
    myPointer.getChildAt(0).setBackgroundResource(R.drawable.mypointer_active);
}}
```

上面这段代码在 MyCode\MySample730\app\src\main\java\com\bin\luo\mysample\MainActivity.java 文件中。在这段代码中，myViewPager.setAdapter(myAdapter)表示使用自定义适配器 MyAdapter 的实例管理 myViewPager 的所有图像，关于 MyAdapter 的详细内容请参考 MyCode\MySample730\app\src\main\java\com\bin\luo\mysample\MyAdapter.java 文件。需要说明的是，使用此实例的相关类需要在 gradle 中引入 compile 'com.android.support:design:23.3.0' 依赖项。此实例的完整项目在 MyCode\MySample730 文件夹中。

022　自定义 selector 以透明前景切换控件

此实例主要通过设置 CardView 控件的 android:foreground 属性为自定义 selector，从而实现在单击 CardView 控件时新增透明前景层的效果。当实例运行之后，单击 CardView 控件，在按下该控件时，该控件新增浅绿色的（半）透明层，如图 022-1 的右图所示；在离开该控件时，移除该层（恢复）至正常状态，如图 022-1 的左图所示。

主要代码如下：

```
<android.support.v7.widget.CardView
    xmlns:card_view = "http://schemas.android.com/apk/res-auto"
    android:layout_width = "300dp"
    android:layout_height = "wrap_content"
    android:layout_centerInParent = "true"
    android:clickable = "true"
    android:foreground = "@drawable/myselector"
    card_view:cardBackgroundColor = "#00EEEE"
```

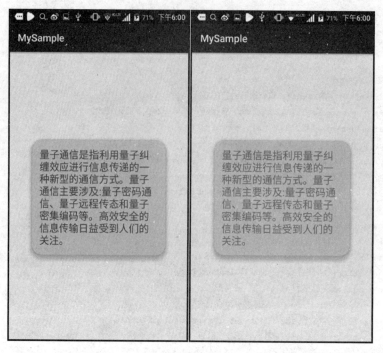

图 022-1

```
        card_view:cardCornerRadius = "20dp"
        card_view:cardElevation = "10dp"
        card_view:cardPreventCornerOverlap = "true"
        card_view:cardUseCompatPadding = "true"
        card_view:contentPadding = "15dp">
    < TextView android:layout_width = "match_parent"
            android:layout_height = "wrap_content"
            android:text = "量子通信是指利用量子纠缠效应进行信息传递的一种新型的通信方式.量子通信
主要涉及:量子密码通信、量子远程传态和量子密集编码等.高效安全的信息传输日益受到人们的关注."
            android:textSize = "20dp"/>
</android.support.v7.widget.CardView>
```

上面这段代码在 MyCode\MySample572\app\src\main\res\layout\activity_main.xml 文件中。在这段代码中,android:foreground="@drawable/myselector"用于设置具有半透明浅绿色前景层的自定义 selector。自定义 selector 的主要代码如下：

```
<?xml version = "1.0" encoding = "utf - 8"?>
< selector xmlns:android = "http://schemas.android.com/apk/res/android">
  < item android:state_pressed = "true">
    < shape >< solid android:color = "@color/myColor" /></ shape >
  </ item >
  < item >
    < shape >< solid android:color = "@android:color/transparent" /></ shape >
  </ item >
</ selector >
```

上面这段代码在 MyCode\MySample572\app\src\main\res\drawable\myselector.xml 文件中。在这段代码中,< item android:state_pressed="true">表示在控件被按下时添加此标签中的半透明浅

绿色前景层。< solid android:color = "@color/myColor"/>用于指定前景的颜色是 myColor,myColor 的颜色值在 colors. xml 文件中定义,主要代码如下:

```
<?xml version = "1.0" encoding = "utf - 8"?>
< resources >
    < color name = "colorPrimary"> # 3F51B5 </color>
    < color name = "colorPrimaryDark"> # 303F9F </color>
    < color name = "colorAccent"> # FF4081 </color>
    < color name = "myColor"> # 5500FF00 </color>
</resources>
```

上面这段代码在 MyCode\MySample572\app\src\main\res\values\colors. xml 文件中。此外,使用此实例的相关控件需要在 gradle 中引入 compile 'com. android. support:cardview-v7:25.3.1'依赖项。此实例的完整项目在 MyCode\MySample572 文件夹中。

023　使用 ViewOutlineProvider 裁剪控件外形

此实例主要通过创建圆形风格的 ViewOutlineProvider 自定义样式,并使用 setOutlineProvider()方法和 setClipToOutline()方法,从而实现将 TextView 控件裁剪成圆形。当实例运行之后,单击"显示裁剪控件"按钮,则 TextView 控件呈现圆形裁剪效果,如图 023-1 的左图所示;单击"显示原始控件"按钮,则 TextView 控件呈现正常显示效果,如图 023-1 的右图所示。

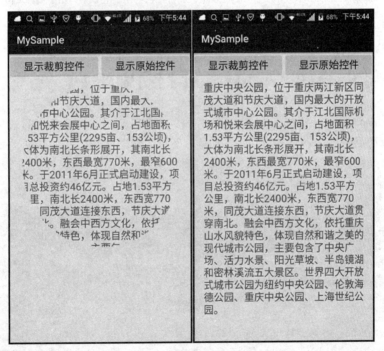

图　023-1

主要代码如下:

```
public class MainActivity extends Activity {
    TextView myTextView;
    ViewOutlineProvider myViewOutlineProvider;
```

```
@Override
protected void onCreate(Bundle savedInstanceState) {
 super.onCreate(savedInstanceState);
 setContentView(R.layout.activity_main);
 myTextView = (TextView)findViewById(R.id.myTextView);
}
public void onClickmyBtn1(View v) {            //响应单击"显示裁剪控件"按钮
 //根据控件宽度设置圆形裁剪
 myViewOutlineProvider = new ViewOutlineProvider() {
  @Override
  public void getOutline(View view, Outline outline) {
   outline.setOval(20, 20, view.getWidth(),view.getWidth());
  } };
 myTextView.setOutlineProvider(myViewOutlineProvider);
 myTextView.setClipToOutline(true);
}
public void onClickmyBtn2(View v) {            //响应单击"显示原始控件"按钮
 myTextView.setOutlineProvider(null);
 //myTextView.setClipToOutline(false);
} }
```

上面这段代码在 MyCode\MySample560\app\src\main\java\com\bin\luo\mysample\MainActivity.java 文件中。在这段代码中，myViewOutlineProvider = new ViewOutlineProvider(){}用于创建圆形样式的 ViewOutlineProvider。myTextView.setOutlineProvider(myViewOutlineProvider) 和 myTextView.setClipToOutline（true）表示使用 myViewOutlineProvider 提供的圆形裁剪 myTextView 控件。此实例的完整项目在 MyCode\MySample560 文件夹中。

024 使用 SwipeRefreshLayout 扫动刷新 UI

此实例主要通过使用 SwipeRefreshLayout 控件，实现扫动手势刷新 UI 的效果。当实例运行之后，手指按住屏幕从上向下滑动再松开，即会出现一个网络数据更新风格的转圈动画；间隔 3s 之后即会显示当前时间，效果分别如图 024-1 的左图和右图所示。

主要代码如下：

```
<android.support.v4.widget.SwipeRefreshLayout
    android:layout_width="match_parent"
    android:layout_height="match_parent"
    xmlns:android="http://schemas.android.com/apk/res/android"
android:id="@+id/mySwipeRefreshLayout">
</android.support.v4.widget.SwipeRefreshLayout>
```

上面这段代码在 MyCode\MySample502\app\src\main\res\layout\activity_main.xml 文件中。在这段代码中，SwipeRefreshLayout 控件通常只接收一个子控件，即需要刷新的控件。它使用一个侦听机制来通知拥有该控件的监听器有刷新事件发生，一旦监听器接收到该事件，就处理刷新动作对应的代码，主要代码如下：

```
public class MainActivity extends AppCompatActivity {
 @Override
 protected void onCreate(@Nullable Bundle savedInstanceState) {
```

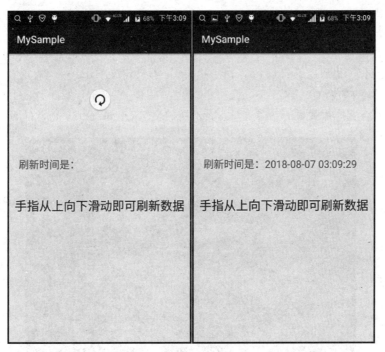

图 024-1

```
super.onCreate(savedInstanceState);
setContentView(R.layout.activity_main);
final SwipeRefreshLayout mySwipeRefreshLayout =
        (SwipeRefreshLayout) findViewById(R.id.mySwipeRefreshLayout);
final TextView myTextView = (TextView) findViewById(R.id.myTextView);
mySwipeRefreshLayout.setOnRefreshListener(
                    new SwipeRefreshLayout.OnRefreshListener(){
 @Override
 public void onRefresh() {
  mySwipeRefreshLayout.setRefreshing(true);
  (new Handler()).postDelayed(new Runnable() {
   @Override
   public void run() {
    mySwipeRefreshLayout.setRefreshing(false);
    //设置日期时间格式
    SimpleDateFormat myDateFormat =
                    new SimpleDateFormat("yyyy-MM-dd hh:mm:ss");
    //获取当前日期时间
    String myDateTime = myDateFormat.format(new java.util.Date());
    myTextView.setText(myDateTime);
  } }, 3000); } });
} }
```

上面这段代码在 MyCode\MySample502\app\src\main\java\com\bin\luo\mysample\MainActivity.java 文件中。在这段代码中，mySwipeRefreshLayout.setRefreshing(true)表示启动转圈刷新动画。mySwipeRefreshLayout.setRefreshing(false)表示停止转圈刷新动画。需要说明的是，使用此实例的相关控件需要在 gradle 中引入 compile 'com.android.support:design:25.0.1'依赖项。此实例的完整项目在 MyCode\MySample502 文件夹中。

025　使用自定义 Behavior 实现滑动遮盖效果

此实例主要通过使用 CoordinatorLayout.Behavior 类自定义 Behavior，实现滑动遮盖效果。当实例运行之后，TextView 控件（文本）在 ImageView 控件（图像）的下面，如图 025-1 的左图所示。向上滑动 TextView 控件（文本），则将遮盖 ImageView 控件（图像），如图 025-1 的右图所示。

图　025-1

主要代码如下：

```java
public class MyCustomBehavior extends CoordinatorLayout.Behavior<View> {
    int myHeaderHeight = 600;                    //垂直可遮盖范围
    public MyCustomBehavior(Context context, AttributeSet attributeSet){
        super(context,attributeSet);
    }
    @Override
    public boolean onLayoutChild(CoordinatorLayout parent,
                                    View child, int layoutDirection) {
        CoordinatorLayout.LayoutParams myLayoutParams =
                    (CoordinatorLayout.LayoutParams) child.getLayoutParams();
        if(myLayoutParams!= null && myLayoutParams.height ==
                        CoordinatorLayout.LayoutParams.MATCH_PARENT){
            child.layout(0,0,parent.getWidth(),parent.getHeight());
            child.setTranslationY(myHeaderHeight);
            return true;
        }
        return super.onLayoutChild(parent, child, layoutDirection);
    }
    @Override
    public boolean onStartNestedScroll(CoordinatorLayout coordinatorLayout,
        View child, View directTargetChild, View target, int nestedScrollAxes) {
        return (nestedScrollAxes & ViewCompat.SCROLL_AXIS_VERTICAL) != 0;
    }
```

```
@Override
public void onNestedPreScroll(CoordinatorLayout coordinatorLayout,
        View child, View target, int dx, int dy, int[] consumed) {
    super.onNestedPreScroll(coordinatorLayout, child, target, dx, dy, consumed);
    //此方法只处理向上滑动
    if(dy < 0){ return; }
    float transY = child.getTranslationY() - dy;
    if(transY > 0){
        child.setTranslationY(transY);
        consumed[1] = dy;
    }
}
@Override
public void onNestedScroll(CoordinatorLayout coordinatorLayout,
        View child, View target, int dxConsumed,
        int dyConsumed, int dxUnconsumed, int dyUnconsumed) {
    super.onNestedScroll(coordinatorLayout, child, target, dxConsumed, dyConsumed, dxUnconsumed, dyUnconsumed);
    //此方法只处理向下滑动
    if(dyUnconsumed > 0){ return; }
    float transY = child.getTranslationY() - dyUnconsumed;
    if(transY > 0 && transY < myHeaderHeight){
        child.setTranslationY(transY);
    }
}}
```

上面这段代码在 MyCode \ MySample496 \ app \ src \ main \ java \ com \ bin \ luo \ mysample \ MyCustomBehavior.java 文件中。此外，使用此实例的相关控件需要在 gradle 中引入 compile 'com.android.support:design：25.0.1'依赖项。此实例的完整项目在 MyCode\MySample496 文件夹中。

026　创建百分比数字跟随进度改变的进度条

此实例主要通过创建自定义进度条 myProgressView，实现进度条的百分比数字位置随着进度的改变而改变。当实例运行之后，进度条马上开始执行进度改变的动作，效果分别如图 026-1 的左图和右图所示。

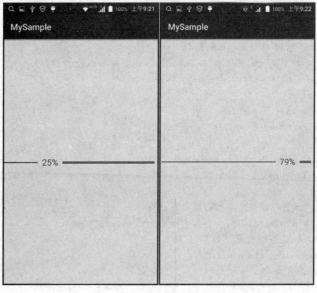

图　026-1

主要代码如下：

```java
public class myProgressView extends ProgressBar {
    protected Paint myPaint = new Paint();
    protected int myTextColor = 0XFF008000;              //百分比数字颜色
    protected int myTextSize = 60;                        //百分比字体大小
    protected int myTextOffset = 60;                      //百分比偏移值
    protected int myLeftHeight = 5;                       //百分比左边的进度条高度
    protected int myLeftColor = 0XFF0000FF;               //百分比左边的进度条颜色
    protected int myRightColor = 0xFFFF0000;              //百分比右边的进度条颜色
    protected int myRightHeight = 15;                     //百分比右边的进度条高度
    protected int mRealWidth;
    protected boolean isDraw = true;
    public myProgressView(Context context,AttributeSet attrs) {
        this(context, attrs, 0);
    }
    public myProgressView(Context context,
                          AttributeSet attrs, int defStyle) {
        super(context, attrs, defStyle);
        setHorizontalScrollBarEnabled(true);
        myPaint.setTextSize(myTextSize);
        myPaint.setColor(myTextColor);
    }
    @Override
    protected synchronized void onMeasure(int widthMeasureSpec,
                                          int heightMeasureSpec){
        int myMode = MeasureSpec.getMode(heightMeasureSpec);
        if (myMode != MeasureSpec.EXACTLY){
            float myHeight = (myPaint.descent() + myPaint.ascent());
            int myExceptHeight = (int)(getPaddingTop()
                    + getPaddingBottom() + Math.max(Math.max(myLeftHeight,
                    myRightHeight), Math.abs(myHeight)));
            heightMeasureSpec = MeasureSpec.makeMeasureSpec(myExceptHeight,
                    MeasureSpec.EXACTLY);
        }
        super.onMeasure(widthMeasureSpec, heightMeasureSpec);
    }
    @Override
    protected synchronized void onDraw(Canvas canvas){
        //平移到(getPaddingLeft(), getHeight()/2)位置,以后坐标都以此为(0,0)
        canvas.translate(getPaddingLeft(), getHeight()/2);
        boolean isNeed = false;
        //当前进度和总值的比例
        float myRatio = getProgress() * 1.0f / getMax();
        //已到达的进度
        float myPosX = (int)(mRealWidth * myRatio);
        String myText = getProgress() + "%";
        //字体的宽度和高度
        float myTextWidth = myPaint.measureText(myText);
        float myTextHeight = (myPaint.descent() + myPaint.ascent()) / 2;
        //如果到达最后,则未到达的进度条不需要绘制
        if (myPosX + myTextWidth > mRealWidth){
            myPosX = mRealWidth - myTextWidth;
```

```
     isNeed = true;
   }
   float myEndX = myPosX - myTextOffset / 2;
   if (myEndX > 0){                    //绘制已完成的进度条
    myPaint.setColor(myLeftColor);
    myPaint.setStrokeWidth(myLeftHeight);
    canvas.drawLine(0, 0, myEndX, 0, myPaint);
   }
   if (isDraw){
    myPaint.setColor(myTextColor);
    canvas.drawText(myText, myPosX, - myTextHeight, myPaint);
   }
   if (!isNeed){                       //绘制未完成的进度条
    float start = myPosX + myTextOffset / 2 + myTextWidth;
    myPaint.setColor(myRightColor);
    myPaint.setStrokeWidth(myRightHeight);
    canvas.drawLine(start, 0, mRealWidth, 0, myPaint);
   } }
  @Override
  protected void onSizeChanged(int w, int h, int oldw, int oldh){
   super.onSizeChanged(w, h, oldw, oldh);
   mRealWidth = w - getPaddingRight() - getPaddingLeft();
} }
```

上面这段代码在 MyCode \ MySample184 \ app \ src \ main \ java \ com \ bin \ luo \ mysample \ myProgressView.java 文件中。由于自定义进度条 myProgressView 继承自 ProgressBar，因此它的进度改变操作仍然与默认的 ProgressBar 相同，主要代码如下：

```
public class MainActivity extends Activity {
 private myProgressView myProgress;
 private int MAX_PROGRESS = 100;
 @Override
 protected void onCreate(Bundle savedInstanceState) {
  super.onCreate(savedInstanceState);
  setContentView(R.layout.activity_main);
  myProgress = (myProgressView)findViewById(R.id.myProgress);
  myProgress.setMax(MAX_PROGRESS);                //设置进度条最大值
  new Thread() {                                  //创建动态更新进度条的线程
   @Override
   public void run() {
    int Progress = 0;
    while (Progress < MAX_PROGRESS) {
     try {
      Thread.sleep(MAX_PROGRESS);
      Progress++;
      myProgress.incrementProgressBy(1);
     } catch (InterruptedException e) { e.printStackTrace(); } }
  }.start();
} }
```

上面这段代码在 MyCode \ MySample184 \ app \ src \ main \ java \ com \ bin \ luo \ mysample \ MainActivity.java 文件中。此实例的完整项目在 MyCode\MySample184 文件夹中。

027 使用 BottomNavigationBar 实现底部导航

此实例主要演示了使用 BottomNavigationBar 控件,从而在屏幕底部实现菜单风格的导航效果。当实例运行之后,在屏幕底部将显示 5 个导航菜单项,并且每个菜单项同时显示图标和文本;单击第一个菜单项,将显示该菜单项对应的电影海报图像,如图 027-1 的左图所示;单击第二个菜单项,也将显示该菜单项对应的电影海报图像,如图 027-1 的右图所示。单击其他菜单项将实现类似的功能。

图 027-1

主要代码如下:

```java
public class MainActivity extends AppCompatActivity implements
BottomNavigationBar.OnTabSelectedListener {
 private ArrayList<Fragment> myFragments;
 @Override
 protected void onCreate(Bundle savedInstanceState) {
  super.onCreate(savedInstanceState);
  setContentView(R.layout.activity_main);
  BottomNavigationBar myBottomNavigationBar =
      (BottomNavigationBar) findViewById(R.id.myBottomNavigationBar);
  myBottomNavigationBar.setMode(BottomNavigationBar.MODE_FIXED);
  myBottomNavigationBar.setBackgroundStyle(
                 BottomNavigationBar.BACKGROUND_STYLE_RIPPLE);
  //在底部导航栏上添加菜单选项
  myBottomNavigationBar.addItem(new BottomNavigationItem(R.mipmap.mymovie,
         "我是传奇").setActiveColorResource(R.color.colorPrimaryDark))
      .addItem(new BottomNavigationItem(R.mipmap.mymovie,
         "银翼杀手").setActiveColorResource(R.color.colorPrimaryDark))
      .addItem(new BottomNavigationItem(R.mipmap.mymovie,
         "三百勇士").setActiveColorResource(R.color.colorPrimaryDark))
      .addItem(new BottomNavigationItem(R.mipmap.mymovie,
         "功夫熊猫").setActiveColorResource(R.color.colorPrimaryDark))
      .addItem(new BottomNavigationItem(R.mipmap.mymovie,
```

```
            "怒火救援").setActiveColorResource(R.color.colorPrimaryDark))
            .setFirstSelectedPosition(0)
            .initialise();
    myFragments = getMyFragments();
    setDefaultFragment();
    myBottomNavigationBar.setTabSelectedListener(this);
}
private void setDefaultFragment() {                    //设置默认的 Fragment
    FragmentManager myHomeFragment = getSupportFragmentManager();
    FragmentTransaction myTransaction = myHomeFragment.beginTransaction();
    myTransaction.replace(R.id.myFrameLayout,
                    myFragment1.newInstance("我是传奇"));
    myTransaction.commit();
}
private ArrayList<Fragment> getMyFragments() {         //向集合中添加 Fragment
    ArrayList<Fragment> myFragments = new ArrayList<>();
    myFragments.add(myFragment1.newInstance("我是传奇"));
    myFragments.add(myFragment2.newInstance("银翼杀手"));
    myFragments.add(myFragment3.newInstance("三百勇士"));
    myFragments.add(myFragment4.newInstance("功夫熊猫"));
    myFragments.add(myFragment5.newInstance("怒火救援"));
    return myFragments;
}
@Override
public void onTabSelected(int position) {              //设置选中的 Fragment
    if (myFragments != null) {
        if (position < myFragments.size()) {
            FragmentManager myManager = getSupportFragmentManager();
            FragmentTransaction myTransaction = myManager.beginTransaction();
            Fragment myFragment = myFragments.get(position);
            if (myFragment.isAdded()) {
                myTransaction.replace(R.id.myFrameLayout, myFragment);
            } else {
                myTransaction.add(R.id.myFrameLayout, myFragment);
            }
            myTransaction.commitAllowingStateLoss();
}}}
@Override
public void onTabUnselected(int position) {            //设置未选中的 Fragment
    if (myFragments != null) {
        if (position < myFragments.size()) {
            FragmentManager myManager = getSupportFragmentManager();
            FragmentTransaction myTransaction = myManager.beginTransaction();
            Fragment myFragment = myFragments.get(position);
            myTransaction.remove(myFragment);
            myTransaction.commitAllowingStateLoss();
}}}
@Override
public void onTabReselected(int position) { }
}
```

上面这段代码在 MyCode\MySample434\app\src\main\java\com\bin\luo\mysample\MainActivity.java 文件中。在这段代码中,myTransaction.replace(R.id.myFrameLayout,myFragment)用于在 myFrameLayout 布局管理器中添加 Frame,即在选择不同的菜单时,在 myFrameLayout 布局管理器中添加不同的 Frame。主要代码如下:

```
< LinearLayout android:layout_width = "match_parent"
               android:layout_height = "match_parent"
               android:orientation = "vertical" >
  < FrameLayout android:background = "#000"
                android:layout_gravity = "bottom"
                android:id = "@ + id/myFrameLayout"
                android:layout_width = "match_parent"
                android:layout_height = "0dp"
                android:layout_weight = "1"/>
  < com.ashokvarma.bottomnavigation.BottomNavigationBar
                android:id = "@ + id/myBottomNavigationBar"
                android:layout_width = "match_parent"
                android:layout_height = "wrap_content"
                android:background = "#00FFFF"
                android:layout_gravity = "bottom"/>
</LinearLayout >
```

上面这段代码在 MyCode\MySample434\app\src\main\res\layout\activity_main.xml 文件中。BottomNavigationBar 是一个底部导航菜单控件，它通过在其中添加菜单项实现导航效果；通常每个菜单项对应一个 Fragment；即 myFragment1、myFragment2、myFragment3、myFragment4、myFragment5。需要说明的是，使用 BottomNavigationBar 控件需要在 gradle 中引入 compile 'com.ashokvarma.android：bottom-navigation-bar：1.2.0' 依赖项。此实例的完整项目在 MyCode\MySample434 文件夹中。

028　使用两个 Spinner 实现省市两级联动

此实例主要通过使用两个 Spinner 控件实现省市两级列表框的联动。当实例运行之后，在第一个列表框中选择"山西省"，将在第二个列表框中显示"山西省"所辖的地区，选择其他选项将实现类似的功能，效果分别如图 028-1 的左图和右图所示。

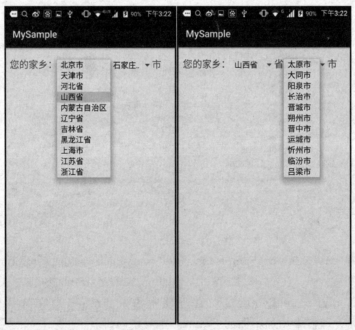

图　028-1

主要代码如下：

```java
public class MainActivity extends Activity {
    List<String> myList1 = new ArrayList<String>();
    List<String> myList2 = new ArrayList<String>();
    JSONObject myJson;
    static String myData = "{\"北京市\":[\"北京市\"],\"天津市\":[\"天津市\"],\"河北省\":[\"石家庄市\",\"唐山市\",\"秦皇岛市\",\"邯郸市\",\"邢台市\",\"保定市\",\"张家口市\",\"承德市\",\"沧州市\",\"廊坊市\",\"衡水市\"],\"山西省\":[\"太原市\",\"大同市\",\"阳泉市\",\"长治市\",\"晋城市\",\"朔州市\",\"晋中市\",\"运城市\",\"忻州市\",\"临汾市\",\"吕梁市\"],\"内蒙古自治区\":[\"呼和浩特市\",\"包头市\",\"乌海市\",\"赤峰市\",\"通辽市\",\"鄂尔多斯市\",\"呼伦贝尔市\",\"巴彦淖尔市\",\"乌兰察布市\",\"兴安盟\",\"锡林郭勒盟\",\"阿拉善盟\"],\"辽宁省\":[\"沈阳市\",\"大连市\",\"鞍山市\",\"抚顺市\",\"本溪市\",\"丹东市\",\"锦州市\",\"营口市\",\"阜新市\",\"辽阳市\",\"盘锦市\",\"铁岭市\",\"朝阳市\",\"葫芦岛市\"],\"吉林省\":[\"长春市\",\"吉林市\",\"四平市\",\"辽源市\",\"通化市\",\"白山市\",\"松原市\",\"白城市\",\"延边朝鲜族自治州\"],\"黑龙江省\":[\"哈尔滨市\",\"齐齐哈尔市\",\"鸡西市\",\"鹤岗市\",\"双鸭山市\",\"大庆市\",\"伊春市\",\"佳木斯市\",\"七台河市\",\"牡丹江市\",\"黑河市\",\"绥化市\",\"大兴安岭地区\"],\"上海市\":[\"上海市\"],\"江苏省\":[\"南京市\",\"无锡市\",\"徐州市\",\"常州市\",\"苏州市\",\"南通市\",\"连云港市\",\"淮安市\",\"盐城市\",\"扬州市\",\"镇江市\",\"泰州市\",\"宿迁市\"],\"浙江省\":[\"杭州市\",\"宁波市\",\"温州市\",\"嘉兴市\",\"湖州市\",\"绍兴市\",\"金华市\",\"衢州市\",\"舟山市\",\"台州市\",\"丽水市\"]}";
    @Override
    protected void onCreate(Bundle savedInstanceState) {
        super.onCreate(savedInstanceState);
        setContentView(R.layout.activity_main);
        try {
            //将 JSON 格式数据转换为 JSONObject 对象
            myJson = new JSONObject(myData);
            Spinner mySpinner1 = (Spinner) findViewById(R.id.mySpinner1);
            Spinner mySpinner2 = (Spinner) findViewById(R.id.mySpinner2);
            InitProvinceList(myList1);
            mySpinner1.setAdapter(new ArrayAdapter<String>(this,
                                android.R.layout.simple_spinner_item, myList1));
            final ArrayAdapter<String> myAdapter = new ArrayAdapter<String>(this,
                                android.R.layout.simple_spinner_item, myList2);
            mySpinner2.setAdapter(myAdapter);
            //对一级 Spinner 进行 Item 选中监听
            mySpinner1.setOnItemSelectedListener(
                                new AdapterView.OnItemSelectedListener() {
                @Override
                public void onItemSelected(AdapterView<?> parent,
                                View view, int position, long id) {
                    try {
                        myList2 = new ArrayList<String>();
                        //根据所选一级 Spinner 名称加载二级 Spinner 数据
                        InitDistrictList(myList1.get(position), myList2);
                        //将二级 Spinner 数据清空并重新加载
                        myAdapter.clear();
                        myAdapter.addAll(myList2);
                    } catch (Exception e) { }
                }
                @Override
                public void onNothingSelected(AdapterView<?> parent) { }
            });
        } catch (Exception e) { }
```

```
}
//初始化一级 Spinner 数据
public List < String > InitProvinceList(List < String > provinceList)
                                                    throws Exception {
    Iterator < String > myArray1 = myJson.keys();
    while (myArray1.hasNext()) { provinceList.add(myArray1.next()); }
    return provinceList;
}
//初始化二级 Spinner 数据
public List < String > InitDistrictList(String province,
                    List < String > districtList) throws Exception {
    //根据所选一级 Spinner 名称加载对应的二级 Spinner 数据
    JSONArray myArray2 = myJson.getJSONArray(province);
    for (int i = 0; i < myArray2.length(); i++) {
        districtList.add(myArray2.getString(i));
    }
    return districtList;
} }
```

上面这段代码在 MyCode \ MySample798 \ app \ src \ main \ java \ com \ bin \ luo \ mysample \ MainActivity.java 文件中。此实例的完整项目在 MyCode\MySample798 文件夹中。

029 使用 BackgroundColorSpan 定制背景

此实例主要通过使用 ForegroundColorSpan 和 BackgroundColorSpan，实现在 TextView 中定制部分内容的前景颜色和背景颜色。当实例运行之后，默认的文本显示效果如图 029-1 的左图所示。单击"显示定制的文本颜色和背景颜色"按钮，则显示经过定制的文本颜色和背景颜色，如图 029-1 的右图所示。

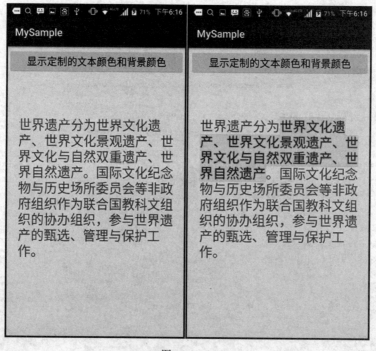

图 029-1

主要代码如下：

```java
public class MainActivity extends Activity {
    SpannableString mySpannableString;
    TextView myTextView;
    @Override
    protected void onCreate(Bundle savedInstanceState) {
        super.onCreate(savedInstanceState);
        setContentView(R.layout.activity_main);
        myTextView = (TextView) findViewById(R.id.myTextView);
        mySpannableString = new SpannableString("世界遗产分为世界文化遗产、" +
                "世界文化景观遗产、世界文化与自然双重遗产、世界自然遗产." +
                "国际文化纪念物与历史场所委员会等非政府组织作为联" +
                "合国教科文组织的协办组织,参与世界遗产的甄选、管理与保护工作.");
        myTextView.setText(mySpannableString);
    }
    public void onClickBtn1(View v) { //响应单击"显示定制的文本颜色和背景颜色"按钮
        //查找"世界文化遗产"在字符串中的索引位置
        int myStart = mySpannableString.toString().indexOf("世界文化遗产");
        int myEnd = myStart + ("世界文化遗产、世界文化景观遗产、" +
                "世界文化与自然双重遗产、世界自然遗产").toString().length();
        mySpannableString.setSpan(new ForegroundColorSpan(Color.BLUE),
                myStart, myEnd, Spanned.SPAN_EXCLUSIVE_EXCLUSIVE); //设置文本颜色
        mySpannableString.setSpan(new BackgroundColorSpan(Color.CYAN),
                myStart, myEnd, Spanned.SPAN_EXCLUSIVE_EXCLUSIVE); //设置背景颜色
        myTextView.setText(mySpannableString);
    }
}
```

上面这段代码在 MyCode \ MySample032 \ app \ src \ main \ java \ com \ bin \ luo \ mysample \ MainActivity.java 文件中。在这段代码中，设置文本前景颜色的 mySpannableString.setSpan(new ForegroundColorSpan(Color.BLUE), myStart, myEnd, Spanned.SPAN_EXCLUSIVE_EXCLUSIVE)的 Color.BLUE 表示文本前景颜色，myStart 表示该颜色作用范围的开始位置，myEnd 表示该颜色作用范围的结束位置。设置文本背景颜色的 mySpannableString.setSpan(new BackgroundColorSpan(Color.CYAN), myStart, myEnd, Spanned.SPAN_EXCLUSIVE_EXCLUSIVE)的 Color.CYAN 表示背景颜色，myStart 表示该颜色作用范围的开始位置，myEnd 表示该颜色作用范围的结束位置。此实例的完整项目在 MyCode\MySample032 文件夹中。

第 2 章

通知栏

030 动态设置当前应用的标题栏文本

此实例主要通过使用 ActionBar 的 setTitle()方法,实现动态设置应用的标题栏文本。当实例运行之后,单击"设置标题栏标题文本一"按钮,则标题栏的标题文本如图 030-1 的左图所示。单击"设置标题栏标题文本二"按钮,则标题栏的标题文本如图 030-1 的右图所示。

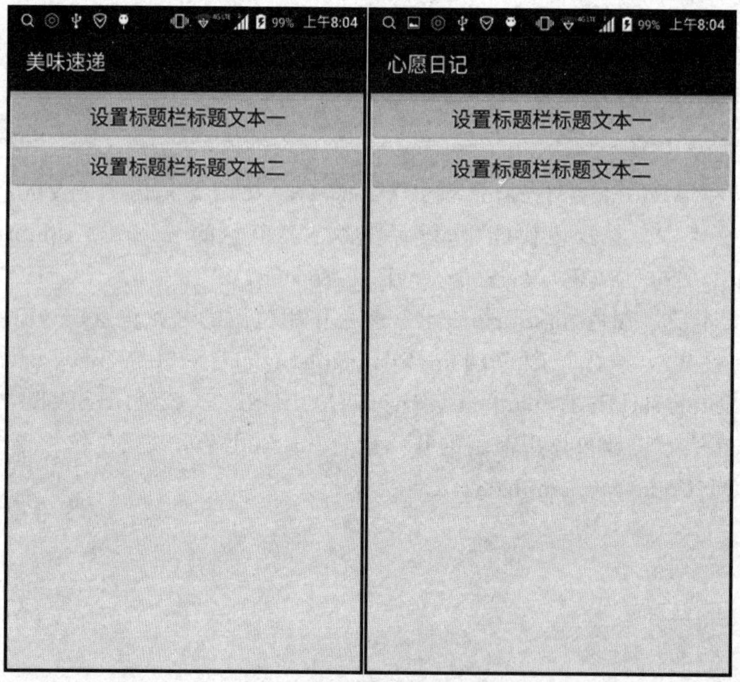

图　030-1

主要代码如下:

```
public void onClickmyBtn1(View v) {      //响应单击"设置标题栏标题文本一"按钮
  ActionBar myActionBar = getSupportActionBar();
  myActionBar.setTitle("美味速递");
}
public void onClickmyBtn2(View v) {      //响应单击"设置标题栏标题文本二"按钮
```

```
ActionBar myActionBar = getSupportActionBar();
myActionBar.setTitle("心愿日记");
}
```

上面这段代码在 MyCode \ MySample517 \ app \ src \ main \ java \ com \ bin \ luo \ mysample \ MainActivity.java 文件中。此外,使用此实例的相关类需要在 gradle 中引入 compile 'com.android.support:design:23.3.0' 依赖项。此实例的完整项目在 MyCode\MySample517 文件夹中。

031　动态设置当前应用的标题栏背景

此实例主要通过使用 ActionBar 的 setBackgroundDrawable()方法,实现动态设置当前应用的标题栏背景图像。当实例运行之后,单击"设置标题栏背景图像一"按钮,标题栏的背景如图 031-1 的左图所示。单击"设置标题栏背景图像二"按钮,标题栏的背景如图 031-1 的右图所示。

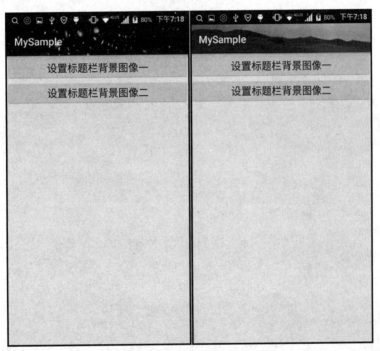

图　031-1

主要代码如下:

```
public void onClickmyBtn1(View v) {      //响应单击"设置标题栏背景图像一"按钮
    ActionBar myActionBar = getSupportActionBar();
    myActionBar.setBackgroundDrawable(getDrawable(R.drawable.myimage1));
}
public void onClickmyBtn2(View v) {      //响应单击"设置标题栏背景图像二"按钮
    ActionBar myActionBar = getSupportActionBar();
    myActionBar.setBackgroundDrawable(getDrawable(R.drawable.myimage2));
}
```

上面这段代码在 MyCode \ MySample516 \ app \ src \ main \ java \ com \ bin \ luo \ mysample \ MainActivity.java 文件中。在这段代码中,setBackgroundDrawable()方法用于根据指定的图像设置

标题栏的背景,也可以使用 setBackgroundDrawable()方法设置透明的标题栏,如下面的代码:

```
myActionBar.setBackgroundDrawable(
            new ColorDrawable(Color.parseColor("#33000000")))
```

此外,使用此实例的相关类需要在 gradle 中引入 compile 'com.android.support:design:23.3.0' 依赖项。此实例的完整项目在 MyCode\MySample516 文件夹中。

032　动态隐藏或显示当前应用标题栏

此实例主要通过使用 getSupportActionBar()方法获取当前应用的 ActionBar,并使用 ActionBar 的 hide()方法和 show()方法,从而实现动态隐藏或显示当前应用的标题栏。当实例运行之后,单击"隐藏标题栏"按钮,标题栏消失,如图 032-1 的左图所示。单击"显示标题栏"按钮,则标题栏出现,如图 032-1 的右图所示。

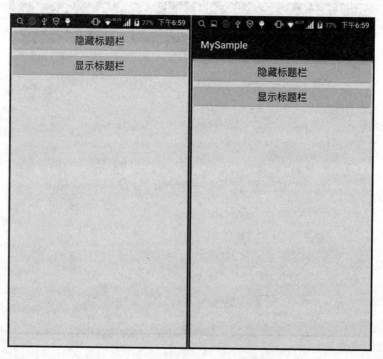

图　032-1

主要代码如下:

```
public void onClickmyBtn1(View v) {          //响应单击"隐藏标题栏"按钮
    ActionBar myActionBar = getSupportActionBar();
    myActionBar.hide();
}
public void onClickmyBtn2(View v) {          //响应单击"显示标题栏"按钮
    ActionBar myActionBar = getSupportActionBar();
    myActionBar.show();
}
```

上面这段代码在 MyCode\MySample515\app\src\main\java\com\bin\luo\mysample\MainActivity.java 文件中。在这段代码中,getSupportActionBar()方法用于获取当前应用的标题栏

ActionBar，ActionBar 的 hide()方法用于隐藏标题栏，ActionBar 的 show()方法用于显示标题栏。此外，使用此实例的相关类需要在 gradle 中引入 compile 'com.android.support:design:23.3.0'依赖项。此实例的完整项目在 MyCode\MySample515 文件夹中。

033　自定义 TextView 创建渐变标题栏

此实例主要通过以 TextView 为基类创建自定义 MyTextView，从而使标题栏的文本呈现从左向右移动的动态渐变效果。当实例运行之后，标题栏文本呈现渐变色，并且渐变色从左向右动态移动，效果分别如图 033-1 的左图和右图所示。

图　033-1

主要代码如下：

```
public class MainActivity extends Activity {
  @Override
  protected void onCreate(Bundle savedInstanceState) {
    super.onCreate(savedInstanceState);
    setContentView(R.layout.activity_main);
    final ActionBar myActionBar = getActionBar();            //获取 ActionBar 对象
myActionBar.setDisplayOptions(ActionBar.DISPLAY_SHOW_CUSTOM);
    myActionBar.setCustomView(R.layout.mylayout);            //自定义 ActionBar 布局
myActionBar.setBackgroundDrawable(
    new ColorDrawable(Color.parseColor("#708090")));        //设置 ActionBar 背景
} }
```

上面这段代码在 MyCode\MySample702\app\src\main\java\com\bin\luo\mysample\MainActivity.java 文件中。在这段代码中，myActionBar.setCustomView(R.layout.mylayout)表示使用 mylayout 布局设置标题栏，mylayout 布局的主要内容如下：

```xml
< com.bin.luo.mysample.MyTextView
    android:layout_width = "match_parent"
    android:layout_height = "wrap_content"
    android:gravity = "center_horizontal"
    android:padding = "2dp"
    android:text = "Android 精彩实例集锦"
    android:textSize = "26dp"/>
```

上面这段代码在 MyCode\MySample702\app\src\main\res\layout\mylayout.xml 文件中。在这段代码中，com.bin.luo.mysample 是实例的包名，在使用自定义 MyTextView 时，一定要修改为自己应用的包名。关于自定义类 MyTextView 的详细内容请参考 MyCode\MySample702\app\src\main\java\com\bin\luo\mysample\MyTextView.java 文件。此实例的完整项目在 MyCode\MySample702 文件夹中。

034　使用自定义布局创建个性化标题栏

此实例主要通过创建自定义布局文件，实现定制应用的标题栏。当实例运行之后，经过定制的应用标题栏的效果如图 034-1 所示。

图　034-1

主要代码如下：

```java
public class MainActivity extends Activity {
    @Override
    protected void onCreate(Bundle savedInstanceState) {
        super.onCreate(savedInstanceState);
        requestWindowFeature(Window.FEATURE_CUSTOM_TITLE);    //指定使用自定义标题
```

```
            setContentView(R.layout.activity_main);
            getWindow().setFeatureInt(Window.FEATURE_CUSTOM_TITLE,
                            R.layout.mylayout);         //设置窗口的自定义标题布局文件
    } }
```

上面这段代码在 MyCode\MySample246\app\src\main\java\com\bin\luo\mysample\MainActivity.java 文件中。在这段代码中，mylayout 表示自定义标题栏布局，该布局的主要内容如下：

```xml
<?xml version="1.0" encoding="utf-8"?>
<LinearLayout xmlns:android="http://schemas.android.com/apk/res/android"
            android:layout_width="match_parent"
            android:layout_height="match_parent"
            android:gravity="center"
            android:orientation="horizontal">
    <ImageView android:layout_width="wrap_content"
            android:layout_height="wrap_content"
            android:layout_marginLeft="10dp"
            android:src="@drawable/myimage2"/>
    <TextView android:layout_width="wrap_content"
            android:layout_height="wrap_content"
            android:layout_gravity="center_vertical"
            android:text="这是自定义的标题栏"
            android:textColor="#000"
            android:textSize="20sp"/>
    <ImageView android:layout_width="wrap_content"
            android:layout_height="wrap_content"
            android:layout_marginLeft="10dp"
            android:src="@drawable/myimage2"/>
</LinearLayout>
```

上面这段代码在 MyCode\MySample246\app\src\main\res\layout\mylayout.xml 文件中。此布局文件与普通的布局文件没有什么不同，但是如果需要使自定义布局文件在标题栏上产生作用，通常应该在 AndroidManifest.xml 文件中修改应用主题，即 android:theme="@style/myTheme"，此 myTheme 在 styles.xml 文件中定义，主要代码如下：

```xml
<resources>
    <!-- Base application theme. -->
    <style name="AppTheme" parent="android:Theme.Material.Light.DarkActionBar">
        <!-- Customize your theme here. -->
    </style>
    <style name="CustomWindowTitleBackground">
        <!-- 设置标题栏背景图像 -->
        <item name="android:background">@drawable/myimage1</item>
    </style>
    <!-- 该样式继承系统的默认样式 -->
    <style name="myTheme" parent="android:Theme">
        <!-- 设置标题栏高度为 50dp -->
        <item name="android:windowTitleSize">50dp</item>
        <item name="android:windowTitleBackgroundStyle">@style/CustomWindowTitleBackground</item>
```

```
</style>
</resources>
```

上面这段代码在 MyCode\MySample246\app\src\main\res\values\styles.xml 文件中。此实例的完整项目在 MyCode\MySample246 文件夹中。

035　在标题栏左侧添加默认的后退按钮

此实例主要通过调用 ActionBar 的 setDisplayHomeAsUpEnabled()方法，并重写 Activity 的 onOptionsItemSelected()方法，实现在 Activity 标题栏的左侧添加后退按钮。当实例运行之后，单击 MainActivity 显示上海夜景图像，将跳转到 SubActivity，此时 SubActivity 标题栏的左侧将有一个后退按钮（左箭头），单击该后退按钮，则返回到 MainActivity，效果分别如图 035-1 的左图和右图所示。

图　035-1

主要代码如下：

```
public class MainActivity extends Activity {
  @Override
  protected void onCreate(Bundle savedInstanceState) {
    super.onCreate(savedInstanceState);
    setContentView(R.layout.activity_main);
    getActionBar().setTitle("MainActivity");
  }
  public void onClickScreen(View v) { //响应单击屏幕,跳转到SubActivity
    Intent myIntent = new Intent(MainActivity.this, SubActivity.class);
    startActivity(myIntent);
  }
}
```

上面这段代码在 MyCode\MySample886\app\src\main\java\com\bin\luo\mysample\MainActivity.java 文件中。在这段代码中，SubActivity 是自定义 Activity，该 Activity 的标题栏左上角包含后退按钮。SubActivity 类的主要代码如下：

```java
public class SubActivity extends Activity {
    @Override
    protected void onCreate(Bundle savedInstanceState) {
        super.onCreate(savedInstanceState);
        setContentView(R.layout.activity_sub);
        ActionBar myActionBar = getActionBar();
        //在标题栏左侧显示后退按钮
        myActionBar.setDisplayHomeAsUpEnabled(true);
        myActionBar.setTitle("SubActivity");
    }
    @Override
    //响应 ActionBar 按钮单击事件
    public boolean onOptionsItemSelected(MenuItem item) {
        //响应单击后退按钮
        if (item.getItemId() == android.R.id.home)
            onBackPressed();
        return false;
    } }
```

上面这段代码在 MyCode\MySample886\app\src\main\java\com\bin\luo\mysample\SubActivity.java 文件中。此外，当新增了 SubActivity，则应该在 AndroidManifest.xml 文件中添加注册代码< activity android:name=".SubActivity"/>。此实例的完整项目在 MyCode\MySample886 文件夹中。

036　在标题栏右侧添加分享按钮分享文本

此实例主要通过使用 ShareActionProvider，实现在标题栏的右侧添加分享功能的按钮。当实例运行之后，在"分享主题："和"分享内容："输入框中输入分享主题和内容，单击右上角的 QQ 按钮（也可以单击右上角的分享按钮，在弹出的下拉菜单中选择其他分享工具），如图 036-1 的左图所示，将弹出"发送到"窗口。在"发送到"窗口选择接收文本的 QQ 好友，如"渝北万博"，将显示"发送给渝北万博"窗口，"发送给渝北万博"窗口将自动显示分享内容，当然也可以在此修改分享内容，如图 036-1 的右图所示，单击"发送"按钮，则将在 QQ 的聊天界面中显示分享内容。

主要代码如下：

```xml
<?xml version = "1.0" encoding = "utf - 8"?>
<menu xmlns:android = "http://schemas.android.com/apk/res/android" >
  < item android:id = "@ + id/myShareMenu"
      android:actionProviderClass = "android.widget.ShareActionProvider"
      android:enabled = "true"
      android:showAsAction = "ifRoom"
      android:title = "分享菜单"/>
</menu >
```

上面这段代码在 MyCode\MySample290\app\src\main\res\menu\mymenu.xml 文件中。在 Java 代码中加载该文件的菜单则可以使用 MenuInflater 的 inflate()方法实现，主要代码如下：

图 036-1

```java
public class MainActivity extends Activity {
  EditText myEditSubject;
  EditText myEditText;
  private ShareActionProvider mShareActionProvider;
  @Override
  protected void onCreate(Bundle savedInstanceState) {
    super.onCreate(savedInstanceState);
    setContentView(R.layout.activity_main);
    myEditSubject = (EditText)findViewById(R.id.myEditSubject);
    myEditText = (EditText)findViewById(R.id.myEditText);
  }
  public boolean onCreateOptionsMenu(Menu menu) {
    MenuInflater myMenuInflater = getMenuInflater();
    myMenuInflater.inflate(R.menu.mymenu, menu);
    //添加分享按钮及其响应事件
    MenuItem myShareMenu = menu.findItem(R.id.myShareMenu);
    mShareActionProvider = (ShareActionProvider) myShareMenu.getActionProvider();
    mShareActionProvider.setShareIntent(getDefaultShareIntent());
    return super.onCreateOptionsMenu(menu);
  }
  //通过 Intent 发送分享信息
  private Intent getDefaultShareIntent() {
    Intent myIntent = new Intent(Intent.ACTION_SEND);
    myIntent.setType("text/plain");
    myIntent.putExtra(Intent.EXTRA_SUBJECT, myEditSubject.getText().toString());
    myIntent.putExtra(Intent.EXTRA_TEXT, myEditText.getText().toString());
    myIntent.setFlags(Intent.FLAG_ACTIVITY_NEW_TASK);
    return myIntent;
  }
}
```

上面这段代码在 MyCode\MySample290\app\src\main\java\com\bin\luo\mysample\MainActivity.java 文件中。在这段代码中，mShareActionProvider =（ShareActionProvider） myShareMenu.getActionProvider()用于设置菜单实现分享的功能。mShareActionProvider.setShareIntent(getDefaultShareIntent())用于在单击分享菜单时通过 Intent 发送分享信息。此实例的完整项目在 MyCode\MySample290 文件夹中。

037　使用 SearchView 在标题栏添加搜索框

此实例主要通过在重写的 onCreateOptionsMenu()方法中，将 menu 中的 item 转化成 SearchView，实现在 Activity 的标题栏上添加搜索框。当实例运行之后，在标题栏的右端将出现一个搜索按钮，单击该搜索按钮，将显示搜索（输入）框，在搜索框中输入查询条件，如"今日头条"，如图 037-1 的左图所示；然后单击右下角软键盘上的"搜索"按钮，则将在百度中显示搜索结果，如图 037-1 的右图所示。

图　037-1

主要代码如下：

```
@Override
public boolean onCreateOptionsMenu(Menu menu) {
    //显示自定义 menu
    getMenuInflater().inflate(R.menu.myactionbar_menu, menu);
    //获取 ActionBar 内部搜索框对象
    mySearchView = (SearchView) menu.findItem(R.id.myMenuSearch).getActionView();
    //为搜索框动态添加输入监听
    mySearchView.setOnQueryTextListener(new SearchView.OnQueryTextListener() {
        @Override
        //在进行提交操作时,获取输入的搜索关键词,并重定向至指定网页
        public boolean onQueryTextSubmit(String query) {
            myWebView.loadUrl("http://www.baidu.com/s?wd=" + query);
            return false;
        }
        @Override
        public boolean onQueryTextChange(String newText) { return false; }
```

```
    });
    return true;
}
```

上面这段代码在 MyCode\MySample887\app\src\main\java\com\bin\luo\mysample\MainActivity.java 文件中。在这段代码中，getMenuInflater().inflate(R.menu.myactionbar_menu, menu)用于从 myactionbar_menu 布局中加载搜索框，myactionbar_menu 布局的主要内容如下：

```
<?xml version = "1.0" encoding = "utf - 8"?>
<menu xmlns:android = "http://schemas.android.com/apk/res/android">
<item android:id = "@ + id/myMenuSearch"
      android:actionViewClass = "android.widget.SearchView"
      android:showAsAction = "ifRoom"
      android:title = "搜索"/>
</menu>
```

上面这段代码在 MyCode\MySample887\app\src\main\res\menu\myactionbar_menu.xml 文件中。此外，使用 WebView 控件访问网络需要在 AndroidManifest.xml 文件中添加<uses-permission android:name = "android.permission.INTERNET"/>权限。此实例的完整项目在 MyCode\MySample887 文件夹中。

038　使用 SearchManager 实现标题栏搜索

此实例主要通过使用 SearchManager、SearchableInfo 等搜索服务，实现在两个不同的 Activity 之间实现搜索功能。当实例运行之后，单击 MainActivity 标题栏右端的搜索按钮，将在标题栏上显示搜索框，在搜索框中输入搜索内容，如"3.1415926"，如图 038-1 的左图所示；然后单击输入法软键盘上的搜索按钮，将从 MainActivity 跳转到 SearchActivity，并通过百度显示搜索结果，如图 038-1 的右图所示。

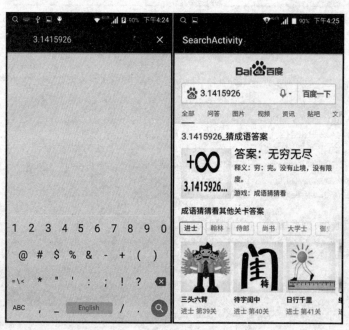

图　038-1

主要代码如下:

```java
@Override
public boolean onCreateOptionsMenu(Menu menu) {
    getMenuInflater().inflate(R.menu.myactionbar_menu, menu);
    //获取 ActionBar 内部搜索框对象
    mySearchView = (SearchView) menu.findItem(R.id.myMenuSearch).getActionView();
    //为搜索框动态添加输入监听
    mySearchView.setOnQueryTextListener(new SearchView.OnQueryTextListener() {
        @Override
        //在进行提交操作时,获取输入的搜索关键词
        public boolean onQueryTextSubmit(String query) {
            //获取搜索服务管理器
            SearchManager mySearchManager =
                    (SearchManager)getSystemService(Context.SEARCH_SERVICE);
            //指定搜索内容要传递到 SearchActivity
            SearchableInfo mySearchableInfo = mySearchManager.getSearchableInfo(
                    new ComponentName(MainActivity.this,SearchActivity.class));
            //向搜索服务传递搜索内容
            mySearchView.setSearchableInfo(mySearchableInfo);
            return false;
        }
        @Override
        public boolean onQueryTextChange(String newText) { return false; }
    });
    return true;
}
```

上面这段代码在 MyCode\MySample888\app\src\main\java\com\bin\luo\mysample\MainActivity.java 文件中。在这段代码中,getMenuInflater().inflate(R.menu.myactionbar_menu,menu)用于从 myactionbar_menu 布局中加载搜索框,myactionbar_menu 布局的主要内容如下:

```xml
<?xml version = "1.0" encoding = "utf-8"?>
<menu xmlns:android = "http://schemas.android.com/apk/res/android">
  <item android:id = "@ + id/myMenuSearch"
      android:actionViewClass = "android.widget.SearchView"
      android:showAsAction = "ifRoom"
      android:title = "搜索"/>
</menu>
```

上面这段代码在 MyCode\MySample888\app\src\main\res\menu\myactionbar_menu.xml 文件中。在 MainActivity.java 中,SearchManager mySearchManager =(SearchManager) getSystemService(Context.SEARCH_SERVICE)用于获取系统的搜索服务,使该服务可实现多种搜索功能。SearchableInfo mySearchableInfo = mySearchManager.getSearchableInfo(new ComponentName(MainActivity.this,SearchActivity.class))中的 SearchActivity 是此实例实现搜索功能的 Activity,SearchActivity 类的主要代码如下:

```java
public class SearchActivity extends Activity {
    @Override
    protected void onCreate(Bundle savedInstanceState) {
```

```
        super.onCreate(savedInstanceState);
        setContentView(R.layout.activity_search);
        getActionBar().setTitle("SearchActivity");
        WebView myWebView = (WebView) findViewById(R.id.myWebView);
        myWebView.getSettings().setJavaScriptEnabled(true);
        myWebView.setWebViewClient(new WebViewClient());
        //判断是否匹配搜索服务对应的 Action 字符串
        if ((Intent.ACTION_SEARCH).equals(getIntent().getAction())) {
         //获取传递的搜索内容
         String myQueryString = getIntent().getStringExtra(SearchManager.QUERY);
         //在百度中搜索指定关键字,并显示搜索结果页面
         myWebView.loadUrl("http://www.baidu.com/s?wd=" + myQueryString);
} } }
```

上面这段代码在 MyCode\MySample888\app\src\main\java\com\bin\luo\mysample\SearchActivity.java 文件中。此外,当新增 SearchActivity 之后,通常应该在 AndroidManifest.xml 文件中进行注册,主要代码如下:

```
<activity android:name=".SearchActivity">
  <intent-filter>
    <action android:name="android.intent.action.SEARCH"/>
  </intent-filter>
  <meta-data android:name="android.app.searchable"
             android:resource="@xml/searchable"/>
</activity>
```

上面这段代码在 MyCode\MySample888\app\src\main\AndroidManifest.xml 文件中。在这段代码中,android:resource="@xml/searchable"主要用于配置搜索选项 searchable,searchable 的主要内容如下:

```
<?xml version="1.0" encoding="utf-8"?>
<searchable xmlns:android="http://schemas.android.com/apk/res/android"
            android:label="@string/app_name"
            android:imeOptions="actionSearch"/>
```

上面这段代码在 MyCode\MySample888\app\src\main\res\xml\searchable.xml 文件中。此外,使用 WebView 控件访问网络需要在 AndroidManifest.xml 文件中添加 <uses-permission android:name="android.permission.INTERNET"/> 权限。此实例的完整项目在 MyCode\MySample888 文件夹中。

039 使用 ActionBar 通过布局定制标题栏

此实例主要通过使用 ActionBar 的 setDisplayOptions()方法和 setCustomView()方法,从而实现在当前应用的标题栏上显示自定义布局。当实例运行之后,在标题栏上显示的不是普通的标题文本,而是两个按钮;单击"机器人总动员"按钮,ImageView 控件显示《机器人总动员》的电影海报图像,如图 039-1 的左图所示。单击"华尔街之狼"按钮,ImageView 控件显示《华尔街之狼》的电影海报图像,如图 039-1 的右图所示。

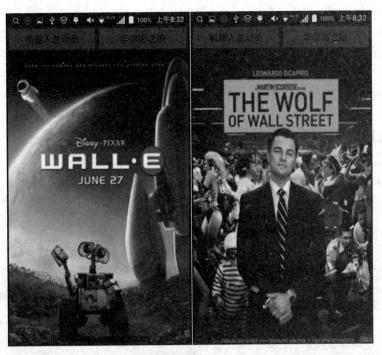

图 039-1

主要代码如下：

```
public class MainActivity extends Activity {
 ImageView myImageView;
 @Override
 protected void onCreate(Bundle savedInstanceState) {
  super.onCreate(savedInstanceState);
  setContentView(R.layout.activity_main);
  ViewGroup myViewGroup =
      (ViewGroup)LayoutInflater.from(this).inflate(R.layout.mylayout,
      new LinearLayout(this),false);
  ActionBar myActionBar = getActionBar();
  myActionBar.setDisplayOptions(ActionBar.DISPLAY_SHOW_CUSTOM);
  myActionBar.setCustomView(myViewGroup);
  myImageView = (ImageView)findViewById(R.id.myImageView);
 }
 public void onClickmyBtn1(View v) {        //响应单击"机器人总动员"按钮
  myImageView.setImageResource(R.mipmap.myimage1);
 }
 public void onClickmyBtn2(View v) {        //响应单击"华尔街之狼"按钮
  myImageView.setImageResource(R.mipmap.myimage2);
} }
```

上面这段代码在 MyCode＼MySample518＼app＼src＼main＼java＼com＼bin＼luo＼mysample＼MainActivity.java 文件中。在这段代码中，myActionBar.setDisplayOptions（ActionBar.DISPLAY_SHOW_CUSTOM）表示标题栏使用定制的布局，如果没有这行代码，标题栏仍然使用默认的风格，即 myActionBar.setCustomView（myViewGroup）这行代码单独调用不起作用；当调用了 myActionBar.setDisplayOptions（ActionBar.DISPLAY_SHOW_CUSTOM）方法，使用

myActionBar.setCustomView(myViewGroup)这行代码才能正常加载并显示自定义布局创建的myViewGroup。ViewGroup myViewGroup=(ViewGroup)LayoutInflater.from(this).inflate(R.layout.mylayout,new LinearLayout(this),false)用于加载包含两个按钮的自定义布局mylayout,关于mylayout布局文件的详细内容请参考 MyCode\MySample518\app\src\main\res\layout\mylayout.xml文件。此实例的完整项目在 MyCode\MySample518 文件夹中。

040 使用 UI 标志动态隐藏或显示通知栏

此实例主要通过使用 RelativeLayout 的 setSystemUiVisibility()方法设置 UI 标志参数,从而实现在应用中动态隐藏或显示通知栏。当实例运行之后,单击"隐藏通知栏"按钮,屏幕顶端的通知栏消失,如图 040-1 的右图所示;单击"显示通知栏"按钮,在屏幕顶端显示通知栏,如图 040-1 的左图所示。

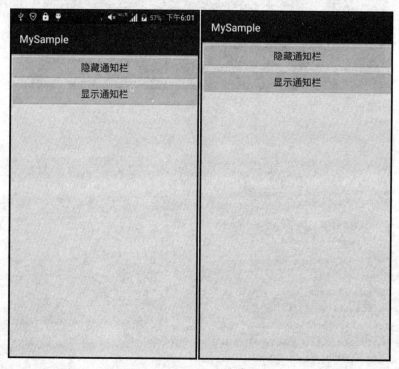

图 040-1

主要代码如下:

```
public void onClickmyBtn1(View v) {        //响应单击"隐藏通知栏"按钮
    myLayout.setSystemUiVisibility(View.SYSTEM_UI_FLAG_FULLSCREEN);
}
public void onClickmyBtn2(View v) {        //响应单击"显示通知栏"按钮
    myLayout.setSystemUiVisibility(View.SYSTEM_UI_FLAG_VISIBLE);
}
```

上面这段代码在 MyCode\MySample453\app\src\main\java\com\bin\luo\mysample\MainActivity.java 文件中。在这段代码中,myLayout.setSystemUiVisibility(View.SYSTEM_UI_FLAG_FULLSCREEN)中的 View.SYSTEM_UI_FLAG_FULLSCREEN 标志参数表示全屏显示当

前应用,因此通知栏隐藏。myLayout. setSystemUiVisibility (View. SYSTEM _ UI _ FLAG _ VISIBLE)中的 View. SYSTEM_UI_FLAG_VISIBLE 标志参数表示正常显示当前应用。此实例的完整项目在 MyCode\MySample453 文件夹中。

041　使用 UI 标志动态隐藏或显示通知栏图标

此实例主要通过使用 RelativeLayout 的 setSystemUiVisibility()方法设置 UI 标志参数,从而实现在应用中动态隐藏或显示通知栏的部分图标。当实例运行之后,单击"隐藏通知栏部分图标"按钮,通知栏左半部分的应用图标消失,仅在右端存留电量和时间图标,如图 041-1 的右图所示;单击"显示通知栏全部图标"按钮,通知栏恢复正常状态显示,如图 041-1 的左图所示。

主要代码如下:

```
public void onClickmyBtn1(View v) {         //响应单击"隐藏通知栏部分图标"按钮
    myLayout.setSystemUiVisibility(View.SYSTEM_UI_FLAG_LOW_PROFILE);
}
public void onClickmyBtn2(View v) {         //响应单击"显示通知栏全部图标"按钮
    myLayout.setSystemUiVisibility(View.SYSTEM_UI_FLAG_VISIBLE);
}
```

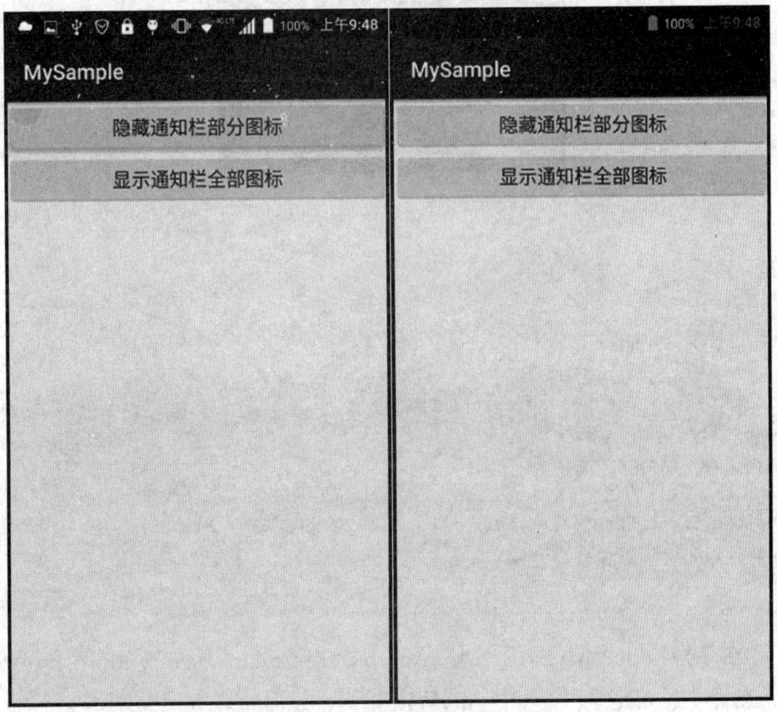

图　041-1

上面这段代码在 MyCode\MySample454\app\src\main\java\com\bin\luo\mysample\MainActivity.java 文件中。在这段代码中,myLayout. setSystemUiVisibility(View. SYSTEM_UI_FLAG_LOW_PROFILE)用于隐藏通知栏的部分图标,并变暗,它主要是为了降低电量消耗。myLayout. setSystemUiVisibility(View. SYSTEM_UI_FLAG_VISIBLE)用于正常显示通知栏图标。此实例的完整项目在 MyCode\MySample454 文件夹中。

042 使用窗口标志实现半透明显示通知栏

此实例主要通过在 Window 的 addFlags()方法中使用 WindowManager. LayoutParams. FLAG_TRANSLUCENT_STATUS 设置窗口标志,从而实现以半透明的方式显示通知栏。当实例运行之后,通知栏以半透明方式显示的效果如图 042-1 所示。

图 042-1

主要代码如下:

```java
public class MainActivity extends Activity {
    @Override
    protected void onCreate(Bundle savedInstanceState) {
        super.onCreate(savedInstanceState);
        Window myWindow = getWindow();
        //以半透明方式显示应用的通知栏
        myWindow.addFlags(WindowManager.LayoutParams.FLAG_TRANSLUCENT_STATUS);
        setContentView(R.layout.activity_main);
    }
}
```

上面这段代码在 MyCode\MySample456\app\src\main\java\com\bin\luo\mysample\MainActivity.java 文件中。在这段代码中,myWindow. addFlags(WindowManager. LayoutParams. FLAG_TRANSLUCENT_STATUS)用于以半透明方式显示应用的通知栏。此实例的完整项目在 MyCode\MySample456 文件夹中。

043 使用指定颜色动态设置通知栏背景色

此实例主要通过使用 Window 的 setStatusBarColor()方法,实现以不同的颜色设置通知栏的背景颜色。当实例运行之后,单击屏幕,通知栏的背景呈现红色,如图 043-1 的左图所示;再次单击屏幕,通知栏的背景呈现黑色,如图 043-1 的右图所示。

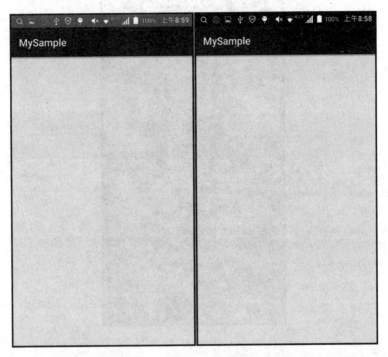

图 043-1

主要代码如下:

```
void onClickScreen(View view) {                    //响应手指单击屏幕
    if (bCheck) {                                  //设置通知栏背景颜色为红色
        Window myWindow = getWindow();
        myWindow.setStatusBarColor(Color.RED);
    } else {                                       //设置通知栏背景颜色为黑色
        Window myWindow = getWindow();
        myWindow.setStatusBarColor(Color.BLACK);
    }
    bCheck = !bCheck;
}
```

上面这段代码在 MyCode \ MySample637 \ app \ src \ main \ java \ com \ bin \ luo \ mysample \ MainActivity.java 文件中。此实例的完整项目在 MyCode\MySample637 文件夹中。

044 将应用的背景图像扩展至通知栏

此实例主要演示了使用 setSystemUiVisibility() 方法和 setStatusBarColor() 方法,从而实现将应用的背景图像扩展至通知(状态)栏。当实例运行之后,应用的背景图像扩展至通知(状态)栏之后的效果如图 044-1 所示。

主要代码如下:

```
public class MainActivity extends Activity {
    @Override
    protected void onCreate(Bundle savedInstanceState) {
```

图 044-1

```
super.onCreate(savedInstanceState);
setContentView(R.layout.activity_main);
//将应用背景图像扩展至通知(状态)栏
View myDecorView = getWindow().getDecorView();
int myOption = View.SYSTEM_UI_FLAG_LAYOUT_FULLSCREEN
                    |View.SYSTEM_UI_FLAG_LAYOUT_STABLE;
myDecorView.setSystemUiVisibility(myOption);
//设置通知(状态)栏的背景颜色为透明色
getWindow().setStatusBarColor(Color.TRANSPARENT);
}}
```

上面这段代码在 MyCode\MySample438\app\src\main\java\com\bin\luo\mysample\MainActivity.java 文件中。此实例的完整项目在 MyCode\MySample438 文件夹中。

045 在通知栏上添加通知并实现跳转功能

此实例主要通过使用 PendingIntent、NotificationManager 和 Notification 等，实现在通知栏上添加通知并实现跳转功能。当实例运行之后，单击"在通知栏上显示图标并实现跳转功能"按钮，将在通知栏上显示一个机器人图标，向下滑动通知栏，将滑出通知列表，在该列表中选择实例刚添加的"这是自定义通知(去百度搜索)"，将通过浏览器显示百度搜索页面，效果分别如图 045-1 的左图和右图所示。

主要代码如下：

```
//响应单击"在通知栏上显示图标并实现跳转功能"按钮
public void onClickmyBtn1(View v) {
    Intent myIntent = new Intent(Intent.ACTION_VIEW,
```

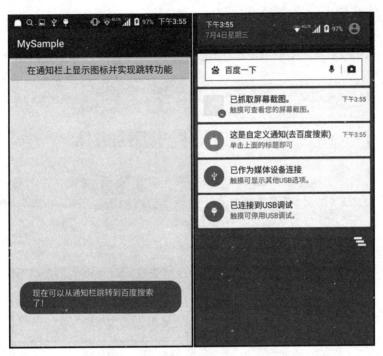

图 045-1

```
                    Uri.parse("https://www.baidu.com/"));  //打开百度搜索
PendingIntent myPendingIntent =
        PendingIntent.getActivity(MainActivity.this, 0, myIntent, 0);
Notification myNotification = new Notification.Builder(this)
        .setAutoCancel(true)
        .setContentTitle("这是自定义通知(去百度搜索)")
        .setContentText("单击上面的标题即可")
        .setContentIntent(myPendingIntent)
        .setDefaults(Notification.DEFAULT_SOUND)
        .setSmallIcon(R.mipmap.ic_launcher)
        .setWhen(System.currentTimeMillis())
        .build();
NotificationManager myNotificationManager =
        (NotificationManager) getSystemService(NOTIFICATION_SERVICE);
myNotificationManager.notify(110, myNotification);
Toast.makeText(getApplicationContext(),
        "现在可以从通知栏跳转到百度搜索了!", Toast.LENGTH_SHORT).show();
}
```

上面这段代码在 MyCode \ MySample439 \ app \ src \ main \ java \ com \ bin \ luo \ mysample \ MainActivity.java 文件中。在这段代码中，myIntent = new Intent(Intent.ACTION_VIEW, Uri.parse("https://www.baidu.com/"))用于创建显示网页的 Intent（即显示百度搜索）。myPendingIntent = PendingIntent.getActivity(MainActivity.this, 0, myIntent, 0)用于根据 myIntent 创建 PendingIntent 的实例，PendingIntent 主要用来在某个事件完成后执行特定的 Action。PendingIntent 包含了 Intent 及 Context，所以即使 Intent 所属应用结束，PendingIntent 依然有效，可以在其他应用中使用。PendingIntent 一般作为参数传给某个实例，在该实例完成某个操作后自动执行 PendingIntent 上的 Action，也可以通过 PendingIntent 的 send()方法手动执行，并可以在 send()方

法中设置 OnFinished 表示 send 成功后执行的动作。myNotification = new Notification.Builder (this)用于创建 Notification 的实例，在 Notification 实例中可以设置图标、提示文字、提示声音等参数。myNotificationManager =（NotificationManager）getSystemService（NOTIFICATION_SERVICE）用于创建 NotificationManager 的实例，NotificationManager 实例用于向通知栏发送通知。此实例的完整项目在 MyCode\MySample439 文件夹中。

046 使用 Notification 在通知栏上添加图标

此实例主要通过使用 Notification、PendingIntent 等类的实例，实现在手机通知栏上显示当前应用图标。当实例运行之后，单击"在手机通知栏上显示当前应用图标"按钮，当前应用图标将显示在通知栏的左上角，如图 046-1 的左图所示；打开通知栏的通知列表，第一个通知即是当前应用通知，如图 046-1 的右图所示。

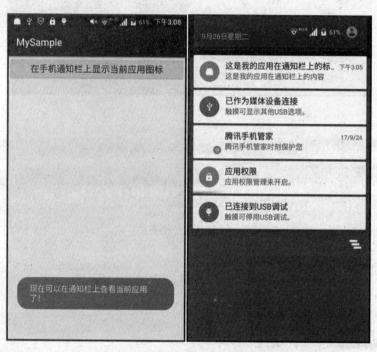

图 046-1

主要代码如下：

```
//响应单击"在手机通知栏上显示当前应用图标"按钮
public void onClickmyBtn1(View v) {
  Intent myIntent = new Intent(this,MainActivity.class);
  PendingIntent myPending = PendingIntent.getActivity(MainActivity.this, 0, myIntent, 0);
  Notification myNotification = new Notification.Builder(this)
        .setAutoCancel(true)
        .setContentTitle("这是我的应用在通知栏上的标题")
        .setContentText("这是我的应用在通知栏上的内容")
        .setContentIntent(myPending)
        .setSmallIcon(R.mipmap.ic_launcher)
        .setWhen(System.currentTimeMillis())
        .build();
```

```
NotificationManager myManager =
        (NotificationManager) getSystemService(NOTIFICATION_SERVICE);
myManager.notify(110, myNotification);
Toast.makeText(getApplicationContext(),
        "现在可以在通知栏上查看当前应用了!", Toast.LENGTH_SHORT).show();
}
```

上面这段代码在 MyCode\MySample277\app\src\main\java\com\bin\luo\mysample\MainActivity.java 文件中。在这段代码中，myPending = PendingIntent.getActivity（MainActivity.this, 0, myIntent, 0）用于创建 PendingIntent 的实例，PendingIntent 可以看作是对 Intent 的一个封装，但它不是立刻执行某个行为，而是在满足某些条件或触发某些事件后才执行指定的行为。myNotification = new Notification.Builder（this）用于创建 Notification 的实例，在 Notification 实例中可以设置图标、提示文字、提示声音等参数。myManager =（NotificationManager）getSystemService（NOTIFICATION_SERVICE）用于创建 NotificationManager 的实例，NotificationManager 实例用于向通知栏发送通知。此实例的完整项目在 MyCode\MySample277 文件夹中。

047　使用 RemoteViews 自定义通知栏视图

此实例主要通过设置 Notification 的 bigContentView 属性为 RemoteViews 创建的自定义视图，实现在通知栏的下拉列表中显示自定义视图。当实例运行之后，单击"在通知栏的下拉列表中显示自定义视图"按钮，将在通知栏上显示一个机器人图标，向下滑动通知栏，将在滑出的通知下拉列表中显示自定义视图（电影《勇闯夺命岛》的海报图像），如图 047-1 的左图所示；在下拉列表中单击该自定义视图（电影海报图像），将跳转到百度百科对应的网页，如图 047-1 的右图所示。

图　047-1

主要代码如下：

```java
//响应单击"在通知栏的下拉列表中显示自定义视图"按钮
public void onClickmyBtn1(View v) {
    Intent myIntent = new Intent(Intent.ACTION_VIEW,
            Uri.parse("https://baike.baidu.com/item/"
    +"%E5%8B%87%E9%97%AF%E5%A4%BA%E5%91%BD%E5%B2%9B/376814?fr=aladdin"));
    PendingIntent myPendingIntent = PendingIntent.getActivity(this, 0, myIntent, 0);
    NotificationManager myNotificationManager =
            (NotificationManager) getSystemService(Context.NOTIFICATION_SERVICE);
    Notification myNotification = new Notification.Builder(this)
            .setContentTitle("折叠式通知")
            .setContentText("This is content text")
            .setWhen(System.currentTimeMillis())
            .setSmallIcon(R.mipmap.ic_launcher)
            .setLargeIcon(BitmapFactory.decodeResource(getResources(),
                                    R.mipmap.ic_launcher))
            .setContentIntent(myPendingIntent)
            .setAutoCancel(true)
            .build();
    RemoteViews myRemoteViews = new RemoteViews(getPackageName(),
                    R.layout.mylayout);           //使用 RemoteViews 创建自定义视图
    myNotification.bigContentView = myRemoteViews;  //指定展开视图
    myNotification.contentView = myRemoteViews;     //指定普通视图
    myNotificationManager.notify(2, myNotification);
}
```

上面这段代码在 MyCode\MySample440\app\src\main\java\com\bin\luo\mysample\MainActivity.java 文件中。在这段代码中，myNotification.bigContentView = myRemoteViews 用于设置在通知栏的下拉列表展开时所显示的自定义视图，该自定义视图通常用 RemoteViews 创建，如 myRemoteViews = new RemoteViews(getPackageName(), R.layout.mylayout)。mylayout 即是该自定义视图的布局，mylayout 布局的主要内容如下：

```xml
<?xml version="1.0" encoding="utf-8"?>
<LinearLayout xmlns:android="http://schemas.android.com/apk/res/android"
            android:layout_width="match_parent"
            android:layout_height="match_parent"
            android:orientation="vertical">
<ImageView android:layout_width="match_parent"
            android:layout_height="match_parent"
            android:layout_margin="10dp"
            android:src="@mipmap/myimage"/>
</LinearLayout>
```

上面这段代码在 MyCode\MySample440\app\src\main\res\layout\mylayout.xml 文件中。此实例的完整项目在 MyCode\MySample440 文件夹中。

048 以悬挂式窗口显示新增的通知栏任务

此实例主要通过使用 Notification 的 setContentIntent() 和 setFullScreenIntent() 方法，实现以悬挂式窗口显示新增的通知栏任务。当实例运行之后，单击"以悬挂式窗口显示新增的通知栏任务"按钮,将在屏幕顶部从上向下滑出一个悬挂式窗口,显示刚添加的通知栏任务("去百度搜索"),如图 048-1 的左图所示;此时如果单击在窗口中的"去百度搜索"任务,将跳转到百度搜索主页面;如果不予理会,则悬挂式窗口在停留几秒后,将自动隐藏,并且在通知栏中添加一个机器人图标,显示新增的任务"去百度搜索"。如果向下滑动通知栏,将在通知栏列表中显示"去百度搜索",单击列表中的"去百度搜索"任务,将跳转到百度搜索主页面,如图 048-1 的右图所示。

图 048-1

主要代码如下:

```java
//响应单击"以悬挂式窗口显示新增的通知栏任务"按钮
public void onClickmyBtn1(View v) {
    Intent myIntent = new Intent(Intent.ACTION_VIEW,
                                 Uri.parse("https://www.baidu.com/"));
    PendingIntent myPendingIntent =
                    PendingIntent.getActivity(this, 0, myIntent, 0);
    Intent myIntentView = new Intent(Intent.ACTION_VIEW);
    myIntentView.setFlags(Intent.FLAG_ACTIVITY_NEW_TASK);
    myIntentView.setClass(this, MainActivity.class);
    PendingIntent myPendingIntentView = PendingIntent.getActivity(this,
            0, myIntentView, PendingIntent.FLAG_CANCEL_CURRENT);
    NotificationManager myNotificationManager =
        (NotificationManager) getSystemService(Context.NOTIFICATION_SERVICE);
    Notification myNotification = new Notification.Builder(this)
            .setContentTitle("去百度搜索")
            .setContentText("单击上面的文本即可实现跳转")
            .setWhen(System.currentTimeMillis())
            .setSmallIcon(R.mipmap.ic_launcher)
            .setLargeIcon(BitmapFactory.decodeResource(getResources(),
                                    R.mipmap.ic_launcher))
            .setContentIntent(myPendingIntent)
```

```
            .setAutoCancel(true)
            .setFullScreenIntent(myPendingIntentView, true)
            .build();
    myNotificationManager.notify(3, myNotification);
}
```

上面这段代码在 MyCode\MySample441\app\src\main\java\com\bin\luo\mysample\MainActivity.java 文件中。在这段代码中，setFullScreenIntent（myPendingIntentView，true）用于实现悬挂式窗口，如果不调用 setFullScreenIntent（）方法，Notification 将仅实现默认的通知栏任务，不显示悬挂式窗口。此实例的完整项目在 MyCode\MySample441 文件夹中。

049　允许直接在通知栏上显示消息内容

此实例主要通过使用 Notification.Builder 的 setTicker（）方法，实现允许或禁止直接在通知栏上显示消息文本的功能。当实例运行之后，单击"允许直接显示文本"按钮，将在通知栏上显示消息文本"APP 又有更新版本了，还在等什么？"，如图 049-1 的左图所示。单击"禁止直接显示文本"按钮，则在通知栏上只显示通知图标（机器人图标），不显示消息文本，如图 049-1 的右图所示。

图　049-1

主要代码如下：

```
public class MainActivity extends Activity {
    private NotificationManager myManager;
    private Notification myNotification;
    private Notification.Builder myBuilder;
    @Override
    protected void onCreate(Bundle savedInstanceState) {
```

```java
    super.onCreate(savedInstanceState);
    setContentView(R.layout.activity_main);
    myBuilder = new Notification.Builder(this);
    myBuilder.setLargeIcon(BitmapFactory.decodeResource(
                            getResources(),R.mipmap.ic_launcher));
    myBuilder.setContentTitle("Android精彩实例集锦");
    myBuilder.setContentText("APP又有更新版本了,还在等什么?");
    myBuilder.setTicker("APP又有更新版本了,还在等什么?");
    myBuilder.setSmallIcon(R.drawable.myicon);
    myManager = (NotificationManager)getSystemService(NOTIFICATION_SERVICE);
    myNotification = myBuilder.build();
    myNotification.icon = R.mipmap.ic_launcher;
    myManager.notify(1, myNotification);
}
public void onClickmyBtn1(View v) {    //响应单击"允许直接显示文本"按钮
    myBuilder.setTicker("APP又有更新版本了,还在等什么?");
    myNotification = myBuilder.build();
    myNotification.icon = R.mipmap.ic_launcher;
    myManager.notify(1,myNotification);
}
public void onClickmyBtn2(View v) {    //响应单击"禁止直接显示文本"按钮
    myBuilder.setTicker(null);
    myNotification = myBuilder.build();
    myNotification.icon = R.mipmap.ic_launcher;
    myManager.notify(1,myNotification);
}}
```

上面这段代码在 MyCode\MySample694\app\src\main\java\com\bin\luo\mysample\MainActivity.java 文件中。在这段代码中,myBuilder.setTicker("APP又有更新版本了,还在等什么?")用于设置在通知栏上直接显示消息的文本,当 setTicker()方法的参数为 null 时,只显示通知图标。注意:此实例在部分模拟器中没有测试成功,因此请直接在手机上测试。此实例的完整项目在 MyCode\MySample694 文件夹中。

050 禁止在通知栏上以右滑方式移除通知

此实例主要通过调用 Notification.Builder 的 setOngoing(true)方法,实现禁止用户在通知栏上以右滑方式移除通知。当实例运行之后,将自动在手机的通知栏中增加"Android精彩应用实例集锦"列表项(通知),向右滑动"Android精彩应用实例集锦"列表项,则该列表项无法从通知栏移除,如图 050-1 的左图所示;单击"Android精彩应用实例集锦"列表项,将启动或跳转到 QQ 应用,如图 050-1 的右图所示。

主要代码如下:

```java
public class MainActivity extends Activity {
    @Override
    protected void onCreate(Bundle savedInstanceState) {
        super.onCreate(savedInstanceState);
        setContentView(R.layout.activity_main);
        Notification.Builder myBuilder = new Notification.Builder(this);
        myBuilder.setLargeIcon(BitmapFactory.decodeResource(getResources(),
```

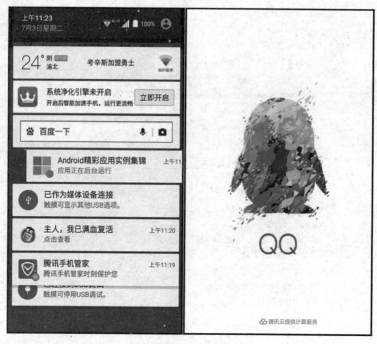

图 050-1

```
                                        R.mipmap.myicon_large));
myBuilder.setContentTitle("Android精彩应用实例集锦");
myBuilder.setContentText("应用正在后台运行");
myBuilder.setOngoing(true);          //禁止通知被滑动删除掉
Intent myIntent = new Intent(this, MyNotificationReceiver.class);
myIntent.setAction("onclick");
PendingIntent myPendingIntent = PendingIntent.getBroadcast(this,
                    0, myIntent, PendingIntent.FLAG_ONE_SHOT);
myBuilder.setContentIntent(myPendingIntent);
NotificationManager myNotificationManager =
        (NotificationManager) getSystemService(NOTIFICATION_SERVICE);
Notification myNotification = myBuilder.build();
myNotification.icon = R.mipmap.myicon_large;
myNotificationManager.notify(1, myNotification);
} }
```

上面这段代码在 MyCode\MySample758\app\src\main\java\com\bin\luo\mysample\MainActivity.java 文件中。在这段代码中，Intent myIntent = new Intent(this, MyNotificationReceiver.class) 用于以 Intent 方式调用自定义广播接收者，自定义广播接收者 MyNotificationReceiver 的内容如下：

```
public class MyNotificationReceiver extends BroadcastReceiver {
  @Override
  public void onReceive(Context context, Intent intent) {
    String myAction = intent.getAction();
    if (myAction.equals("onclick")) {
      Intent myIntent = context.getPackageManager().
                  getLaunchIntentForPackage("com.tencent.mobileqq");
```

```
        context.startActivity(myIntent);                    //启动或跳转到 QQ
}}}
```

上面这段代码在 MyCode \ MySample758 \ app \ src \ main \ java \ com \ bin \ luo \ mysample \ MyNotificationReceiver.java 文件中。需要注意的是,MyNotificationReceiver 广播接收者必须在 AndroidManifest.xml 文件中注册,主要代码如下:

```
<receiver android:name=".MyNotificationReceiver">
    <intent-filter>
        <action android:name="onclick"/>
    </intent-filter>
</receiver>
```

此实例的完整项目在 MyCode\MySample758 文件夹中。

051　在向通知栏发送消息时同时振动手机

此实例主要通过使用 getSystemService(Context.VIBRATOR_SERVICE)方法获取手机的振动服务,从而实现在向通知栏发送通知时同时振动手机。当实例运行之后,单击"发送通知并启用振动"按钮,将向通知栏发送一条消息("Android 精彩应用实例集锦"),同时振动手机;单击该消息,将打开百度网页。单击"发送通知并禁用振动"按钮,也将向通知栏发送一条消息("Android 精彩应用实例集锦"),但是没有振动;单击该消息,将打开百度网页,效果分别如图 051-1 的左图和右图所示。

图　051-1

主要代码如下:

```
public class MainActivity extends Activity {
    private NotificationManager myNotificationManager;
```

```java
private Notification myNotification;
@Override
protected void onCreate(Bundle savedInstanceState) {
    super.onCreate(savedInstanceState);
    setContentView(R.layout.activity_main);
    Intent myIntent = new Intent(Intent.ACTION_VIEW,
                        Uri.parse("https://www.baidu.com/"));
    PendingIntent myPendingIntent =
            PendingIntent.getActivity(MainActivity.this, 0, myIntent, 0);
    Notification.Builder myBuilder = new Notification.Builder(this);
    myBuilder.setLargeIcon(BitmapFactory.decodeResource(getResources(),
                        R.mipmap.ic_launcher));
    myBuilder.setContentTitle("Android精彩应用实例集锦");
    myBuilder.setContentText("马上去百度搜索……");
    myBuilder.setSmallIcon(R.mipmap.ic_launcher);
    myBuilder.setContentIntent(myPendingIntent);
    myBuilder.setAutoCancel(true);
    myNotificationManager =
            (NotificationManager) getSystemService(NOTIFICATION_SERVICE);
    myNotification = myBuilder.build();
    myNotification.icon = R.mipmap.ic_launcher;
}
public void onClickmyBtn1(View v) {              //响应单击"发送通知并启用振动"按钮
    myNotificationManager.notify(1, myNotification);    //发送通知
    Vibrator myVibrator = (Vibrator)getSystemService(Context.VIBRATOR_SERVICE);
    long[] myPattern = new long[]{100, 400, 100, 400};  //设置振动间隔
    myVibrator.vibrate(myPattern, -1);               //启动指定模式振动
}
public void onClickmyBtn2(View v) {              //响应单击"发送通知并禁用振动"按钮
    myNotificationManager.notify(1, myNotification);    //发送通知
} }
```

上面这段代码在 MyCode\MySample654\app\src\main\java\com\bin\luo\mysample\MainActivity.java 文件中。此外，操作振动器需要在 AndroidManifest.xml 文件中添加＜uses-permission android:name="android.permission.VIBRATE"/＞权限。此实例的完整项目在 MyCode\MySample654 文件夹中。

052 在滚动文本时自动隐藏或显示工具栏

此实例主要通过使用 AppBarLayout.LayoutParams 的 setScrollFlags()方法设置滚动标志，从而实现在滚动文本时自动决定隐藏或显示工具栏。在实例运行之后，如果下滑滚动文本，则滑出工具栏，如图 052-1 的左图所示；如果上滑滚动文本，则隐藏工具栏，如图 052-1 的右图所示。

主要代码如下：

```java
public class MainActivity extends Activity{
@Override
protected void onCreate(Bundle savedInstanceState){
    super.onCreate(savedInstanceState);
    setContentView(R.layout.activity_main);
    Toolbar myToolBar = (Toolbar)findViewById(R.id.myToolBar);
```

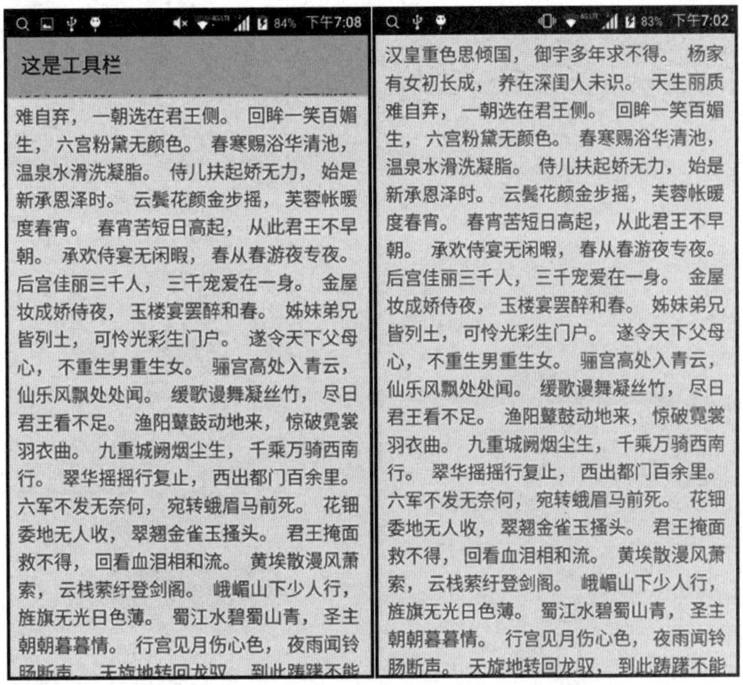

图 052-1

```
myToolBar.setTitle("这是工具栏");
AppBarLayout.LayoutParams myLayoutParams =
            (AppBarLayout.LayoutParams)myToolBar.getLayoutParams();
//为 ToolBar 添加动态显示或隐藏功能
myLayoutParams.setScrollFlags(AppBarLayout.LayoutParams.SCROLL_FLAG_SCROLL|
        AppBarLayout.LayoutParams.SCROLL_FLAG_ENTER_ALWAYS);
TextView myTextView = (TextView)findViewById(R.id.myTextView);
myTextView.setLineSpacing(1,1.25f);        //设置文本内容行间距
} }
```

上面这段代码在 MyCode\MySample909\app\src\main\java\com\bin\luo\mysample\MainActivity.java 文件中。此外，使用 AppBarLayout 等控件需要在 gradle 中引入 compile 'com.android.support:design:24.2.0'依赖项。如果应用在运行时崩溃，请参考源代码中的 app\src\main\res\values\styles.xml 文件修改与主题相关的内容。此实例的完整项目在 MyCode\MySample909 文件夹中。

053 使用 Toolbar 在工具栏上添加查找按钮

此实例主要通过在 Toolbar 上添加菜单的方式，实现在工具栏上添加查找按钮。当实例运行之后，在工具栏的输入框中输入查找条件，如"重庆"，然后单击右上角的放大镜（开始查找）按钮，将在弹出的 Toast 中显示查找条件，如图 053-1 的左图和右图所示。

主要代码如下：

```
public class MainActivity extends AppCompatActivity {
  private Toolbar myToolbar;
```

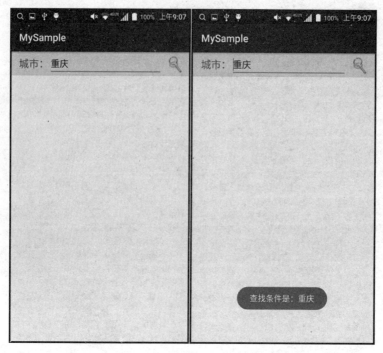

图 053-1

```
protected void onCreate(Bundle savedInstanceState) {
    super.onCreate(savedInstanceState);
    setContentView(R.layout.activity_main);
    myToolbar = (Toolbar)findViewById(R.id.myToolbar);
    myToolbar.inflateMenu(R.menu.mymenu);
    myToolbar.setOnMenuItemClickListener(new Toolbar.OnMenuItemClickListener(){
        @Override
        public boolean onMenuItemClick(MenuItem item) {
            if(item.getItemId() == R.id.myFindButton){          //响应单击开始查找按钮(菜单)
                EditText myEditText = (EditText) myToolbar.findViewById(R.id.myEditText);
                String myCity = myEditText.getText().toString();    //获取ToolBar输入框内容
                Toast.makeText(getApplicationContext(),
                            "查找条件是: " + myCity,Toast.LENGTH_SHORT).show();
            }
            return false;
        }});;}}
```

上面这段代码在 MyCode\MySample484\app\src\main\java\com\bin\luo\mysample\MainActivity.java 文件中。在这段代码中，EditText 控件用于输入查找条件，它是 Toolbar 的子控件，对应的布局如下：

```xml
<?xml version = "1.0" encoding = "utf-8"?>
<RelativeLayout xmlns:android = "http://schemas.android.com/apk/res/android"
    xmlns:tools = "http://schemas.android.com/tools"
    android:id = "@ + id/activity_main"
    android:layout_width = "match_parent"
    android:layout_height = "match_parent"
    tools:context = "com.bin.luo.mysample.MainActivity">
```

```xml
<android.support.v7.widget.Toolbar android:id="@+id/myToolbar"
                                    android:layout_width="match_parent"
                                    android:layout_height="46dp"
                                    android:background="#8EE5EE">
    <TextView android:layout_width="wrap_content"
              android:layout_height="wrap_content"
              android:textSize="20dp"
              android:text="城市:"/>
    <EditText android:text="重庆"
              android:id="@+id/myEditText"
              android:layout_width="match_parent"
              android:layout_height="match_parent"/>
</android.support.v7.widget.Toolbar>
</RelativeLayout>
```

上面这段代码在 MyCode\MySample484\app\src\main\res\layout\activity_main.xml 文件中。在这段代码中,并没有"开始查找"按钮的内容,"开始查找"按钮是通过加载菜单文件实现的。菜单文件的主要内容如下:

```xml
<?xml version="1.0" encoding="utf-8"?>
<menu xmlns:android="http://schemas.android.com/apk/res/android"
      xmlns:app="http://schemas.android.com/apk/res-auto">
    <item android:id="@+id/myFindButton"
          android:title="开始查找"
          android:icon="@mipmap/myimage1"
          app:showAsAction="always"/>
</menu>
```

上面这段代码在 MyCode\MySample484\app\src\main\res\menu\mymenu.xml 文件中。在这段代码中,app:showAsAction="always" 表示"开始查找"菜单在工具栏上以按钮的形式显示。在 MainActivity.java 中,Toolbar 则使用 inflateMenu() 方法加载该菜单文件,即 myToolbar.inflateMenu(R.menu.mymenu),从而显示放大镜查找按钮;默认情况下,当菜单以按钮的风格显示时,标题文本不显示。此外,使用 Toolbar 需要在 gradle 中引入 compile 'com.android.support:design:25.0.1' 依赖项。此实例的完整项目在 MyCode\MySample484 文件夹中。

054 使用 Toolbar 为导航图标添加关闭功能

此实例主要通过使用 Toolbar 的 setNavigationOnClickListener() 方法,实现为工具栏导航图标添加关闭功能。当实例运行之后,在工具栏的左上角将会出现红色的关闭(导航)图标,如图 054-1 的左图所示。单击该关闭(导航)图标,将出现水波状的扩散效果,然后关闭应用,如图 054-1 的右图所示。

主要代码如下:

```java
public class MainActivity extends AppCompatActivity {
    private Toolbar myToolbar;
    protected void onCreate(Bundle savedInstanceState) {
        super.onCreate(savedInstanceState);
        setContentView(R.layout.activity_main);
```

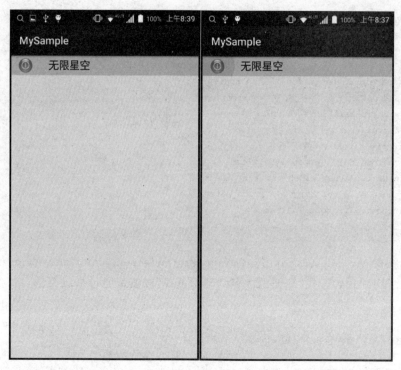

图 054-1

```
myToolbar = (Toolbar)findViewById(R.id.myToolbar);
myToolbar.setTitle("无限星空");                    //设置工具栏标题
myToolbar.setNavigationIcon(R.mipmap.myimage);   //设置导航图标
//设置导航图标单击事件响应
myToolbar.setNavigationOnClickListener(new View.OnClickListener() {
  @Override
  public void onClick(View v) {
    finish();                                   //关闭当前应用
  } });
} }
```

上面这段代码在 MyCode\MySample483\app\src\main\java\com\bin\luo\mysample\MainActivity.java 文件中。在这段代码中,Toolbar 是工具栏控件,它实际上相当于容器,里面可以容纳多种不同类型的控件,通常在布局文件中添加,主要代码如下:

```
< android.support.v7.widget.Toolbar
    android:id = "@ + id/myToolbar"
    android:layout_width = "match_parent"
    android:layout_height = "36dp"
    android:background = "#8EE5EE">
</android.support.v7.widget.Toolbar>
```

上面这段代码在 MyCode\MySample483\app\src\main\res\layout\activity_main.xml 文件中。此外,使用 Toolbar 控件需要在 gradle 中引入 compile 'com.android.support:appcompat-v7:25.2.0' 依赖项。此实例的完整项目在 MyCode\MySample483 文件夹中。

055 在拖动改变控件大小时实现工具栏跟随

此实例主要通过设置 Toolbar 控件的 layout_gravity 属性为 bottom,实现在拖动图像、文本、工具栏等控件使其改变在垂直方向的相对大小时,工具栏与文本控件始终跟随的效果。当实例运行之后,拖住文本上下移动,蓝色的工具栏和文本始终相随,如图 055-1 的左图所示。如果未设置 Toolbar 控件的 layout_gravity 属性为 bottom,当在拖住文本上下移动时,工具栏始终位于屏幕顶部,如图 055-1 的右图所示。

图 055-1

主要代码如下:

```xml
<?xml version = "1.0" encoding = "utf-8"?>
<android.support.design.widget.CoordinatorLayout
    xmlns:android = "http://schemas.android.com/apk/res/android"
    xmlns:app = "http://schemas.android.com/apk/res-auto"
    android:orientation = "vertical"
    android:layout_width = "match_parent"
    android:layout_height = "match_parent">
    <android.support.design.widget.AppBarLayout
        android:layout_width = "match_parent"
        android:layout_height = "wrap_content">
        <android.support.design.widget.CollapsingToolbarLayout
            android:layout_width = "match_parent"
            android:layout_height = "wrap_content"
            app:contentScrim = "@color/colorPrimary"
            app:layout_scrollFlags = "scroll|exitUntilCollapsed" >
            <ImageView android:layout_width = "match_parent"
                android:layout_height = "match_parent"
```

```xml
            android:scaleType="centerCrop"
            android:src="@mipmap/myimage2"
            app:layout_collapseMode="parallax"/>
        <android.support.v7.widget.Toolbar android:layout_width="match_parent"
            android:layout_height="48dp"
            android:layout_gravity="bottom"
            android:background="#0000FF"
            app:navigationIcon="@mipmap/myimage1"
            app:layout_collapseMode="pin">
            <TextView android:layout_width="wrap_content"
                android:layout_height="wrap_content"
                android:textSize="18dp"
                android:textColor="#fff"
                android:text="姓名："/>
            <EditText android:textColor="#fff"
                android:text="罗帅"
                android:layout_width="match_parent"
                android:layout_height="match_parent"/>
        </android.support.v7.widget.Toolbar>
    </android.support.design.widget.CollapsingToolbarLayout>
</android.support.design.widget.AppBarLayout>
<android.support.v4.widget.NestedScrollView
    android:layout_width="match_parent"
    android:layout_height="match_parent"
    app:layout_behavior="@string/appbar_scrolling_view_behavior">
    <TextView android:layout_width="wrap_content"
        android:layout_height="wrap_content"
        android:layout_margin="20dp"
        android:textSize="20dp"
        android:text="Android是一个以 Linux 为基础的开源操作系统，主要用于移动设备，由 Google
和开放手持设备联盟开发与领导。Android 系统最初由安迪·鲁宾(Andy Rubin)制作，最初主要支持手机。2005 年
8 月 17 日被 Google 收购。2007 年 11 月 5 日，Google 与 84 家硬件制造商、软件开发商及电信营运商组成开放手
持设备联盟(Open Handset Alliance)来共同研发改良 Android 系统并生产搭载 Android 的智慧型手机，然后逐渐
拓展到平板电脑及其他领域上。接着，Google 以 Apache 免费开源许可证的授权方式，发布了 Android 的源代
码。"/>
</android.support.v4.widget.NestedScrollView>
</android.support.design.widget.CoordinatorLayout>
```

上面这段代码在 MyCode\MySample486\app\src\main\res\layout\activity_main.xml 文件中。在这段代码中，Toolbar 是 CollapsingToolbarLayout 的子控件，该子控件的 android:layout_gravity="bottom"表示 Toolbar 始终与 CollapsingToolbarLayout 的底部靠齐。此外，使用此实例的相关控件需要在 gradle 中引入 compile 'com.android.support:design：25.0.1'依赖项。此实例的完整项目在 MyCode\MySample486 文件夹中。

056 创建 CollapsingToolbarLayout 工具栏

此实例主要通过使用 CollapsingToolbarLayout 等控件，创建可在垂直方向折叠拉伸的工具栏。当实例运行之后，向上（在图像上）拖动工具栏，则工具栏将从图 056-1 左图所示的效果改变为图 056-1 右图所示的效果；向下（在图像上）拖动工具栏，则工具栏将从图 056-1 右图所示的效果改变为图 056-1 左图所示的效果。

第2章 通知栏

图 056-1

主要代码如下:

```xml
<?xml version = "1.0" encoding = "utf-8"?>
<android.support.design.widget.CoordinatorLayout
    xmlns:android = "http://schemas.android.com/apk/res/android"
    xmlns:app = "http://schemas.android.com/apk/res-auto"
    android:layout_width = "match_parent"
    android:layout_height = "match_parent"
    android:fitsSystemWindows = "true">
    <android.support.design.widget.AppBarLayout
        android:layout_width = "match_parent"
        android:layout_height = "wrap_content"
        android:fitsSystemWindows = "true"
        android:theme = "@style/ThemeOverlay.AppCompat.Dark.ActionBar">
    <android.support.design.widget.CollapsingToolbarLayout
        android:layout_width = "match_parent"
        android:layout_height = "wrap_content"
        android:fitsSystemWindows = "true"
        app:contentScrim = "?attr/colorPrimary"
        app:expandedTitleMarginEnd = "64dp"
        app:expandedTitleMarginStart = "48dp"
        app:layout_scrollFlags = "scroll|exitUntilCollapsed"
        app:title = "折叠效果的工具栏">
    <ImageView android:layout_width = "match_parent"
            android:layout_height = "wrap_content"
            android:adjustViewBounds = "true"
            android:fitsSystemWindows = "true"
            android:src = "@mipmap/myimage1"
            app:layout_collapseMode = "parallax"
            app:layout_collapseParallaxMultiplier = "0.7"/>
```

```
    <android.support.v7.widget.Toolbar
        android:layout_width = "match_parent"
        android:layout_height = "?attr/actionBarSize"
        app:layout_collapseMode = "pin"
        app:popupTheme = "@style/ThemeOverlay.AppCompat.Light"
        app:title = "@string/app_name"/>
   </android.support.design.widget.CollapsingToolbarLayout>
  </android.support.design.widget.AppBarLayout>
  <android.support.v4.widget.SwipeRefreshLayout
      android:layout_width = "match_parent"
      android:layout_height = "match_parent"
      app:layout_behavior = "@string/appbar_scrolling_view_behavior">
   <android.support.v7.widget.CardView
              xmlns:card_view = "http://schemas.android.com/apk/res-auto"
              android:layout_width = "match_parent"
              android:layout_height = "wrap_content"
              card_view:cardBackgroundColor = "#00EEEE"
              card_view:cardCornerRadius = "20dp"
              card_view:cardElevation = "5dp"
              card_view:cardPreventCornerOverlap = "true"
              card_view:cardUseCompatPadding = "true"
              card_view:contentPadding = "15dp">
     <TextView android:layout_width = "wrap_content"
         android:layout_height = "wrap_content"
         android:text = "Android 是一个以 Linux 为基础的开源操作系统,主要用于移动设备,由 Google 和开放手持设备联盟开发与领导。Android 系统最初由安迪·鲁宾(Andy Rubin)制作,最初主要支持手机。2005 年 8 月 17 日被 Google 收购。2007 年 11 月 5 日,Google 与 84 家硬件制造商、软件开发商及电信营运商组成开放手持设备联盟(Open Handset Alliance)来共同研发改良 Android 系统并生产搭载 Android 的智慧型手机,然后逐渐拓展到平板电脑及其他领域上。接着,Google 以 Apache 免费开源许可证的授权方式,发布了 Android 的源代码。"
         android:textSize = "18dp"/>
   </android.support.v7.widget.CardView>
  </android.support.v4.widget.SwipeRefreshLayout>
 </android.support.design.widget.CoordinatorLayout>
```

上面这段代码在 MyCode\MySample480\app\src\main\res\layout\activity_main.xml 文件中。在这段代码中,android.support.design.widget.CollapsingToolbarLayout 控件的 app:layout_scrollFlags = "scroll|exitUntilCollapsed"用于指定工具栏在被上下拖动时,自动折叠或拉伸。此外,使用 CollapsingToolbarLayout 控件需要在 gradle 中引入 compile 'com.android.support:design:25.0.1'依赖项,使用 CardView 控件需要在 gradle 中引入 compile 'com.android.support:cardview-v7:25.2.0'依赖项。此实例的完整项目在 MyCode\MySample480 文件夹中。

057 使用 Snackbar 在底部创建浮出信息栏

此实例主要通过使用 Snackbar 实现在屏幕底部浮出具有黏滞效果的信息栏。当实例运行之后,单击"显示最简单的 Snackbar"按钮,将在屏幕底部浮出一个信息栏"这是最简单的 Snackbar",如图 057-1 的左图所示。单击信息栏上的"单击显示提示窗口"按钮,将在弹出的 Toast 中提示"一个时刻只能有唯一的 Snackbar 显示",如图 057-1 的右图所示。Snackbar 的显示风格也被人称为沉浸式状态栏。

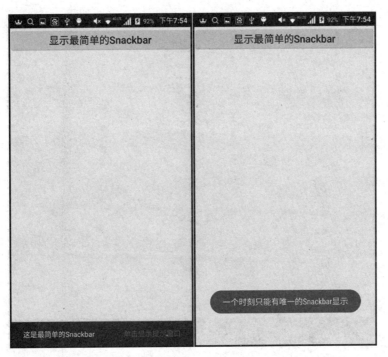

图 057-1

主要代码如下:

```
public void onClickmyBtn1(View v) {    //响应单击"显示最简单的 Snackbar"按钮
    Snackbar.make(myCoordinatorLayout, "这是最简单的 Snackbar",
            Snackbar.LENGTH_LONG).setAction("单击显示提示窗口",
            new View.OnClickListener() {
        @Override
        public void onClick(View v) {
            Toast.makeText(getApplicationContext(),"一个时刻只能有唯一的 Snackbar 显示",
            Toast.LENGTH_SHORT).show();
        } }).show();
}
```

上面这段代码在 MyCode\MySample360\app\src\main\java\com\bin\luo\mysample\MainActivity.java 文件中。在这段代码中,Snackbar 的 make()方法用于创建 Snackbar,setAction()方法用于设置 Snackbar 的按钮单击等事件,show()方法用于显示 Snackbar。此外,使用 Snackbar 需要在 gradle 中引入 compile 'com.android.support:support-v4:24.2.1'和 compile 'com.android.support:design:24.2.1'依赖项。此实例的完整项目在 MyCode\MySample360 文件夹中。

058 自定义 Snackbar 文本颜色和字体大小

此实例主要通过设置 Snackbar 的文本控件 R.id.snackbar_text 的字体大小属性和颜色属性,从而实现自定义 Snackbar 的文本颜色和字体大小。当实例运行之后,单击"显示文本颜色可变化的 Snackbar"按钮,将在屏幕底部浮出红色文字的 Snackbar,如图 058-1 的左图所示。单击 Snackbar 右端的"下一条"按钮,将在屏幕底部浮出绿色的 Snackbar,如图 058-1 的右图所示。

图 058-1

主要代码如下：

```
//响应单击"显示文本颜色可变化的 Snackbar"按钮
public void onClickmyBtn1(View v) {
    Snackbar mySnackbar1 =
            Snackbar.make(myCoordinatorLayout, "中国向菲马拉维捐重建机械",
Snackbar.LENGTH_LONG).setAction("下一条", new View.OnClickListener() {
            @Override
            public void onClick(View v) {
              Snackbar mySnackbar2 =
                      Snackbar.make(myCoordinatorLayout,
                      "特斯拉在华启动校招规模近500人", Snackbar.LENGTH_LONG);
              setSnackbarMessageText(mySnackbar2, Color.GREEN);
              mySnackbar2.show();
            } });
    setSnackbarMessageText(mySnackbar1, Color.parseColor("#FF0000"));
    mySnackbar1.show();
}
public static void setSnackbarMessageText(Snackbar snackbar,
                    int color) {    //设置 Snackbar 的文本颜色和大小
    View view = snackbar.getView();
    TextView myTextView = (TextView) view.findViewById(R.id.snackbar_text);
    myTextView.setTextColor(color);          //设置 Snackbar 的文本颜色
    myTextView.setTextSize(20);              //设置 Snackbar 的字体大小
}
```

上面这段代码在 MyCode\MySample361\app\src\main\java\com\bin\luo\mysample\MainActivity.java 文件中。在这段代码中，myTextView =（TextView）view.findViewById（R.id.snackbar_text）用于根据 R.id.snackbar_text 获取 Snackbar 内置的显示文本的 TextView 控件，然后

可以按照普通的 TextView 控件自定义在 Snackbar 中显示的文本。此外，使用 Snackbar 需要在 gradle 中引入 compile 'com.android.support:support-v4:24.2.1'和 compile 'com.android.support:design:24.2.1'依赖项。此实例的完整项目在 MyCode\MySample361 文件夹中。

059　在 Snackbar 上新增自定义风格布局

此实例主要通过使用 Snackbar.SnackbarLayout 的 addView()方法，实现在 Snackbar 上新增带图像的自定义布局。当实例运行之后，单击"在 Snackbar 上新增自定义布局"按钮，将在屏幕底部浮出带图像的 Snackbar，如图 059-1 的左图所示。单击 Snackbar 右端的"购买"按钮，将弹出一个 Toast，如图 059-1 的右图所示。

图　059-1

主要代码如下：

```
//响应单击"在Snackbar上新增自定义布局"按钮
public void onClickmyBtn1(View v) {
    Snackbar mySnackbar =
            Snackbar.make(myCoordinatorLayout, "", Snackbar.LENGTH_LONG)
                    .setAction("购买", new View.OnClickListener() {
                        @Override
                        public void onClick(View view) {
                            Toast.makeText(MainActivity.this,
                                    "成功加入购物车", Toast.LENGTH_LONG).show();} });
    View mySnackView = mySnackbar.getView();
    Snackbar.SnackbarLayout mySnackLayout = (Snackbar.SnackbarLayout)mySnackView;
    //设置 Snackbar 的背景色
    mySnackView.setBackgroundColor(Color.parseColor("#66CDAA"));
    //加载自定义图像布局 mylayout
```

```
        View myView = LayoutInflater.from(mySnackView.getContext()).inflate
                                            (R.layout.mylayout,null);
        //设置 Snackbar 的按钮(购买)颜色
        ((Button)mySnackView.findViewById(R.id.snackbar_action)).
                            setTextColor(Color.parseColor("#ff0000"));
        LinearLayout.LayoutParams myLayoutParams =
            new LinearLayout.LayoutParams(LinearLayout.LayoutParams.WRAP_CONTENT,
            LinearLayout.LayoutParams.WRAP_CONTENT);
        myLayoutParams.gravity = Gravity.CENTER_VERTICAL;
        mySnackLayout.addView(myView, 0, myLayoutParams);   //添加自定义图像布局
        mySnackbar.show();
    }
```

上面这段代码在 MyCode\MySample362\app\src\main\java\com\bin\luo\mysample\MainActivity.java 文件中。在这段代码中，myView = LayoutInflater.from(mySnackView.getContext()).inflate(R.layout.mylayout,null)用于加载自定义布局 mylayout，mylayout 布局的主要内容如下：

```xml
<?xml version = "1.0" encoding = "utf-8"?>
<LinearLayout xmlns:android = "http://schemas.android.com/apk/res/android"
            android:layout_width = "match_parent"
            android:layout_height = "match_parent"
            android:orientation = "horizontal" >
    <ImageView  android:layout_margin = "2dp"
            android:layout_width = "60dp"
            android:layout_height = "60dp"
            android:src = "@mipmap/myimage1"/>
    <TextView   android:gravity = "center_vertical"
            android:layout_margin = "1dp"
            android:layout_width = "match_parent"
            android:layout_height = "match_parent"
            android:text = "我的成功你也可以复制"
            android:textColor = "#9A32CD"
            android:textSize = "20dp"/>
</LinearLayout>
```

上面这段代码在 MyCode\MySample362\app\src\main\res\layout\mylayout.xml 文件中。此外，使用 Snackbar 需要在 gradle 中引入 compile 'com.android.support:support-v4：24.2.1' 和 compile 'com.android.support：design：24.2.1' 依赖项。此实例的完整项目在 MyCode\MySample362 文件夹中。

第3章

菜单

060 在 ActionBar 上以按钮风格显示菜单

此实例主要通过在 MenuItem 的 setShowAsAction()方法中使用 MenuItem. SHOW_AS_ACTION_ALWAYS 参数,从而使 ActionBar 的菜单以按钮形式显示。当实例运行之后,单击 ActionBar 上的"黑曼巴"按钮(菜单),效果如图 060-1 的左图所示。单击 ActionBar 上的"神战"按钮(菜单),效果如图 060-1 的右图所示。

图 060-1

主要代码如下:

```
public boolean onCreateOptionsMenu(Menu myMenu) {
    super.onCreateOptionsMenu(myMenu);
    //添加菜单项
    MenuItem myItem1 = myMenu.add(0,ITEM1,0,"黑曼巴");
```

```
    MenuItem myItem2 = myMenu.add(0,ITEM2,1,"神战");
    MenuItem myItem3 = myMenu.add(0,ITEM3,2,"吹梦巨人");
    //在 ActionBar 上以按钮形式显示菜单
    myItem1.setShowAsAction(MenuItem.SHOW_AS_ACTION_ALWAYS);
    myItem2.setShowAsAction(MenuItem.SHOW_AS_ACTION_ALWAYS);
    myItem3.setShowAsAction(MenuItem.SHOW_AS_ACTION_ALWAYS);
    return true;
}
public boolean onOptionsItemSelected(MenuItem item) { //响应单击菜单
    switch (item.getItemId()) {
      case ITEM1:
        myImageView.setImageResource(R.mipmap.myimage1);
        break;
      case ITEM2:
        myImageView.setImageResource(R.mipmap.myimage2);
        break;
      case ITEM3:
        myImageView.setImageResource(R.mipmap.myimage3);
        break;
      default:
        break;
    }
    return super.onOptionsItemSelected(item);
}
```

上面这段代码在 MyCode\MySample288\app\src\main\java\com\bin\luo\mysample\MainActivity.java 文件中。在这段代码中，myItem1.setShowAsAction(MenuItem.SHOW_AS_ACTION_ALWAYS)表示 myItem1 菜单始终以按钮的形式显示，如果这行代码被修改为 myItem1.setShowAsAction(MenuItem.SHOW_AS_ACTION_NEVER)，则 myItem1 菜单将以菜单的形式显示。此实例的完整项目在 MyCode\MySample288 文件夹中。

061　使用 ActionBar 在标题栏添加下拉菜单

此实例主要通过使用 ActionBar 的 setListNavigationCallbacks()等方法，实现在标题栏上添加下拉菜单。当实例运行之后，单击标题栏右端的下拉箭头，将弹出下拉菜单，如图 061-1 的左图所示。在下拉菜单中选择"当幸福来敲门"菜单项，将在弹出的 Toast 中显示选择结果，如图 061-1 的右图所示。

主要代码如下：

```
public class MainActivity extends Activity {
  @Override
  protected void onCreate(Bundle savedInstanceState) {
    super.onCreate(savedInstanceState);
    setContentView(R.layout.activity_main);
    SpinnerAdapter mySpinnerAdapter = ArrayAdapter.createFromResource(this,
        R.array.myItems, android.R.layout.simple_spinner_dropdown_item);
    ActionBar myActionBar = getActionBar();
    //将 ActionBar 的操作模式设置为 NAVIGATION_MODE_LIST 下拉模式
    myActionBar.setNavigationMode(ActionBar.NAVIGATION_MODE_LIST);
```

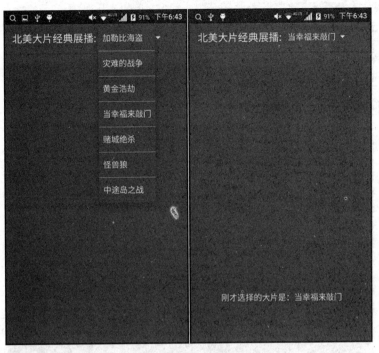

图 061-1

```
//为 ActionBar 设置下拉菜单和监听器
myActionBar.setListNavigationCallbacks(mySpinnerAdapter,
                                       new DropDownListenser());
}
//实现 ActionBar.OnNavigationListener 接口
class DropDownListenser implements ActionBar.OnNavigationListener {
    String[] myItems = getResources().getStringArray(R.array.myItems);
    //当选择下拉菜单项的时候响应
    public boolean onNavigationItemSelected(int myPosition, long itemId) {
        Toast.makeText(getApplicationContext(),
            "刚才选择的大片是:" + myItems[myPosition], Toast.LENGTH_SHORT).show();
        return true;
}}}
```

上面这段代码在 MyCode \ MySample291 \ app \ src \ main \ java \ com \ bin \ luo \ mysample \ MainActivity.java 文件中。在这段代码中，SpinnerAdapter mySpinnerAdapter ＝ ArrayAdapter. createFromResource(this，R. array. myItems, android. R. layout. simple_ spinner_ dropdown_item) 用于从资源 myItems 中加载下拉菜单的菜单项，关于 myItems 资源的详细内容请参考 MyCode\ MySample291\app \ src \ main \ res \ values \ strings. xml 文件。此实例的完整项目在 MyCode \ MySample291 文件夹中。

062 在 ActionBar 上使用 XML 文件创建菜单

此实例主要通过使用 MenuInflater 的 inflate()方法，实现在 ActionBar 上加载在 XML 文件中定义的菜单。当实例运行之后，单击 ActionBar(标题栏)右端的菜单图标，将弹出菜单列表，在菜单列表中选择"好莱坞影片"，如图 062-1 的左图所示，将弹出该菜单的二级菜单列表，在二级菜单列表中选择

"吹梦巨人",将显示《吹梦巨人》的电影海报图像,如图 062-1 的右图所示。

图 062-1

主要代码如下：

```java
public boolean onCreateOptionsMenu(Menu myMenu) {
    super.onCreateOptionsMenu(myMenu);
    MenuInflater myMenuInflater = getMenuInflater();
    myMenuInflater.inflate(R.menu.mymenu, myMenu);
    return true;
}
public boolean onOptionsItemSelected(MenuItem item) {        //响应选择菜单项
    switch (item.getItemId()) {
     case R.id.item21:
      myImageView.setImageResource(R.mipmap.myimage1);
      break;
     case R.id.item22:
      myImageView.setImageResource(R.mipmap.myimage2);
      break;
     case R.id.item23:
      myImageView.setImageResource(R.mipmap.myimage3);
      break;
     default:
      break;
    }
    return super.onOptionsItemSelected(item);
}
```

上面这段代码在 MyCode\MySample289\app\src\main\java\com\bin\luo\mysample\MainActivity.java 文件中。在这段代码中,myMenuInflater.inflate(R.menu.mymenu,myMenu)用

于加载mymenu菜单资源,关于mymenu菜单资源的详细内容请参考MyCode\MySample289\app\src\main\res\menu\mymenu.xml文件。此实例的完整项目在MyCode\MySample289文件夹中。

063 使用Toolbar在工具栏上添加下拉菜单

此实例主要通过使用Toolbar的inflateMenu()方法,实现在工具栏的右端添加下拉菜单。当实例运行之后,在工具栏的右端将会出现一个菜单图标,单击该图标,会弹出下拉菜单;单击"技术前沿"菜单,则在弹出的Toast中显示"刚才单击的菜单是:技术前沿";单击"经营管理"菜单,在弹出的Toast中显示"刚才单击的菜单是:经营管理",效果分别如图063-1的左图和右图所示。单击其他菜单将实现类似的功能。

图 063-1

主要代码如下:

```
public class MainActivity extends AppCompatActivity {
  private Toolbar myToolbar;
  protected void onCreate(Bundle savedInstanceState) {
    super.onCreate(savedInstanceState);
    setContentView(R.layout.mytoolbar);
    myToolbar = (Toolbar) findViewById(R.id.myToolbar);
    myToolbar.setNavigationIcon(R.mipmap.navigation);      //设置导航图标
    myToolbar.inflateMenu(R.menu.mymenu);                  //设置右上角的菜单
    myToolbar.setOnMenuItemClickListener(new Toolbar.OnMenuItemClickListener() {
      @Override
      public boolean onMenuItemClick(MenuItem item) {      //响应单击菜单
        int menuItemId = item.getItemId();
        if (menuItemId == R.id.myMenu1) {
          Toast.makeText(MainActivity.this,
```

```
                    "刚才单击的菜单是：经营管理", Toast.LENGTH_SHORT).show();
        } else if (menuItemId == R.id.myMenu2) {
            Toast.makeText(MainActivity.this,
                    "刚才单击的菜单是：文学艺术", Toast.LENGTH_SHORT).show();
        } else if (menuItemId == R.id.myMenu3) {
            Toast.makeText(MainActivity.this,
                    "刚才单击的菜单是：职场励志", Toast.LENGTH_SHORT).show();
        } else if (menuItemId == R.id.myMenu4) {
            Toast.makeText(MainActivity.this,
                    "刚才单击的菜单是：技术前沿", Toast.LENGTH_SHORT).show();
        }
        return true;
    }});}}
```

上面这段代码在 MyCode\MySample294\app\src\main\java\com\bin\luo\mysample\MainActivity.java 文件中。在这段代码中，myToolbar.inflateMenu(R.menu.mymenu)用于在工具栏上加载 mymenu 资源。mymenu 资源的主要内容如下：

```xml
<?xml version = "1.0" encoding = "utf-8"?>
<menu xmlns:android = "http://schemas.android.com/apk/res/android"
      xmlns:app = "http://schemas.android.com/apk/res-auto">
    <item android:id = "@+id/myMenu1"
          android:title = "经营管理"
          app:showAsAction = "never" />
    <item android:id = "@+id/myMenu2"
          android:title = "文学艺术"
          app:showAsAction = "never" />
    <item android:id = "@+id/myMenu3"
          android:title = "职场励志"
          app:showAsAction = "never" />
    <item android:id = "@+id/myMenu4"
          android:title = "技术前沿"
          app:showAsAction = "never" />
</menu>
```

上面这段代码在 MyCode\MySample294\app\src\main\res\menu\mymenu.xml 文件中。在这段代码中，app:showAsAction = "never" 表示该 item 仅以菜单的形式显示，如果是 app:showAsAction = "ifRoom"，则在工具栏有足够空间的情况下，菜单以按钮的形式显示。此外，使用 Toolbar 需要在 gradle 中引入 compile 'com.android.support:appcompat-v7:25.2.0' 依赖项。此实例的完整项目在 MyCode\MySample294 文件夹中。

064 使用 ActionProvider 创建二级菜单

此实例主要通过使用自定义 ActionProvider 类作为 setActionProvider()方法的参数，实现在标题栏上添加二级菜单的功能。当实例运行之后，单击标题栏右端的菜单图标，将弹出一级菜单，在一级菜单中选择"购物网站"，如图 064-1 的左图所示，将弹出二级菜单；在二级菜单中选择"京东商城"，如图 064-1 的右图所示，将在 WebView 中显示京东商城的主页。

图 064-1

主要代码如下：

```java
@Override
public boolean onCreateOptionsMenu(Menu myMenu) {
    //动态创建一级菜单,并指定 MyActionProvider 对象来动态创建二级菜单
    myMenu.add("购物网站").setActionProvider(new MyActionProvider(this)).
            setShowAsAction(MenuItem.SHOW_AS_ACTION_NEVER);
    myMenu.add("新浪网").setOnMenuItemClickListener(
                        new MenuItem.OnMenuItemClickListener(){
        @Override
        public boolean onMenuItemClick(MenuItem item) {
            MainActivity.myWebView.loadUrl("http://www.sina.com.cn/");
            return true;
        } });
    myMenu.add("搜狐网").setOnMenuItemClickListener(
                        new MenuItem.OnMenuItemClickListener() {
        @Override
        public boolean onMenuItemClick(MenuItem item) {
            MainActivity.myWebView.loadUrl("http://www.sohu.com/");
            return true;
        } });
    myMenu.add("安居客").setOnMenuItemClickListener(
                        new MenuItem.OnMenuItemClickListener() {
        @Override
        public boolean onMenuItemClick(MenuItem item) {
            MainActivity.myWebView.loadUrl("https://chongqing.anjuke.com");
            return true;
        } });
    return super.onCreateOptionsMenu(myMenu);
}
```

上面这段代码在 MyCode \ MySample897 \ app \ src \ main \ java \ com \ bin \ luo \ mysample \ MainActivity.java 文件中。在这段代码中，myMenu.add("购物网站").setActionProvider（new MyActionProvider(this)).setShowAsAction(MenuItem.SHOW_AS_ACTION_NEVER)用于创建一级菜单"购物网站"，并在该菜单下使用自定义 MyActionProvider 创建二级菜单。自定义 MyActionProvider 的主要代码如下：

```java
public class MyActionProvider extends ActionProvider {
    public MyActionProvider(Context context) { super(context); }
    @Override
    public View onCreateActionView() { return null; }
    @Override
    //创建二级菜单
    public void onPrepareSubMenu(SubMenu subMenu) {
        subMenu.clear();
        //通过 subMenu 创建二级菜单，并设置菜单图标以及响应单击事件
        subMenu.add("京东商城").setIcon(R.mipmap.ic_launcher)
            .setOnMenuItemClickListener(new MenuItem.OnMenuItemClickListener() {
                @Override
                public boolean onMenuItemClick(MenuItem item) {
                    MainActivity.myWebView.loadUrl("https://www.jd.com");
                    return true;
                }
            });
        subMenu.add("天猫商城").setIcon(R.mipmap.ic_launcher)
            .setOnMenuItemClickListener(new MenuItem.OnMenuItemClickListener() {
                @Override
                public boolean onMenuItemClick(MenuItem item) {
                    MainActivity.myWebView.loadUrl("https://www.tmall.com");
                    return true;
                }
            });
        subMenu.add("苏宁易购").setIcon(R.mipmap.ic_launcher)
            .setOnMenuItemClickListener(new MenuItem.OnMenuItemClickListener() {
                @Override
                public boolean onMenuItemClick(MenuItem item) {
                    MainActivity.myWebView.loadUrl("https://www.suning.com");
                    return true;
                }
            });
    }
    @Override
    //返回 true 表示创建二级菜单
    public boolean hasSubMenu() { return true; }
}
```

上面这段代码在 MyCode \ MySample897 \ app \ src \ main \ java \ com \ bin \ luo \ mysample \ MyActionProvider.java 文件中。此外，访问网络需要在 AndroidManifest.xml 文件中添加< uses-permission android:name = "android.permission.INTERNET"/>权限。此实例的完整项目在 MyCode\MySample897 文件夹中。

065　在右上角二级菜单中实现单选按钮风格

此实例主要通过使用 setGroupCheckable()方法，实现二级菜单项以单选按钮的形式显示。当实例运行之后，单击应用右上角的菜单图标，将滑出一级菜单，如图 065-1 的左图所示。在一级菜单中选择"报考单位"，将滑出"报考单位"的二级菜单，并且以单选按钮的风格显示，选择"长寿区云集镇人民

政府"单选按钮,如图 065-1 的右图所示,将在弹出的 Toast 中显示选择结果。

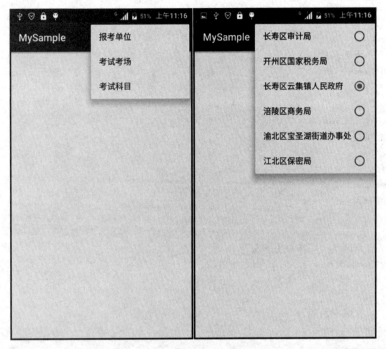

图 065-1

主要代码如下:

```
public boolean onCreateOptionsMenu(Menu myMenu){
  //创建一级菜单
  SubMenu newMenu = myMenu.addSubMenu("报考单位");
  //newMenu.setHeaderTitle("选择您的报考单位");
  //创建"报考单位"的二级菜单
  newMenu.add(0, ITEM1, 0, "长寿区审计局");
  newMenu.add(0, ITEM2, 1, "开州区国家税务局");
  newMenu.add(0, ITEM3, 2, "长寿区云集镇人民政府");
  newMenu.add(0, ITEM4, 3, "涪陵区商务局");
  newMenu.add(0, ITEM5, 4, "渝北区宝圣湖街道办事处");
  newMenu.add(0, ITEM6, 5, "江北区保密局");
  //设置二级菜单以单选按钮风格显示
  newMenu.setGroupCheckable(0 , true , true);
  //创建一级菜单"考试考场"
  newMenu = myMenu.addSubMenu("考试考场");
  //创建一级菜单考试科目"
  newMenu = myMenu.addSubMenu("考试科目");
  return super.onCreateOptionsMenu(myMenu);
}
@Override
//菜单项被单击后的回调方法
public boolean onOptionsItemSelected(MenuItem myItem){
  String myInfo = "你选择的报考单位是:";
  //判断单击了哪个菜单项
  switch (myItem.getItemId()){
    case ITEM1:
```

```
            myInfo += "长寿区审计局";
            myItem.setChecked(true);
            break;
        case ITEM2:
            myInfo += "开州区国家税务局";
            myItem.setChecked(true);
            break;
        case ITEM3:
            myInfo += "长寿区云集镇人民政府";
            myItem.setChecked(true);
            break;
        case ITEM4:
            myInfo += "涪陵区商务局";
            myItem.setChecked(true);
            break;
        case ITEM5:
            myInfo += "渝北区宝圣湖街道办事处";
            myItem.setChecked(true);
            break;
        case ITEM6:
            myInfo += "江北区保密局";
            myItem.setChecked(true);
            break;
    }
    if(myInfo.length()>10){                         //仅在二级菜单中显示 Toast
        Toast.makeText(getApplicationContext(),
                            myInfo, Toast.LENGTH_SHORT).show();
    }
    return true;
}
```

上面这段代码在 MyCode\MySample279\app\src\main\java\com\bin\luo\mysample\MainActivity.java 文件中。在这段代码中，newMenu.add(0，ITEM1，0，"长寿区审计局")用于在菜单中增加子菜单，add()方法的语法声明如下：

```
add( int groupId, int itemId, int order, CharSequence title)
```

其中，参数 int groupId 表示菜单所属组别。参数 int itemId 表示当前菜单项的标识。参数 int order 表示菜单在组中的序号。参数 CharSequence title 表示菜单标题。

newMenu.setGroupCheckable（0，true，truc)用于使菜单以单选按钮的形式显示，setGroupCheckable()方法的语法声明如下：

```
setGroupCheckable( int group, boolean checkable, boolean exclusive)
```

其中，参数 int group 表示组别。参数 boolean checkable 表示是否允许检查标记，true 表示允许，false 表示禁止。参数 boolean exclusive 表示此菜单是以单选按钮形式显示还是以多选框形式显示，true 表示以单选按钮形式显示，false 表示以多选框形式显示。

此实例的完整项目在 MyCode\MySample279 文件夹中。

066 在右上角二级菜单中实现多选框风格

此实例主要通过使用 setCheckable() 方法实现二级菜单项以多选框的形式显示。当实例运行之后，单击应用右上角的菜单图标，将滑出一级菜单，在一级菜单中选择"考试科目"，将滑出"考试科目"的二级菜单，并且以多选框的风格显示；在其中分别选择"行政职业能力测验""行政法律制度"等菜单，如图 066-1 的左图所示，将在弹出的 Toast 中显示选择结果，如图 066-1 的右图所示。注意：一次只能选择（或取消）一个菜单，同时自动保存以前的选择（或取消）结果。

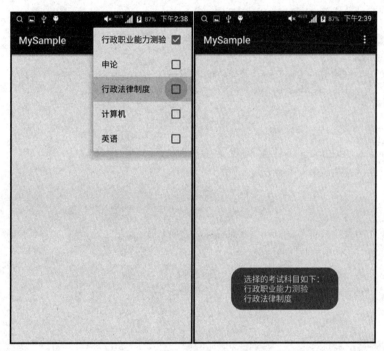

图 066-1

主要代码如下：

```
public boolean onCreateOptionsMenu(Menu myMenu) {
    //创建一级菜单"报考单位"
    SubMenu mySubMenu = myMenu.addSubMenu("报考单位");
    //创建一级菜单"考试考场"
    mySubMenu = myMenu.addSubMenu("考试考场");
    //创建一级菜单"考试科目"
    mySubMenu = myMenu.addSubMenu("考试科目");
    //创建"考试科目"的二级菜单,并设置它为多选框
    myItems[0] = mySubMenu.add(0, ITEM1, 0, myTitles[0]).setCheckable(true);
    myItems[1] = mySubMenu.add(0, ITEM2, 1, myTitles[1]).setCheckable(true);
    myItems[2] = mySubMenu.add(0, ITEM3, 2, myTitles[2]).setCheckable(true);
    myItems[3] = mySubMenu.add(0, ITEM4, 3, myTitles[3]).setCheckable(true);
    myItems[4] = mySubMenu.add(0, ITEM5, 4, myTitles[4]).setCheckable(true);
    return super.onCreateOptionsMenu(myMenu);
}
@Override
//菜单项被单击后的回调方法
```

```java
public boolean onOptionsItemSelected(MenuItem myItem) {
    //判断单击了哪个菜单项
    switch (myItem.getItemId()) {
        case ITEM1:
            if (myItem.isChecked()) { myItem.setChecked(false); }
            else { myItem.setChecked(true); }
            showInfo();
            break;
        case ITEM2:
            if (myItem.isChecked()) { myItem.setChecked(false); }
            else { myItem.setChecked(true); }
            showInfo();
            break;
        case ITEM3:
            if (myItem.isChecked()) { myItem.setChecked(false); }
            else { myItem.setChecked(true); }
            showInfo();
            break;
        case ITEM4:
            if (myItem.isChecked()) { myItem.setChecked(false); }
            else { myItem.setChecked(true); }
            showInfo();
            break;
        case ITEM5:
            if (myItem.isChecked()) { myItem.setChecked(false); }
            else { myItem.setChecked(true); }
            showInfo();
            break;
    }
    return true;
}
private void showInfo() {
    String myInfo = "选择的考试科目如下：";
    for (int i = 0; i < myItems.length; i++) {
        if (myItems[i].isChecked()) { myInfo += "\n" + myTitles[i]; }
    }
    Toast.makeText(this, myInfo, Toast.LENGTH_SHORT).show();
}
```

上面这段代码在 MyCode\MySample280\app\src\main\java\com\bin\luo\mysample\MainActivity.java 文件中。在这段代码中，myItems[0] = mySubMenu.add(0, ITEM1, 0, myTitles[0]).setCheckable(true)用于向二级菜单中添加一个多选框风格的子菜单项，setCheckable()方法的语法声明如下：

setCheckable(boolean checkable)

其中，参数 boolean checkable 用于控制菜单项是否以多选框形式显示，当该参数值为 true 时，菜单项以多选框形式显示。

myItem.setChecked(true)用于设置菜单项以选中状态显示，myItem.setChecked(false)用于设置菜单项以未选中状态显示。此实例的完整项目在 MyCode\MySample280 文件夹中。

067 使用 DrawerLayout 创建抽屉式侧滑菜单

此实例主要通过使用 DrawerLayout 控件和 ListView 控件，从而创建抽屉式的侧滑菜单。当实例运行之后，手指按住屏幕从左向右滑动，将从左侧滑出一个菜单列表，然后单击其中的菜单项，如"三个老枪手"，侧滑菜单消失，屏幕将显示《三个老枪手》的电影海报图像，效果分别如图 067-1 的左图和右图所示。选择其他菜单项将实现类似的功能。

图 067-1

主要代码如下：

```xml
<?xml version = "1.0" encoding = "utf - 8"?>
<android.support.v4.widget.DrawerLayout
    xmlns:android = "http://schemas.android.com/apk/res/android"
    android:id = "@ + id/myDrawerlayout"
    android:layout_width = "match_parent"
    android:layout_height = "match_parent">
    <FrameLayout android:id = "@ + id/v4_drawerlayout_frame"
        android:layout_width = "match_parent"
        android:layout_height = "match_parent" >
        <LinearLayout android:layout_width = "match_parent"
            android:layout_height = "match_parent"
            android:orientation = "vertical">
            <TextView android:layout_width = "match_parent"
                android:layout_height = "wrap_content"
                android:id = "@ + id/myTextView"
                android:text = "蜘蛛侠：英雄归来"
                android:textSize = "22sp"
                android:textColor = "@color/colorAccent"
                android:gravity = "center"/>
```

```xml
<ImageView android:layout_width = "match_parent"
    android:layout_height = "match_parent"
    android:scaleType = "fitXY"
    android:id = "@+id/myImageView"
    android:src = "@mipmap/myimage1"
    android:gravity = "center"/>
    </LinearLayout>
</FrameLayout>
<ListView android:layout_width = "200dp"
    android:layout_height = "match_parent"
    android:layout_gravity = "left"
    android:id = "@+id/myListview"
    android:choiceMode = "singleChoice"
    android:background = "#76EEC6" />
</android.support.v4.widget.DrawerLayout>
```

上面这段代码在 MyCode\MySample358\app\src\main\res\layout\activity_main.xml 文件中。在这段代码中，ListView 控件主要用于管理侧滑菜单的菜单项，FrameLayout 控件则用于承载在单击侧滑菜单之后的显示内容。在 Android 中，DrawerLayout 控件分为侧边菜单和主内容区两部分，侧边菜单可以根据手势展开与隐藏，主内容区的部分可以随着菜单的单击而变化。菜单单击事件则通过 Java 代码响应，主要代码如下：

```java
private void initData() {
    final List<String> myList = new ArrayList<String>();
    myList.add("蜘蛛侠：英雄归来");
    myList.add("乐高蝙蝠侠");
    myList.add("海滩游侠");
    myList.add("三个老枪手");
    ArrayAdapter<String> adapter = new ArrayAdapter<>(this,
                        android.R.layout.simple_list_item_1, myList);
    myListView.setAdapter(adapter);
    myListView.setOnItemClickListener(new AdapterView.OnItemClickListener() {
        @Override
        public void onItemClick(AdapterView<?> parent,
                            View view, int position, long id) {
            switch (position) {                    //显示电影海报图像
            case 0:
                myImageView.setImageResource(R.mipmap.myimage1);
                break;
            case 1:
                myImageView.setImageResource(R.mipmap.myimage2);
                break;
            case 2:
                myImageView.setImageResource(R.mipmap.myimage3);
                break;
            case 3:
                myImageView.setImageResource(R.mipmap.myimage4);
                break;
            }
            myTextView.setText(myList.get(position));  //显示电影名称
            showDrawerLayout();
        }});
```

```
    myDrawerLayout.openDrawer(Gravity.LEFT);        //打开侧滑菜单,默认不会打开
  }
  private void showDrawerLayout() {                 //关闭或显示左侧侧滑菜单
    if (!myDrawerLayout.isDrawerOpen(Gravity.LEFT)) {
      myDrawerLayout.openDrawer(Gravity.LEFT);
    } else {
      myDrawerLayout.closeDrawer(Gravity.LEFT);
    }
  }
```

上面这段代码在 MyCode\MySample358\app\src\main\java\com\bin\luo\mysample\MainActivity.java 文件中。在这段代码中,myDrawerLayout.openDrawer(Gravity.LEFT)用于从左侧打开侧滑菜单。此外,使用 DrawerLayout 控件时需要在 gradle 中添加 compile 'com.android.support：appcompat-v7:24.2.1'依赖项。默认情况下,当在 Android Studio 中添加此依赖项(或其他依赖项)时,在右上角将立即弹出一个黄色的提示框,单击右侧的"Sync Now"超链接,Android Studio 将会自动进行同步操作。此实例的完整项目在 MyCode\MySample358 文件夹中。

068 使用手机菜单键控制侧滑菜单是否显示

此实例主要通过在 onKeyDown()事件响应方法中检测菜单键 KEYCODE_MENU,并使用 DrawerLayout 的 closeDrawer()和 openDrawer()方法,实现使用手机菜单键控制侧滑菜单是否显示。当实例运行之后,单击手机左下角的菜单键,从左侧滑出侧滑菜单,单击侧滑菜单中的任一菜单,如"我的消息",将在弹出的 Toast 中显示对应的内容,如图 068-1 的左图所示;再次单击手机左下角的菜单键,关闭侧滑菜单,如图 068-1 的右图所示。当然,在屏幕上从左向右滑动,也可以拉出侧滑菜单;在屏幕上从右向左滑动,可以关闭侧滑菜单。

图 068-1

主要代码如下：

```java
public class MainActivity extends Activity {
  public DrawerLayout myDrawerLayout;
  public NavigationView myNavigationView;
  public boolean IsMyDrawerOpen = false;                    //判断侧滑菜单的显示状态
  @Override
  protected void onCreate(Bundle savedInstanceState) {
    super.onCreate(savedInstanceState);
    setContentView(R.layout.activity_main);
    myDrawerLayout = (DrawerLayout) findViewById(R.id.myDrawerLayout);
    myNavigationView = (NavigationView) findViewById(R.id.myNavigationView);
    myNavigationView.setItemIconTintList(null);
    myNavigationView.setNavigationItemSelectedListener(
            new NavigationView.OnNavigationItemSelectedListener() {
      @Override
      public boolean onNavigationItemSelected(MenuItem item) {
        switch(item.getItemId()){
        case R.id.myItem1:
          Toast.makeText(getApplicationContext(),
                  "刚才单击的菜单是【我的收藏】", Toast.LENGTH_SHORT).show();
          break;
        case R.id.myItem2:
          Toast.makeText(getApplicationContext(),
                  "刚才单击的菜单是【我的钱包】", Toast.LENGTH_SHORT).show();
          break;
        case R.id.myItem3:
          Toast.makeText(getApplicationContext(),
                  "刚才单击的菜单是【我的相册】", Toast.LENGTH_SHORT).show();
          break;
        case R.id.myItem4:
          Toast.makeText(getApplicationContext(),
                  "刚才单击的菜单是【我的消息】", Toast.LENGTH_SHORT).show();
          break;
        }
        return true;
      } });
  }
  @Override
  //响应单击手机菜单键(部分模拟器没有此键,一定要在手机上测试)
  public boolean onKeyDown(int keyCode, KeyEvent event) {
    if (IsMyDrawerOpen && keyCode == event.KEYCODE_MENU) {
      myDrawerLayout.closeDrawer(myNavigationView);         //关闭侧滑菜单
      IsMyDrawerOpen = false;
    } else if (!IsMyDrawerOpen && keyCode == event.KEYCODE_MENU) {
      myDrawerLayout.openDrawer(myNavigationView);          //打开侧滑菜单
      IsMyDrawerOpen = true;
    }
    return super.onKeyDown(keyCode, event);
  }
}
```

上面这段代码在 MyCode \ MySample652 \ app \ src \ main \ java \ com \ bin \ luo \ mysample \

MainActivity.java 文件中。在这段代码中，keyCode == event.KEYCODE_MENU 表示当前用户按下了手机的菜单键，myDrawerLayout.closeDrawer(myNavigationView)用于关闭侧滑菜单，myDrawerLayout.openDrawer(myNavigationView)用于打开侧滑菜单。一般情况下，承载菜单项的 NavigationView 控件位于 DrawerLayout 控件的内部，主要代码如下：

```xml
<android.support.v4.widget.DrawerLayout
    android:id = "@+id/myDrawerLayout"
    android:layout_width = "match_parent"
    android:layout_height = "match_parent">
<LinearLayout
    android:layout_width = "match_parent"
    android:layout_height = "match_parent"
    android:background = "@mipmap/myimage1"
    android:orientation = "vertical">
</LinearLayout>
<android.support.design.widget.NavigationView
    android:id = "@+id/myNavigationView"
    android:layout_width = "180dp"
    android:layout_height = "match_parent"
    android:layout_gravity = "left"
    android:fitsSystemWindows = "true"
    app:menu = "@menu/mymenu">
</android.support.design.widget.NavigationView>
</android.support.v4.widget.DrawerLayout>
```

上面这段代码在 MyCode\MySample652\app\src\main\res\layout\activity_main.xml 文件中。在这段代码中，app:menu="@menu/mymenu"用于加载菜单资源 mymenu，mymenu 的主要内容如下：

```xml
<?xml version = "1.0" encoding = "utf-8"?>
<menu xmlns:android = "http://schemas.android.com/apk/res/android">
<item android:id = "@+id/myItem1"
    android:icon = "@drawable/myicon"
    android:title = "我的收藏"/>
<item android:id = "@+id/myItem2"
    android:icon = "@drawable/myicon"
    android:title = "我的钱包"/>
<item android:id = "@+id/myItem3"
    android:icon = "@drawable/myicon"
    android:title = "我的相册"/>
<item android:id = "@+id/myItem4"
    android:icon = "@drawable/myicon"
    android:title = "我的消息"/>
</menu>
```

上面这段代码在 MyCode\MySample652\app\src\main\res\menu\mymenu.xml 文件中。此外，使用此实例的相关控件需要在 gradle 中引入 compile 'com.android.support:design:24.2.0'依赖项。此实例的完整项目在 MyCode\MySample652 文件夹中。

069　在侧滑菜单中使用 NavigationView 导航

此实例主要通过使用 NavigationView 控件加载菜单项，实现对侧滑菜单进行导航。当实例运行之后，如果手指在屏幕上从最左端向右滑动，将从左向右拉出侧滑菜单，任意选择一个菜单项，如"我是传奇"，将显示该菜单项对应的电影名字和海报，如图 069-1 的左图所示；如果手指在侧滑菜单上从右向左滑动，将关闭侧滑菜单，满屏显示对应的电影海报，如图 069-1 的右图所示。选择其他菜单将实现类似的功能。

图　069-1

主要代码如下：

```
public class MainActivity extends Activity {
 NavigationView myNavigationView;
 TextView myTextView;
 ImageView myImageView;
 @Override
 protected void onCreate(Bundle savedInstanceState) {
  super.onCreate(savedInstanceState);
  setContentView(R.layout.activity_main);
  myTextView = (TextView) findViewById(R.id.myTextView);
  myImageView = (ImageView) findViewById(R.id.myImageView);
  myNavigationView = (NavigationView) findViewById(R.id.myNavigationView);
  //显示图标的原始颜色
  myNavigationView.setItemIconTintList(null);
  //在这里处理侧滑菜单的单击事件
  myNavigationView.setNavigationItemSelectedListener(
          new NavigationView.OnNavigationItemSelectedListener() {
   @Override
```

```
    public boolean onNavigationItemSelected(MenuItem item) {
     switch(item.getItemId()){
      case R.id.myItem1:
       myTextView.setText("我是传奇");
       myImageView.setImageResource(R.mipmap.myimage1);
      break;
      case R.id.myItem2:
       myTextView.setText("银翼杀手");
       myImageView.setImageResource(R.mipmap.myimage2);
      break;
      case R.id.myItem3:
       myTextView.setText("三百勇士");
       myImageView.setImageResource(R.mipmap.myimage3);
      break;
      case R.id.myItem4:
       myTextView.setText("功夫熊猫");
       myImageView.setImageResource(R.mipmap.myimage4);
      break;
      case R.id.myItem5:
       myTextView.setText("怒火救援");
       myImageView.setImageResource(R.mipmap.myimage5);
      break;
     }
     return true;
    } });
} }
```

上面这段代码在 MyCode\MySample436\app\src\main\java\com\bin\luo\mysample\MainActivity.java 文件中。在这段代码中，myNavigationView.setItemIconTintList(null)用于强制菜单左边的图标以原色显示，否则以黑白风格显示。onNavigationItemSelected()方法用于响应单击菜单项事件。NavigationView 控件通常在 DrawerLayout 控件的里面，主要代码如下：

```
<android.support.v4.widget.DrawerLayout android:layout_width="match_parent"
                                        android:layout_height="match_parent">
    <LinearLayout android:layout_width="match_parent"
                  android:layout_height="match_parent"
                  android:orientation="vertical">
        <TextView android:id="@+id/myTextView"
                  android:layout_width="match_parent"
                  android:layout_height="wrap_content"
                  android:gravity="end"
                  android:padding="5dp"
                  android:background="#000"
                  android:textColor="#fff"
                  android:textSize="24dp"
                  android:text="我是传奇"/>
        <ImageView android:id="@+id/myImageView"
                   android:layout_width="match_parent"
                   android:layout_height="match_parent"
                   android:scaleType="centerCrop"
                   android:src="@mipmap/myimage1"/>
    </LinearLayout>
```

```xml
<android.support.design.widget.NavigationView
    android:id="@+id/myNavigationView"
    android:layout_width="165dp"
    android:layout_height="match_parent"
    android:layout_gravity="left"
    android:fitsSystemWindows="true"
    app:headerLayout="@layout/mylayout"
    app:menu="@menu/mymenu">
</android.support.design.widget.NavigationView>
</android.support.v4.widget.DrawerLayout>
```

上面这段代码在 MyCode\MySample436\app\src\main\res\layout\activity_main.xml 文件中。在这段代码中，app:headerLayout="@layout/mylayout"用于设置侧滑菜单上面的视图（此实例仅为一幅图像，电影《末日迷踪》的海报图像）。mylayout 布局的主要内容如下：

```xml
<?xml version="1.0" encoding="utf-8"?>
<LinearLayout xmlns:android="http://schemas.android.com/apk/res/android"
    android:layout_width="match_parent"
    android:layout_height="280dp"
    android:orientation="vertical">
<ImageView
    android:layout_width="match_parent"
    android:layout_height="match_parent"
    android:scaleType="centerCrop"
    android:src="@mipmap/myimage"/>
</LinearLayout>
```

上面这段代码在 MyCode\MySample436\app\src\main\res\layout\mylayout.xml 文件中。在 activity_main.xml 文件中，app:menu="@menu/mymenu"用于在 NavigationView 控件上加载菜单资源 mymenu，关于 mymenu 资源的详细内容请参考 MyCode\MySample436\app\src\main\res\menu\mymenu.xml 文件。此外，使用此实例的相关控件需要在 gradle 中引入 compile 'com.android.support:design:24.2.0'依赖项。此实例的完整项目在 MyCode\MySample436 文件夹中。

070 使用 TabLayout 高仿微信底部导航菜单

此实例主要通过使用 TabLayout 控件和 ViewPager 控件，从而实现高仿微信底部的导航菜单。当实例运行之后，单击底部导航条上的"微信"菜单，则将显示该菜单对应的选项卡，效果如图 070-1 的左图所示。单击底部导航条上的"朋友圈"菜单，也将显示该菜单对应的选项卡，效果如图 070-1 的右图所示。单击底部导航条上的其他两个菜单，将实现类似的功能。

主要代码如下：

```xml
<?xml version="1.0" encoding="utf-8"?>
<LinearLayout xmlns:android="http://schemas.android.com/apk/res/android"
    android:layout_width="match_parent"
    android:layout_height="match_parent"
    android:orientation="vertical">
<android.support.v4.view.ViewPager
    android:id="@+id/myViewPager"
    android:layout_width="match_parent"
```

第3章 菜单

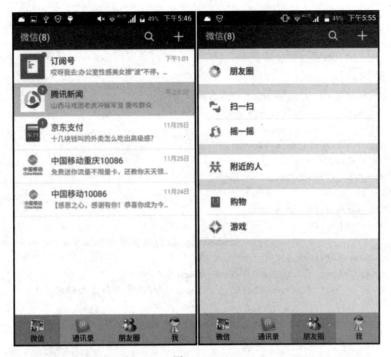

图 070-1

```
        android:layout_height = "0dp"
        android:layout_weight = "1">
</android.support.v4.view.ViewPager>
< android.support.design.widget.TabLayout
        android:id = "@ + id/myTabLayout"
        android:layout_width = "match_parent"
        android:layout_height = "60dp"
        android:padding = "1dp"
        android:background = "#00ccff">
        </android.support.design.widget.TabLayout>
</LinearLayout>
```

上面这段代码在 MyCode\MySample481\app\src\main\res\layout\activity_main.xml 文件中。在这段代码中，TabLayout 控件主要用于管理底部的导航菜单条，ViewPager 控件则用于管理选项卡的内容视图（View）。此外，在此实例中，MainActivity 需要从 FragmentActivity 继承，主要代码如下：

```
public class MainActivity extends FragmentActivity {
  private TabLayout myTabLayout;
  private ViewPager myViewPager;
  private List < String > myItems;
  private List < Fragment > myTabFragments;
  private ContentPagerAdapter myContentPagerAdapter;
  @Override
  protected void onCreate(@Nullable Bundle savedInstanceState) {
    super.onCreate(savedInstanceState);
    setContentView(R.layout.activity_main);
    myTabLayout = (TabLayout) findViewById(R.id.myTabLayout);
```

```java
        myViewPager = (ViewPager) findViewById(R.id.myViewPager);
        initContent();
        initTab();
    }
    private void initTab() {
        myTabLayout.setTabMode(TabLayout.MODE_FIXED);
        myTabLayout.setSelectedTabIndicatorHeight(0);
        ViewCompat.setElevation(myTabLayout, 10);
        myTabLayout.setupWithViewPager(myViewPager);
        for (int i = 0; i < myItems.size(); i++) {
            TabLayout.Tab myTab = myTabLayout.getTabAt(i);
            if (myTab != null) {
                myTab.setCustomView(R.layout.myitem);
                TextView myMenuText = (TextView) myTab.getCustomView().
                                    findViewById(R.id.myMenu);  //设置导航菜单文本
                myMenuText.setText(myItems.get(i));
                ImageView myMenuImage = (ImageView) myTab.getCustomView().
                                    findViewById(R.id.myImage); //设置导航菜单图标
                switch (i) {
                    case 0:
                        myMenuImage.setImageResource(R.mipmap.myimage1);
                        break;
                    case 1:
                        myMenuImage.setImageResource(R.mipmap.myimage2);
                        break;
                    case 2:
                        myMenuImage.setImageResource(R.mipmap.myimage3);
                        break;
                    case 3:
                        myMenuImage.setImageResource(R.mipmap.myimage4);
                        break;
                }
                myMenuText.setText(myItems.get(i));
            } }
        myTabLayout.getTabAt(0).getCustomView().setSelected(true);
    }
    private void initContent() {
        myItems = new ArrayList<>();
        myItems.add("微信");
        myItems.add("通讯录");
        myItems.add("朋友圈");
        myItems.add("我");
        myTabFragments = new ArrayList<>();
        //根据菜单文本同步 ViewPager
        for (String myMenuText : myItems) {
            myTabFragments.add(MyFragment.newInstance(myMenuText));
        }
        myContentPagerAdapter = new ContentPagerAdapter(getSupportFragmentManager());
        myViewPager.setAdapter(myContentPagerAdapter);
    }
    class ContentPagerAdapter extends FragmentPagerAdapter {
        public ContentPagerAdapter(FragmentManager fm) { super(fm); }
        @Override
```

```
    public Fragment getItem(int position) { return myTabFragments.get(position); }
    @Override
    public int getCount() { return myItems.size(); }
    @Override
    public CharSequence getPageTitle(int position) {
      return myItems.get(position);
} } }
```

上面这段代码在 MyCode\MySample481\app\src\main\java\com\bin\luo\mysample\MainActivity.java 文件中。在这段代码中，MyFragment.newInstance(myMenuText)用于创建在 ViewPager 中显示的内容(此实例是一幅图像)。MyFragment 的主要代码如下：

```
public class MyFragment extends Fragment {
  private static final String EXTRA_CONTENT = "content";
  public static MyFragment newInstance(String content) {
    Bundle myBundle = new Bundle();
    myBundle.putString(EXTRA_CONTENT, content);
    MyFragment myFragment = new MyFragment();
    myFragment.setArguments(myBundle);
    return myFragment;
  }
  public View onCreateView(LayoutInflater inflater,
                ViewGroup container, Bundle savedInstanceState) {
    View myContentView = inflater.inflate(R.layout.mylayout, null);
    ImageView myImageView =
            (ImageView) myContentView.findViewById(R.id.myImageView);
    String myMenu = getArguments().getString(EXTRA_CONTENT);
    if (myMenu.contains("微信")) {
      myImageView.setImageResource(R.mipmap.myimage11);
    } else if (myMenu.contains("通讯录")) {
      myImageView.setImageResource(R.mipmap.myimage22);
    }
    else if (myMenu.contains("朋友圈")) {
      myImageView.setImageResource(R.mipmap.myimage33);
    }
    else if (myMenu.contains("我")) {
      myImageView.setImageResource(R.mipmap.myimage44);
    }
    return myContentView;
} } }
```

上面这段代码在 MyCode\MySample481\app\src\main\java\com\bin\luo\mysample\MyFragment.java 文件中。在这段代码中，myContentView = inflater.inflate(R.layout.mylayout, null)用于根据 mylayout 布局的内容创建 myContentView。mylayout 布局的主要内容如下：

```
<?xml version = "1.0" encoding = "utf-8"?>
<LinearLayout xmlns:android = "http://schemas.android.com/apk/res/android"
              android:layout_width = "match_parent"
              android:layout_height = "match_parent"
              android:gravity = "center"
              android:orientation = "vertical">
  <ImageView android:scaleType = "centerCrop"
             android:id = "@ + id/myImageView"
             android:layout_width = "match_parent"
```

```
        android:layout_height = "match_parent"
        android:src = "@mipmap/myimage11"/>
</LinearLayout>
```

上面这段代码在 MyCode\MySample481\app\src\main\res\layout\mylayout.xml 文件中。在这段代码中，ImageView 控件的主要作用就是在单击底部导航栏的不同菜单时，显示对应的图像（微信截图）。在 MainActivity.java 文件中，myTab.setCustomView（R.layout.myitem）用于根据 myitem 布局的内容创建底部导航栏的各个菜单项，myitem 布局的主要内容如下：

```xml
<?xml version = "1.0" encoding = "utf-8"?>
<LinearLayout xmlns:android = "http://schemas.android.com/apk/res/android"
    android:layout_width = "match_parent"
    android:layout_height = "match_parent"
    android:gravity = "center"
    android:orientation = "vertical">
    <ImageView android:id = "@ + id/myImage"
        android:scaleType = "fitCenter"
        android:layout_width = "30dp"
        android:layout_height = "30dp"
        android:src = "@mipmap/myimage1"/>
    <TextView android:id = "@ + id/myMenu"
        android:layout_width = "wrap_content"
        android:layout_height = "wrap_content"
        android:text = "朋友圈"
        android:textColor = "#000"/>
</LinearLayout>
```

上面这段代码在 MyCode\MySample481\app\src\main\res\layout\myitem.xml 文件中。此外，使用此实例的相关控件需要在 gradle 中引入 compile 'com.android.support:design:23.3.0' 依赖项。此实例的完整项目在 MyCode\MySample481 文件夹中。

071 在弹出底部菜单时主窗口立即变暗

此实例主要通过设置 WindowManager.LayoutParams 的 alpha 属性值，实现在弹出底部菜单时主窗口立即变暗。当实例运行之后，单击屏幕（风景图像），将从底部滑出菜单窗口，并且屏幕变暗，如图 071-1 的左图所示。选择其中的菜单，如"显示下一幅图像"，菜单窗口消失，屏幕恢复正常显示，并在弹出的 Toast 中显示刚才选择的菜单，如图 071-1 的右图所示。

主要代码如下：

```java
public class MainActivity extends Activity implements View.OnClickListener {
    private PopupWindow myMenuWindow;
    @Override
    protected void onCreate(Bundle savedInstanceState) {
        super.onCreate(savedInstanceState);
        setContentView(R.layout.activity_main);
        ImageView myImageView = (ImageView) findViewById(R.id.myImageView);
        //单击图像时从屏幕底部滑出菜单窗口
        myImageView.setOnTouchListener(new View.OnTouchListener() {
            @Override
            public boolean onTouch(View v, MotionEvent event) {
                if (event.getAction() == MotionEvent.ACTION_DOWN) {
```

图 071-1

```
        View myWindow = LayoutInflater.from(MainActivity.this).inflate(
                R.layout.mylayout, null, false);
        myMenuWindow = new PopupWindow();
        myMenuWindow.setWidth(ViewGroup.LayoutParams.MATCH_PARENT);
        myMenuWindow.setHeight(ViewGroup.LayoutParams.WRAP_CONTENT);
        myMenuWindow.setContentView(myWindow);
        TextView myMenuLast = (TextView) myWindow.findViewById(R.id.myMenuLast);
        TextView myMenuNext = (TextView) myWindow.findViewById(R.id.myMenuNext);
        myMenuLast.setOnClickListener(MainActivity.this);
        myMenuNext.setOnClickListener(MainActivity.this);
        //设置菜单窗口从屏幕底部滑出
        myMenuWindow.showAtLocation(findViewById(R.id.activity_main),
                            Gravity.BOTTOM | Gravity.CENTER_HORIZONTAL, 0, 0);
        WindowManager.LayoutParams myParams = getWindow().getAttributes();
        myParams.alpha = 0.4f;
        //设置菜单窗口的背景为半透明
        getWindow().setAttributes(myParams);
        return true;
      }
      return false;
    } });
}
@Override
public void onClick(View v) {                //响应单击菜单项
    if (v.getId() == R.id.myMenuLast) {
        Toast.makeText(MainActivity.this,
                "刚才选择的菜单是：显示上一幅图像", Toast.LENGTH_SHORT).show();
    } else if (v.getId() == R.id.myMenuNext) {
        Toast.makeText(MainActivity.this,
                "刚才选择的菜单是：显示下一幅图像", Toast.LENGTH_SHORT).show();
    }
    myMenuWindow.dismiss();
    WindowManager.LayoutParams myParams = getWindow().getAttributes();
    myParams.alpha = 1;
    getWindow().setAttributes(myParams);      //正常显示窗口
} }
```

上面这段代码在 MyCode\MySample747\app\src\main\java\com\bin\luo\mysample\MainActivity.java 文件中。在这段代码中,myParams = getWindow().getAttributes()用于获取当前窗口的参数信息。myParams.alpha = 0.4f 用于设置窗口的透明度为 0.4。myWindow = LayoutInflater.from(MainActivity.this).inflate(R.layout.mylayout, null, false)用于加载菜单窗口的布局 mylayout。关于 mylayout 布局的详细内容请参考 MyCode\MySample747\app\src\main\res\layout\mylayout.xml 文件。此实例的完整项目在 MyCode\MySample747 文件夹中。

072 在长时间按住控件时弹出上下文菜单

此实例主要通过重写 onCreateContextMenu()和 onContextItemSelected()方法,实现在长时间按住控件时弹出上下文菜单。当实例运行之后,长时间按住图像,将弹出上下文菜单,选择其中的"使用扇形裁剪图像"菜单,如图 072-1 的左图所示,将在弹出的 Toast 中显示"你选择了【使用扇形裁剪图像】菜单",如图 072-1 的右图所示,选择其他菜单将实现类似的功能。

图 072-1

主要代码如下:

```
public class MainActivity extends Activity {
 @Override
 protected void onCreate(Bundle savedInstanceState) {
  super.onCreate(savedInstanceState);
  setContentView(R.layout.activity_main);
  ImageView myImageView = (ImageView) findViewById(R.id.myImageView);
  registerForContextMenu(myImageView);                   //在 ImageView 控件上注册上下文菜单
 }
 public void onCreateContextMenu(ContextMenu myMenu, View v,
     ContextMenu.ContextMenuInfo menuInfo) {            //创建上下文菜单选项
  myMenu.add(0, 1, 0, "使用扇形裁剪图像");
```

```
        myMenu.add(0, 2, 0, "使用椭圆裁剪图像");
        myMenu.add(0, 3, 0, "使用菱形裁剪图像");
        myMenu.add(0, 4, 0, "使用矩形裁剪图像");
        super.onCreateContextMenu(myMenu, v, menuInfo);
    }
    @Override
    public boolean onContextItemSelected(MenuItem myItem) {//响应单击菜单
        switch (myItem.getItemId()) {
         case 1:
          Toast.makeText(this, "你选择了【使用扇形裁剪图像】菜单",
                  Toast.LENGTH_SHORT).show();
          break;
         case 2:
          Toast.makeText(this, "你选择了【使用椭圆裁剪图像】菜单",
                  Toast.LENGTH_SHORT).show();
          break;
         case 3:
          Toast.makeText(this, "你选择了【使用菱形裁剪图像】菜单",
                  Toast.LENGTH_SHORT).show();
          break;
         case 4:
          Toast.makeText(this, "你选择了【使用矩形裁剪图像】菜单",
                  Toast.LENGTH_SHORT).show();
          break;
        }
        return super.onContextItemSelected(myItem);
    }
    //当上下文菜单关闭时调用的方法
    public void onContextMenuClosed(Menu menu) {
        super.onContextMenuClosed(menu);
    } }
```

上面这段代码在 MyCode \ MySample188 \ app \ src \ main \ java \ com \ bin \ luo \ mysample \ MainActivity.java 文件中。Android 系统中的 ContextMenu(上下文菜单)类似于 PC 中的右键弹出菜单,上下文菜单与 Options Menu 主要的不同在于,Options Menu 的拥有者是 Activity,而上下文菜单的拥有者是 Activity 中的 View(控件)。每个 Activity 有且只有一个 Options Menu,它为整个 Activity 服务。而一个 Activity 往往有多个 View,然而并不是每个 View 都有一个上下文菜单,因此这就需要通过 registerForContextMenu(View view)方法的参数来指定目标控件。创建上下文菜单,必须重写方法 onCreateContextMenu()和 onContextItemSelected()。在 onCreateContextMenu()方法中,可以通过使用 add()方法添加菜单项,或者通过扩充一个定义在 XML 中的菜单资源(使用 getMenuInflater()方法获取)。onContextItemSelected()方法则用于响应选择的上下文菜单项。此实例的完整项目在 MyCode\MySample188 文件夹中。

第4章

图形和图像

073 通过像素操作在图像上添加马赛克特效

此实例主要通过创建自定义方法 BitmapMosaic(),以操作像素的方式对图像块的所有像素取平均值,从而使图像产生马赛克效果。当实例运行之后,单击"显示原始图像"按钮,则效果如图 073-1 的左图所示;单击"显示马赛克图像"按钮,则效果如图 073-1 的右图所示。

图 073-1

主要代码如下:

```
public void onClickBtn1(View view) {            //响应单击"显示原始图像"按钮
    Bitmap myBitmap =
            BitmapFactory.decodeResource(getResources(),R.mipmap.myimage1);
    myImageView.setImageBitmap(myBitmap);
}
```

```java
public void onClickBtn2(View view) {//响应单击"显示马赛克图像"按钮
    Bitmap myBitmap =
     BitmapMosaic(((BitmapDrawable) myImageView.getDrawable()).getBitmap(),30);
    myImageView.setImageBitmap(myBitmap);
}
public static Bitmap BitmapMosaic(Bitmap bitmap, int size) {
    if (bitmap == null || bitmap.getWidth() == 0 ||
                          bitmap.getHeight() == 0|| bitmap.isRecycled()) {
     return null;
    }
    int myWidth = bitmap.getWidth();
    int myHeight = bitmap.getHeight();
    Bitmap myBitmap = Bitmap.createBitmap(myWidth,
                                      myHeight, Bitmap.Config.ARGB_8888);
    int myRows = myWidth / size;
    int myCols = myHeight / size;
    int[] myBlocks = new int[size * size];
    for (int i = 0; i <= myRows; i++) {
     for (int j = 0; j <= myCols; j++) {
      int myLength = myBlocks.length;
      int myFlag = 0;              //是否到边界标志
      if (i == myRows && j != myCols) {
       myLength = (myWidth - i * size) * size;
       if (myLength == 0) { break; }
       bitmap.getPixels(myBlocks, 0, size, i * size,
                                     j * size, myWidth - i * size,size);
       myFlag = 1;
      } else if (i != myRows && j == myCols) {
       myLength = (myHeight - j * size) * size;
       if (myLength == 0) { break; }
       bitmap.getPixels(myBlocks, 0, size, i * size,
                                     j * size, size, myHeight - j * size);
       myFlag = 2;
      } else if (i == myRows && j == myCols) {
       myLength = (myWidth - i * size) * (myHeight - j * size);
       if (myLength == 0) { break; }
       bitmap.getPixels(myBlocks, 0, size, i * size,j * size,
                                     myWidth - i * size,myHeight - j * size);
       myFlag = 3;
      } else {
       bitmap.getPixels(myBlocks, 0, size, i * size, j * size, size, size);
      }
      int r = 0, g = 0, b = 0, a = 0;
      for (int k = 0; k < myLength; k++) {
       r += Color.red(myBlocks[k]);
       g += Color.green(myBlocks[k]);
       b += Color.blue(myBlocks[k]);
       a += Color.alpha(myBlocks[k]);
      }
      //获取块内所有像素点的颜色平均值
      int color = Color.argb(a / myLength, r / myLength, g / myLength, b/ myLength);
      for (int k = 0; k < myLength; k++) { myBlocks[k] = color; }
      if (myFlag == 1) {
```

```
                myBitmap.setPixels(myBlocks, 0, myWidth - i * size,
                                 i * size, j * size, myWidth - i * size, size);
            } else if (myFlag == 2) {
                myBitmap.setPixels(myBlocks, 0, size, i * size,
                                 j * size, size, myHeight - j * size);
            } else if (myFlag == 3) {
                myBitmap.setPixels(myBlocks, 0, size, i * size,
                                 j * size, myWidth - i * size, myHeight - j * size);
            } else {
                myBitmap.setPixels(myBlocks, 0, size, i * size, j * size, size, size);
            } } }
    return myBitmap;
}
```

上面这段代码在 MyCode\MySample741\app\src\main\java\com\bin\luo\mysample\MainActivity.java 文件中。在这段代码中，BitmapMosaic(Bitmap bitmap, int size)自定义方法用于使图像产生马赛克效果，参数 Bitmap bitmap 表示原始图像，参数 int size 表示马赛克（块）的大小。此实例的完整项目在 MyCode\MySample741 文件夹中。

074 通过像素操作实现为图像添加冰冻效果

此实例主要通过使用 getPixels()方法和 setPixels()方法操作图像的像素，从而为图像添加冰冻效果。当实例运行之后，单击"显示原图"按钮，将在屏幕上显示原始图像，如图 074-1 的左图所示；单击"冰冻图像"按钮，将显示冰冻效果的图像，如图 074-1 的右图所示。

图 074-1

主要代码如下：

```java
public void onClickmyBtn1(View v) {                    //响应单击"显示原图"按钮
    myImageView.setImageBitmap(myBitmap);
}
public void onClickmyBtn2(View v) {                    //响应单击"冰冻图像"按钮
    Bitmap newImage = getNewImage(myBitmap);
    myImageView.setImageBitmap(newImage);
}
public Bitmap getNewImage(Bitmap oldBmp) {
    int myWidth = oldBmp.getWidth();
    int myHeight = oldBmp.getHeight();
    int[] myPixels = new int[myWidth * myHeight];
    oldBmp.getPixels(myPixels, 0, myWidth, 0, 0, myWidth, myHeight);
    for (int i = 1; i < myHeight - 1; i++) {
        for (int j = 1; j < myWidth - 1; j++) {
            int myPixel = myPixels[(i) * myWidth + j];
            int R = Color.red(myPixel);
            int G = Color.green(myPixel);
            int B = Color.blue(myPixel);
            int myColor = R - G - B;
            myColor = myColor * 3 / 2;
            if (myColor < 0)   myColor = - myColor;
            if (myColor > 255) myColor = 255;
            R = myColor;
            myColor = G - B - R;
            myColor = myColor * 3 / 2;
            if (myColor < 0)   myColor = - myColor;
            if (myColor > 255) myColor = 255;
            G = myColor;
            myColor = B - R - G;
            myColor = myColor * 3 / 2;
            if (myColor < 0)   myColor = - myColor;
            if (myColor > 255)  myColor = 255;
            B = myColor;
            R = Math.min(255, Math.max(0, R));
            G = Math.min(255, Math.max(0, G));
            B = Math.min(255, Math.max(0, B));
            myPixels[i * myWidth + j] = Color.argb(255, R, G, B);
        } }
    Bitmap newBmp = Bitmap.createBitmap(myWidth, myHeight,
            Bitmap.Config.RGB_565);                    //根据原始图像尺寸创建新图像
    newBmp.setPixels(myPixels, 0, myWidth, 0, 0,
            myWidth, myHeight);                        //根据新的像素值填充新图像
    return newBmp;
}
```

上面这段代码在 MyCode\MySample129\app\src\main\java\com\bin\luo\mysample\MainActivity.java 文件中。在这段代码中，oldBmp.getPixels(myPixels，0，myWidth，0，0，myWidth，myHeight)用于获取原始图像的像素值，并保存在数组 myPixels 中。myColor＝R－G－B、myColor＝G－B－R 和 myColor＝B－R－G 用于重置像素的 R、G、B 分量，从而使整个图像产生冰冻效果。R＝Math.min(255，Math.max(0，R))、G＝Math.min(255，Math.max(0，G))和 B＝

Math.min(255,Math.max(0,B))用于处理 R、G、B 分量溢出的情况,即禁止 R、G、B 大于 255 或小于 0。newBmp.setPixels(myPixels,0,myWidth,0,0,myWidth,myHeight)用于根据处理之后的像素值在空白图像中填充像素,即生成冰冻图像。此实例的完整项目在 MyCode\MySample129 文件夹中。

075 通过像素操作将彩色图像改变为怀旧风格

此实例主要通过使用 getPixels()方法和 setPixels()方法,根据特定的算法操作图像的像素,实现将彩色图像转化为怀旧风格的图像。当实例运行之后,单击"显示原图"按钮,将在屏幕上显示原始图像,如图 075-1 的左图所示;单击"显示怀旧图像"按钮,将显示怀旧风格的图像,如图 075-1 的右图所示。

图 075-1

主要代码如下:

```
public void onClickmyBtn1(View v) {            //响应单击"显示原图"按钮
    myImageView.setImageBitmap(myBitmap);
}
public void onClickmyBtn2(View v) {            //响应单击"显示怀旧图像"按钮
    myImageView.setImageBitmap(getNewImage(myBitmap));
}
public Bitmap getNewImage(Bitmap myBitmap) {
    int myWidth = myBitmap.getWidth();
    int myHeight = myBitmap.getHeight();
    int oldR = 0, oldG = 0, oldB = 0, newR = 0;
    int newG = 0, newB = 0, oldPixel = 0;
    int[] myPixels = new int[myWidth * myHeight];
    //获取原始图像每个点的像素值
```

```
    myBitmap.getPixels(myPixels, 0, myWidth, 0, 0, myWidth, myHeight);
    for (int i = 0; i < myHeight; i++) {
     for (int k = 0; k < myWidth; k++) {
      //解析旧的像素值
      oldPixel = myPixels[myWidth * i + k];
      oldR = Color.red(oldPixel);
      oldG = Color.green(oldPixel);
      oldB = Color.blue(oldPixel);
      //设置新的像素值
      newR = (int) (0.393 * oldR + 0.769 * oldG + 0.189 * oldB);
      newG = (int) (0.349 * oldR + 0.686 * oldG + 0.168 * oldB);
      newB = (int) (0.272 * oldR + 0.534 * oldG + 0.131 * oldB);
      int newPixel = Color.argb(255, newR > 255 ? 255 : newR,
               newG > 255 ? 255 : newG, newB > 255 ? 255 : newB);
      myPixels[myWidth * i + k] = newPixel;
    } }
    Bitmap myNewBmp = Bitmap.createBitmap(myWidth, myHeight,
              Bitmap.Config.ARGB_8888);         //根据原始图像尺寸创建新位图
    myNewBmp.setPixels(myPixels, 0, myWidth, 0, 0,
              myWidth, myHeight);       //根据新的像素值填充新图像
    return myNewBmp;
}
```

上面这段代码在 MyCode\MySample119\app\src\main\java\com\bin\luo\mysample\MainActivity.java 文件中。在这段代码中，myBitmap.getPixels(myPixels，0，myWidth，0，0，myWidth，myHeight)用于获取原始图像的像素值，并保存在数组中。getPixels()方法的语法声明如下：

```
getPixels(@ColorInt int[] pixels, int offset,
            int stride, int x, int y, int width, int height)
```

其中，参数 int[] pixels 用于保存读取的所有像素值；参数 int offset 表示像素数组的首个索引；参数 int stride 表示位图行之间跳过的颜色个数；参数 int x 表示读取首个像素的 X 坐标；参数 int y 表示读取首个像素的 Y 坐标；参数 int width 表示一行读取的像素数，通常为图像的宽度；参数 int height 表示读取的像素行数，通常为图像的高度。

myNewBmp.setPixels(myPixels，0，myWidth，0，0，myWidth，myHeight)用于根据像素值在空白图像中填充像素。setPixels()方法的语法声明如下：

```
setPixels(@ColorInt int[] pixels, int offset,
            int stride, int x, int y, int width, int height)
```

其中，参数 int[] pixels 用于保存写入的所有像素值；参数 int offset 表示像素数组的首个索引；参数 int stride 表示位图行之间跳过的颜色个数；参数 int x 表示写入首个像素的 X 坐标；参数 int y 表示写入首个像素的 Y 坐标；参数 int width 表示一行写入的像素数，通常为图像的宽度；参数 int height 表示写入的像素行数，通常为图像的高度。

oldR=Color.red(oldPixel)用于获取像素 oldPixel 的红色分量。oldG=Color.green(oldPixel)用于获取像素 oldPixel 的绿色分量。oldB=Color.blue(oldPixel)用于获取像素 oldPixel 的蓝色分量。newR=(int)（0.393 * oldR＋0.769 * oldG＋0.189 * oldB)、newG=(int)（0.349 * oldR＋

0.686*oldG+0.168*oldB)和 newB=(int)(0.272*oldR+0.534*oldG+0.131*oldB)用于根据像素值将彩色图像转化为怀旧风格图像,它的数学算法如下:

$$\begin{cases} R = 0.393r + 0.769g + 0.189b \\ G = 0.349r + 0.686g + 0.168b \\ B = 0.272r + 0.534g + 0.131b \end{cases}$$

int newPixel = Color.argb(255,newR > 255 ? 255 : newR, newG > 255 ? 255 : newG, newB > 255 ? 255 : newB)用于处理 R、G、B 分量溢出。因为当应用上面的数学算法重置 R、G、B 分量时,很可能产生 R、G、B 分量大于 255 的情况;这种情况在像素值中是不可能存在的,所以一旦发生这种情况,则将其强行设置为最大值 255。

此实例的完整项目在 MyCode\MySample119 文件夹中。

076 使用 PorterDuffXfermode 裁剪六边形

此实例主要通过在 PorterDuffXfermode 的构造函数中使用 PorterDuff.Mode.SRC_IN 参数叠加图像和六边形,实现将图像裁剪成六边形的效果。当实例运行之后,单击"显示原图"按钮,将在屏幕上显示原始图像,如图 076-1 的左图所示;单击"显示六边形图像"按钮,将以六边形形状在屏幕上显示裁剪的图像,如图 076-1 的右图所示。

图 076-1

主要代码如下:

```
public void onClickmyBtn1(View v) {            //响应单击"显示原图"按钮
    myImageView.setImageBitmap(myBitmap);
}
public void onClickmyBtn2(View v) {            //响应单击"显示六边形图像"按钮
    myImageView.setImageBitmap(getNewImage(myBitmap));
```

```
}
public  Bitmap getNewImage(Bitmap myBitmap){
    Bitmap myNewBmp = Bitmap.createBitmap(myWidth,myHeight,
                                Bitmap.Config.ARGB_8888);         //创建新位图
    Canvas myCanvas = new Canvas(myNewBmp);                       //根据新位图创建画布
    myCanvas.translate(0,400);
    Paint myPaint = new Paint();
    Path myPath = new Path();
    myPath.moveTo(0, myWidth * 0.5f);
    myPath.lineTo(myWidth * 0.25f,0);
    myPath.lineTo(myWidth * 0.75f, 0);
    myPath.lineTo(myWidth , myWidth * 0.5f);
    myPath.lineTo(myWidth * 0.75f, myWidth);
    myPath.lineTo(myWidth * 0.25f, myWidth);
    myPath.close();
    myCanvas.drawPath(myPath, myPaint);                           //绘制六边形
    myPaint.setXfermode(new
            PorterDuffXfermode(PorterDuff.Mode.SRC_IN));          //设置相交模式
    Rect myRect = new Rect(0, 0, myWidth, myWidth);
    myCanvas.drawBitmap(myBitmap, null,
                        myRect, myPaint);                         //在六边形中绘制图像
    return myNewBmp;
}
```

上面这段代码在 MyCode\MySample109\app\src\main\java\com\bin\luo\mysample\MainActivity.java 文件中。在这段代码中,myPath 表示使用六边形的 6 个顶点的坐标位置用直线连接而成的封闭路径。myCanvas.drawPath(myPath, myPaint)则是根据封闭路径绘制六边形。myPaint.setXfermode(new PorterDuffXfermode(PorterDuff.Mode.SRC_IN))用于设置画笔的 PorterDuffXfermode 模式为 SRC_IN,即相交模式,该模式规定只在源图像和目标图像相交的地方绘制源图像。myCanvas.drawBitmap(myBitmap, null, myRect, myPaint)则表示在六边形中绘制图像。此实例的完整项目在 MyCode\MySample109 文件夹中。

077 使用 PorterDuffXfermode 抠取异形图像

此实例主要通过使用 PorterDuff.Mode.SRC_OUT 混合模式,实现当两幅图像叠加在一起时,根据不规则的 PNG 图像形状在另一幅图像上抠取该 PNG 图像形状。当实例运行之后,单击"显示原始图像"按钮,将在屏幕上显示原始图像,如图 077-1 的左图所示;单击"显示抠图结果"按钮,则原始图像根据不规则的 PNG 图像形状抠图之后的结果如图 077-1 的右图所示,中心的空白部分(两只乌龟)即是不规则的 PNG 图像形状。

主要代码如下:

```
public void onClickBtn1(View v) {                                 //响应单击"显示原始图像"按钮
    myImageView.setImageBitmap(myBitmap);
}
public void onClickBtn2(View v) {                                 //响应单击"显示抠图结果"按钮
    int myWidth = getWindowManager().getDefaultDisplay().getWidth();
    int myHeight = getWindowManager().getDefaultDisplay().getHeight();
    Bitmap myMaskBitmap = BitmapFactory.decodeResource(getResources(),
```

图 077-1

```
                        R.mipmap.myimage2);          //获取不规则的 PNG 图像
Bitmap myScaledMaskBitmap = Bitmap.createScaledBitmap(myMaskBitmap,
            1400, 1400, true);                       //对不规则的 PNG 图像进行缩放处理
Bitmap myNewBitmap = Bitmap.createBitmap(myWidth,
            myHeight, Bitmap.Config.ARGB_8888);      //创建空白位图
Canvas myCanvas = new Canvas(myNewBitmap);
Paint myPaint = new Paint();
myCanvas.drawBitmap(myScaledMaskBitmap, (myNewBitmap.getWidth() -
    myScaledMaskBitmap.getWidth()) / 2 - 7, (myNewBitmap.getHeight() -
    myScaledMaskBitmap.getHeight()) / 2 - 35, myPaint);   //绘制不规则图像
//设置混合模式(提取原图与不规则的 PNG 图像不相交的部分)
myPaint.setXfermode(new PorterDuffXfermode(PorterDuff.Mode.SRC_OUT));
myCanvas.drawBitmap(myBitmap, (myNewBitmap.getWidth() -
        myBitmap.getWidth()) / 2, (myNewBitmap.getHeight() -
        myBitmap.getHeight()) / 2, myPaint);              //按照混合模式绘制原图
//在 ImageView 上显示抠图(不相交的部分)结果
myImageView.setImageBitmap(myNewBitmap);
}
```

上面这段代码在 MyCode\MySample825\app\src\main\java\com\bin\luo\mysample\MainActivity.java 文件中。在这段代码中，PorterDuff.Mode.SRC_OUT 混合模式表示在 myCanvas 上同时绘制两幅图像时，结果是两幅图像的非交集部分。此实例的完整项目在 MyCode\MySample825 文件夹中。

078 使用 ColorMatrix 增强图像颜色对比度

此实例主要通过直接使用 ColorMatrix 自定义颜色矩阵配置画笔，实现增强图像的颜色对比度的功能。当实例运行之后，单击"显示原图"按钮，将在屏幕上显示原始图像，如图 078-1 的左图所示；单击"显示高对比度图像"按钮,显示颜色对比度增强之后的图像,如图 078-1 的右图所示。

图 078-1

主要代码如下:

```
public void onClickmyBtn1(View v) {              //响应单击"显示原图"按钮
    myImageView.setImageBitmap(myBitmap);
}
public void onClickmyBtn2(View v) {              //响应单击"显示高对比度图像"按钮
    Bitmap newImage = getNewImage(myBitmap);
    myImageView.setImageBitmap(newImage);
}
public Bitmap getNewImage(Bitmap oldBmp) {
    int myWidth = oldBmp.getWidth();
    int myHeight = oldBmp.getHeight();
    float[] myMatrix = {5,0,0,0,-254,
                        0,5,0,0,-254,
                        0,0,5,0,-254,
                        0,0,0,1,0};          //高对比度颜色矩阵
    Bitmap newBmp = Bitmap.createBitmap(myWidth,
            myHeight, Bitmap.Config.ARGB_8888);
    Canvas myCanvas = new Canvas(newBmp);
    Paint myPaint = new Paint();
    ColorMatrix myColorMatrix = new ColorMatrix(); //新建颜色矩阵
    myColorMatrix.set(myMatrix);                   //设置高对比度颜色矩阵
    //通过颜色过滤器配置颜色高对比度画笔
    myPaint.setColorFilter(new ColorMatrixColorFilter(myColorMatrix));
    myCanvas.drawBitmap(oldBmp, 0, 0, myPaint);    //绘制颜色高对比度图像
    return newBmp;
}
```

上面这段代码在 MyCode \ MySample140 \ app \ src \ main \ java \ com \ bin \ luo \ mysample \ MainActivity.java 文件中。在这段代码中，float[] myMatrix = {5,0,0,0,−254,0,5,0,0,−254, 0,0,5,0,−254,0,0,0,1,0}是增强图像颜色对比度的矩阵数组表达式，myColorMatrix.set (myMatrix)在使用该数组构建颜色矩阵与图像在画布上进行绘制时（即 myCanvas.drawBitmap (oldBmp, 0, 0, myPaint)），将增强图像的颜色对比度。此实例的完整项目在 MyCode \ MySample140 文件夹中。

079　使用 ColorMatrix 为图像添加加亮效果

此实例主要通过直接使用 ColorMatrix 颜色矩阵配置画笔，实现为图像添加加亮效果。当实例运行之后，单击"显示原图"按钮，将在屏幕上显示原始图像，如图 079-1 的左图所示；单击"加亮图像"按钮，则显示加亮效果的图像，如图 079-1 的右图所示。

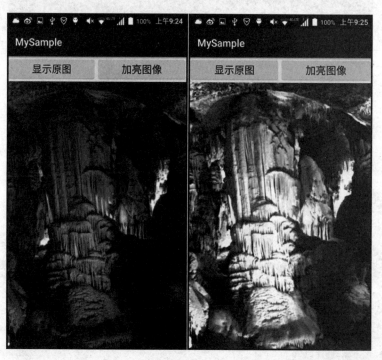

图　079-1

主要代码如下：

```
public void onClickmyBtn1(View v) {                    //响应单击"显示原图"按钮
  myImageView.setImageBitmap(myBitmap);
}
public void onClickmyBtn2(View v) {                    //响应单击"加亮图像"按钮
  Bitmap newImage = getNewImage(myBitmap);             //第 1 次加亮
  newImage = getNewImage(newImage);                    //第 2 次加亮
  newImage = getNewImage(newImage);                    //第 3 次加亮
  myImageView.setImageBitmap(newImage);                //显示最终加亮图像
}
public Bitmap getNewImage(Bitmap oldBmp) {
  int myWidth = oldBmp.getWidth();
  int myHeight = oldBmp.getHeight();
```

```
        float[] myMatrix = {1.438f, -0.122f, -0.016f, 0, -0.03f,
                            -0.062f, 1.378f, -0.016f, 0, 0.05f,
                            -0.062f, -0.122f, 1.483f, 0, -0.02f,
                            0, 0, 0, 1, 0};                    //加亮颜色矩阵
        Bitmap newBmp = Bitmap.createBitmap(myWidth,
              myHeight, Bitmap.Config.ARGB_8888);
        Canvas myCanvas = new Canvas(newBmp);
        Paint myPaint = new Paint();
        ColorMatrix myColorMatrix = new ColorMatrix();       //新建颜色矩阵
        myColorMatrix.set(myMatrix);                         //设置加亮颜色矩阵
        //通过颜色过滤器配置加亮画笔
        myPaint.setColorFilter(new ColorMatrixColorFilter(myColorMatrix));
        myCanvas.drawBitmap(oldBmp, 0, 0, myPaint);          //绘制加亮图像
        return newBmp;
    }
```

上面这段代码在 MyCode\MySample138\app\src\main\java\com\bin\luo\mysample\MainActivity.java 文件中。在这段代码中，float[] myMatrix = {1.438f,－0.122f,－0.016f,0,－0.03f,－0.062f,1.378f,－0.016f,0,0.05f,－0.062f,－0.122f,1.483f,0,－0.02f,0,0,0,1,0}是加亮图像的颜色矩阵的数组表达式，myColorMatrix.set(myMatrix)在使用该数组构建颜色矩阵与图像在画布上进行绘制时（即 myCanvas.drawBitmap(oldBmp,0,0,myPaint)），将使图像产生加亮效果。此实例的完整项目在 MyCode\MySample138 文件夹中。

080　使用 ColorMatrix 调整图像的红色色调

此实例主要通过使用 ColorMatrix 的 setRotate()方法，实现以不同值调整图像的红色色调。当实例运行之后，如果向左拖动滑块，则图像的红色色调减弱，如图 080-1 的左图所示；如果向右拖动滑块，则图像的红色色调增强，如图 080-1 的右图所示。

图　080-1

主要代码如下：

```java
public class MainActivity extends Activity {
    ImageView myImageView;
    Bitmap myBitmap;
    @Override
    protected void onCreate(Bundle savedInstanceState) {
        super.onCreate(savedInstanceState);
        setContentView(R.layout.activity_main);
        myImageView = (ImageView) findViewById(R.id.myImageView);
        myBitmap = BitmapFactory.decodeResource(getResources(), R.mipmap.myimage);
        final SeekBar mySeekBar = (SeekBar) findViewById(R.id.mySeekBar);
        mySeekBar.setOnSeekBarChangeListener(new SeekBar.OnSeekBarChangeListener() {
            //进度改变的时候调用
            public void onProgressChanged(SeekBar arg0, int progress, boolean fromUser) {
                Bitmap myNewBmp = GetNewImage(myBitmap,
                                    progress);               //根据滑块值改变图像的红色色调(相)
                myImageView.setImageBitmap(myNewBmp);         //重置图像
            }
            @Override                                         //开始拖动的时候调用
            public void onStartTrackingTouch(SeekBar seekBar) { }
            @Override                                         //停止拖动的时候调用
            public void onStopTrackingTouch(SeekBar seekBar) { }
        }); }
    public Bitmap GetNewImage(Bitmap myOldBmp, int myValue) {
        Bitmap myNewBmp = Bitmap.createBitmap(myOldBmp.getWidth(),
                myOldBmp.getHeight(), Bitmap.Config.ARGB_8888);
        Canvas myCanvas = new Canvas(myNewBmp);               //根据图像尺寸创建画布
        Paint myPaint = new Paint();
        myPaint.setAntiAlias(true);                           //设置抗锯齿,也即是边缘做平滑处理
        ColorMatrix myColorMatrix = new ColorMatrix();
        float myHueValue = myValue;
        myColorMatrix.reset();
        myColorMatrix.setRotate(0, myHueValue);               //调整图像的红色色调(相)
        //通过颜色过滤器设置新的色调(相)值
        myPaint.setColorFilter(new ColorMatrixColorFilter(myColorMatrix));
        //将重置的图像输出到新创建的位图
        myCanvas.drawBitmap(myOldBmp, 0, 0, myPaint);
        return myNewBmp;                                      //返回新的位图
} }
```

上面这段代码在 MyCode\MySample115\app\src\main\java\com\bin\luo\mysample\MainActivity.java 文件中。在这段代码中，myColorMatrix.setRotate(0，myHueValue)用于设置颜色矩阵的色调，0 表示设置颜色矩阵的红色色调，myHueValue 表示红色色调值。myPaint.setColorFilter(new ColorMatrixColorFilter(myColorMatrix))用于将颜色矩阵中的新色调值传递给画笔。myCanvas.drawBitmap(myOldBmp，0，0，myPaint)用于根据新画笔和原始图像绘制色调改变后的图像。此实例的完整项目在 MyCode\MySample115 文件夹中。

081 使用 ColorMatrix 旋转图像的颜色色相

此实例主要通过直接使用 ColorMatrix 自定义颜色矩阵配置画笔，实现图像颜色发生色相旋转的效果。当实例运行之后，单击"显示原图"按钮，将在屏幕上显示原始图像，如图 081-1 的左图所示；单击"显示色相旋转图像"按钮，则原始图像的红色变为绿色、绿色变为蓝色、蓝色变为红色，即色相旋

转,如图 081-1 的右图所示。

图 081-1

主要代码如下:

```java
public void onClickmyBtn1(View v) {              //响应单击"显示原图"按钮
    myImageView.setImageBitmap(myBitmap);
}
public void onClickmyBtn2(View v) {              //响应单击"显示色相旋转图像"按钮
    Bitmap newImage = getNewImage(myBitmap);
    myImageView.setImageBitmap(newImage);
}
public Bitmap getNewImage(Bitmap oldBmp) {
    int myWidth = oldBmp.getWidth();
    int myHeight = oldBmp.getHeight();
    float[] myMatrix = { -0.41f, 0.539f, 0.873f, 0, 0,
                          0.452f, 0.666f, -0.11f, 0, 0,
                         -0.30f, 1.71f, -0.40f, 0, 0,
                          0, 0, 0, 1, 0};              //色相旋转矩阵
    Bitmap newBmp = Bitmap.createBitmap(myWidth,
                                myHeight, Bitmap.Config.ARGB_8888);
    Canvas myCanvas = new Canvas(newBmp);
    Paint myPaint = new Paint();
    ColorMatrix myColorMatrix = new ColorMatrix();    //新建颜色矩阵
    myColorMatrix.set(myMatrix);                      //设置色相旋转颜色矩阵
    //通过颜色过滤器配置色相旋转画笔
    myPaint.setColorFilter(new ColorMatrixColorFilter(myColorMatrix));
    myCanvas.drawBitmap(oldBmp, 0, 0, myPaint);       //绘制色相旋转图像
    return newBmp;
}
```

上面这段代码在 MyCode\MySample144\app\src\main\java\com\bin\luo\mysample\MainActivity.java 文件中。在这段代码中，float[] myMatrix = {−0.41f, 0.539f, 0.873f, 0, 0, 0.452f, 0.666f, −0.11f, 0, 0, −0.30f, 1.71f, −0.40f, 0, 0, 0, 0, 0, 1, 0}是图像的颜色色相旋转矩阵的数组表达式，myColorMatrix.set(myMatrix)在使用该数组构建颜色矩阵与图像在画布上进行绘制时（即 myCanvas.drawBitmap(oldBmp, 0, 0, myPaint)），将使图像颜色产生色相旋转的效果。此实例的完整项目在 MyCode\MySample144 文件夹中。

082 自定义 ColorMatrix 改变图像对比度

此实例主要通过使用 ColorMatrix 的 set() 方法自定义颜色矩阵，并使用该颜色矩阵创建 ColorMatrixColorFilter，实现以不同的对比度调整图像。当实例运行之后，如果向左拖动滑块，则图像的对比度减弱，如图 082-1 的左图所示；如果向右拖动滑块，则图像的对比度增强，如图 082-1 的右图所示。

图　082-1

主要代码如下：

```java
public class MainActivity extends Activity{
  ImageView myImageView;
  Bitmap myBitmap;
  @Override
  protected void onCreate(Bundle savedInstanceState){
    super.onCreate(savedInstanceState);
    setContentView(R.layout.activity_main);
    myImageView = (ImageView) findViewById(R.id.myImageView);
    myBitmap = BitmapFactory.decodeResource(getResources(), R.mipmap.myimage);
    final SeekBar mySeekBar = (SeekBar) findViewById(R.id.mySeekBar);
    mySeekBar.setOnSeekBarChangeListener(new SeekBar.OnSeekBarChangeListener(){
```

```
    //进度改变的时候调用
    public void onProgressChanged(SeekBar arg0, int progress, boolean fromUser){
      Bitmap myNewBmp = GetNewImage(myBitmap,
                                    progress);           //根据滑块值改变图像的对比度
      myImageView.setImageBitmap(myNewBmp);              //重置图像
    }
    @Override                                            //开始拖动的时候调用
    public void onStartTrackingTouch(SeekBar seekBar){ }
    @Override                                            //停止拖动的时候调用
    public void onStopTrackingTouch(SeekBar seekBar){ }
}); }
public Bitmap GetNewImage(Bitmap myOldBmp, int myValue){
    Bitmap myNewBmp = Bitmap.createBitmap(myOldBmp.getWidth(),
           myOldBmp.getHeight(), Bitmap.Config.ARGB_8888);
    Canvas myCanvas = new Canvas(myNewBmp);              //根据图像大小创建画布
    Paint myPaint = new Paint();
    myPaint.setAntiAlias(true);                          //设置抗锯齿,也即是边缘做平滑处理
    ColorMatrix myColorMatrix = new ColorMatrix();
    float myContrastValue = myValue * 1.0f / 127;
    myColorMatrix.set(new float[] {myContrastValue, 0, 0, 0, 0,
           myContrastValue, 0, 0, 0, 0,
           myContrastValue, 0, 0, 0, 0, 1, 0 });         //自定义 ColorMatrix
    //使用颜色矩阵过滤器调整图像对比度
    myPaint.setColorFilter(new ColorMatrixColorFilter(myColorMatrix));
    myCanvas.drawBitmap(myOldBmp, 0, 0, myPaint);        //将重置图像输出到新位图
    return myNewBmp;                                     //返回新的位图
} }
```

上面这段代码在 MyCode \ MySample113 \ app \ src \ main \ java \ com \ bin \ luo \ mysample \ MainActivity.java 文件中。在这段代码中,myColorMatrix.set(new float[] {myContrastValue, 0, 0, 0, 0, 0, myContrastValue, 0, 0, 0, 0, 0, myContrastValue, 0, 0, 0, 0, 1, 0 })用于在颜色矩阵中设置新对比度。myPaint.setColorFilter(new ColorMatrixColorFilter(myMatrix))用于将颜色矩阵中的对比度传递给画笔。myCanvas.drawBitmap(myOldBmp, 0, 0, myPaint)用于根据新画笔和原始图像绘制对比度改变之后的图像。此实例的完整项目在 MyCode\MySample113 文件夹中。

083 使用 Matrix 实现按照指定角度旋转图像

此实例主要通过使用 Matrix 的 postRotate()方法旋转矩阵,实现按照一定的角度旋转图像的功能。当实例运行之后,单击"显示原图"按钮,将在屏幕上显示原始图像,如图 083-1 的左图所示;单击"90°旋转图像"按钮,图像将会按照顺时针方向旋转 90°,效果如图 083-1 的右图所示。

主要代码如下:

```
public void onClickmyBtn1(View v) {              //响应单击"显示原图"按钮
    myImageView.setImageBitmap(myBitmap);
}
public void onClickmyBtn2(View v) {              //响应单击"90°旋转图像"按钮
    myImageView.setImageBitmap(getNewImage(myBitmap));
}
public Bitmap getNewImage(Bitmap oldBmp) {
```

图 083-1

```
    int myWidth = oldBmp.getWidth();
    int myHeight = oldBmp.getHeight();
    Matrix myMatrix = new Matrix();
    myMatrix.postRotate(90);         //顺时针旋转 90°
    Bitmap newBmp = Bitmap.createBitmap(oldBmp, 0, 0, myWidth, myHeight,
            myMatrix, true);        //根据原始图像和旋转之后的矩阵创建新图像
    return newBmp;
}
```

上面这段代码在 MyCode\MySample126\app\src\main\java\com\bin\luo\mysample\MainActivity.java 文件中。在这段代码中，myMatrix.postRotate(90)表示将 myMatrix 矩阵顺时针旋转 90°。postRotate()方法的语法声明如下：

```
postRotate(float degrees)
```

其中，参数 degrees 表示旋转角度，正数表示沿顺时针方向旋转，负数表示沿逆时针方向旋转。此实例的完整项目在 MyCode\MySample126 文件夹中。

084 通过改变图像透明度重叠显示两幅图像

此实例主要通过调用 ImageView 的 setImageAlpha()方法改变透明度，实现以重叠的方式混合显示 ImageView 控件的背景图像和前景图像。当实例运行之后，SeekBar 控件的滑块处于最右端，ImageView 控件的图像处于完全不透明状态，因此仅显示前景图像，拖动 SeekBar 控件的滑块向左滑动改变图像的透明度，将逐渐显示出背景图像，效果分别如图 084-1 的左图和右图所示。

图 084-1

主要代码如下：

```java
public class MainActivity extends Activity {
 private ImageView myImageView;
 @Override
 protected void onCreate(Bundle savedInstanceState) {
  super.onCreate(savedInstanceState);
  setContentView(R.layout.activity_main);
  myImageView = (ImageView) findViewById(R.id.myImageView);
  myImageView.setBackground(getDrawable(R.mipmap.myimage1));     //设置背景图像
  myImageView.setImageDrawable(getDrawable(R.mipmap.myimage2));  //设置前景图像
  myImageView.setImageAlpha(255);                                //设置透明度
  SeekBar mySeekBar = (SeekBar) findViewById(R.id.mySeekBar);
  mySeekBar.setOnSeekBarChangeListener(new SeekBar.OnSeekBarChangeListener(){
   //进度改变的时候调用
   public void onProgressChanged(SeekBar arg0, int progress, boolean fromUser){
    myImageView.setImageAlpha(progress);                         //改变透明度
   }
   @Override                                                     //开始拖动的时候调用
   public void onStartTrackingTouch(SeekBar seekBar) { }
   @Override                                                     //停止拖动的时候调用
   public void onStopTrackingTouch(SeekBar seekBar) { }
});}}
```

上面这段代码在 MyCode\MySample061\app\src\main\java\com\bin\luo\mysample\MainActivity.java 文件中。在这段代码中，myImageView.setImageAlpha(progress)用于根据 SeekBar 的当前值改变 ImageView 控件的透明度。myImageView.setAlpha(progress)也能够实现 myImageView.setImageAlpha(progress)相同的功能，但不在推荐使用之列。此实例的完整项目在 MyCode\MySample061 文件夹中。

085　根据指定颜色过滤 ImageView 的图像

此实例主要通过使用 ImageView 的 setColorFilter()方法,实现根据指定的颜色和 PorterDuff.Mode.MULTIPLY 一起实现过滤图像。当实例运行之后,单击"绿色过滤"按钮,则图像在经过绿色过滤之后的效果如图 085-1 的左图所示;单击"红色过滤"按钮,则图像在经过红色过滤之后的效果如图 085-1 的右图所示;单击"蓝色过滤"按钮将实现类似的功能。

图　085-1

主要代码如下:

```java
public void onClickmyBtnGreen(View v) {          //响应单击"绿色过滤"按钮
    myImageView.setColorFilter(Color.GREEN, PorterDuff.Mode.MULTIPLY);
}
public void onClickmyBtnRed(View v) {            //响应单击"红色过滤按钮"
    myImageView.setColorFilter(Color.RED, PorterDuff.Mode.MULTIPLY);
}
public void onClickmyBtnBlue(View v) {           //响应单击"蓝色过滤"按钮
    myImageView.setColorFilter(Color.BLUE, PorterDuff.Mode.MULTIPLY);
}
```

上面这段代码在 MyCode \ MySample089 \ app \ src \ main \ java \ com \ bin \ luo \ mysample \ MainActivity.java 文件中。在这段代码中,setColorFilter(int color,Mode mode)方法用于根据指定的颜色对图像进行滤镜特效处理。其中,参数 int color 表示颜色;参数 Mode mode 表示过滤模式,PorterDuff.Mode.MULTIPLY 模式表示将每个位置的两个像素相乘,除以 255,然后使用该值创建一个新的像素进行显示,即,结果颜色＝顶部颜色＊底部颜色/255。此实例的完整项目在 MyCode\MySample089 文件夹中。

086 使用高斯矩阵模板实现图像的柔化特效

此实例主要通过使用高斯矩阵模板重置图像的像素，实现对图像进行柔化特效处理。当实例运行之后，单击"显示原图"按钮，将在屏幕上显示原始图像，如图086-1的左图所示；单击"柔化图像"按钮，则显示经过柔化处理之后的图像，如图086-1的右图所示。

图　086-1

主要代码如下：

```
public void onClickmyBtn1(View v) {                    //响应单击"显示原图"按钮
  myImageView.setImageBitmap(myBitmap);
}
public void onClickmyBtn2(View v) {                    //响应单击"柔化图像"按钮
  Bitmap newImage = getNewImage(myBitmap);
  for(int i = 0;i < 10;i++){ newImage = getNewImage(newImage); }
  myImageView.setImageBitmap(newImage);                //显示经过11次柔化处理之后的图像
}
public Bitmap getNewImage(Bitmap oldBmp) {
  int[] myGauss = new int[]{1, 2, 1, 2, 4, 2, 1, 2, 1};    //高斯矩阵
  int myWidth = oldBmp.getWidth();
  int myHeight = oldBmp.getHeight();
  int[] myPixels = new int[myWidth * myHeight];
  oldBmp.getPixels(myPixels, 0, myWidth, 0, 0, myWidth, myHeight);
  for (int i = 1; i < myHeight - 1; i++) {
   for (int j = 1;j < myWidth - 1; j++) {
    int newR = 0,newG = 0,newB = 0,myDelta = 16,myIndex = 0;
    for (int m = -1; m <= 1; m++) {
```

```
            for (int n = -1; n <= 1; n++) {
                int myPixel = myPixels[(i + m) * myWidth + j + n];
                int oldR = Color.red(myPixel);
                int oldG = Color.green(myPixel);
                int oldB = Color.blue(myPixel);
                newR = newR + (int) (oldR * myGauss[myIndex]);
                newG = newG + (int) (oldG * myGauss[myIndex]);
                newB = newB + (int) (oldB * myGauss[myIndex]);
                myIndex++;
            } }
            newR /= myDelta;
            newG /= myDelta;
            newB /= myDelta;
            newR = Math.min(255, Math.max(0, newR));
            newG = Math.min(255, Math.max(0, newG));
            newB = Math.min(255, Math.max(0, newB));
            myPixels[i * myWidth + j] = Color.argb(255, newR, newG, newB);
        } }
        Bitmap newBmp = Bitmap.createBitmap(myWidth, myHeight,
                        Bitmap.Config.RGB_565);            //根据原始图像大小创建新图像
        newBmp.setPixels(myPixels, 0, myWidth, 0, 0,
                            myWidth, myHeight);            //根据新的像素值填充新图像
        return newBmp;
    }
```

上面这段代码在 MyCode\MySample128\app\src\main\java\com\bin\luo\mysample\MainActivity.java 文件中。在这段代码中，int[] myGauss = new int[]{1, 2, 1, 2, 4, 2, 1, 2, 1} 即是高斯矩阵模板的数组表示方式，它是对图像进行柔化处理的比较优秀的模型；柔化图像的另一个简单算法是取图像上的每一点（图像边缘点忽略），然后计算它周围 8 个点的平均值作为新像素值。oldBmp.getPixels(myPixels, 0, myWidth, 0, 0, myWidth, myHeight) 用于获取原始图像的像素值，并保存在数组 myPixels 中。newR = newR + (int) (oldR * myGauss[myIndex])、newG = newG + (int) (oldG * myGauss[myIndex]) 和 newB = newB + (int) (oldB * myGauss[myIndex]) 即是根据高斯模板对原像素的 RGB 分量进行数学运算从而产生新的像素以形成整幅图像的柔化效果。newR = Math.min(255, Math.max(0, newR))、newG = Math.min(255, Math.max(0, newG)) 和 newB = Math.min(255, Math.max(0, newB)) 用于处理 R、G、B 分量在进行高斯运算时发生溢出的情况，即禁止 R、G、B 分量出现小于 0 或大于 255 的情况。newBmp.setPixels(myPixels, 0, myWidth, 0, 0, myWidth, myHeight) 用于根据处理之后的像素值在空白图像中填充像素，即生成柔化图像。此实例的完整项目在 MyCode\MySample128 文件夹中。

087 使用正弦函数创建波浪起伏风格的图像

此实例主要通过在调用 Canvas 的 drawBitmapMesh() 方法时，根据正弦函数计算偏移量，使普通图像产生波浪起伏的特效。当实例运行之后，单击"显示波浪图像"按钮，则图像在经过特效处理之后的效果如图 087-1 的左图所示。单击"显示原始图像"按钮，则原始图像的显示效果如图 087-1 的右图所示。

图　087-1

主要代码如下：

```
public class MainActivity extends Activity {
 Bitmap myBitmap, myBitmapCopy;
 ImageView myImageView;
 Canvas myCanvas;
 int   myWidth = 100, myHeight = 100,
       myCount = (myWidth + 1) * (myHeight + 1);
 float[] myVertices = new float[myCount * 2],
         myTempVertices = new float[myCount * 2];
 float myRatio;
 @Override
 protected void onCreate(Bundle savedInstanceState) {
  super.onCreate(savedInstanceState);
  setContentView(R.layout.activity_main);
  myImageView = (ImageView) findViewById(R.id.myImageView);
 }
 public void onClickBtn1(View v) {                          //响应单击"显示波浪图像按钮"
  int myIndex = 0;
  myBitmap = BitmapFactory.decodeResource(getResources(), R.mipmap.myimage);
  //预置宽高
  float myBitmapWidth = myBitmap.getWidth() * 0.5f;
  float myBitmapHeight = myBitmap.getHeight() * 1.5f;
  for (int i = 0; i < myHeight + 1; i++) {
   float myYCoordinates = myBitmapWidth / myHeight * i;
   for (int j = 0; j < myWidth + 1; j++) {
    float myXCoordinates = myBitmapHeight / myWidth * j;
    //在奇数和偶数位分别记录X和Y方向的顶点坐标值
    myTempVertices[myIndex * 2 + 0] = myVertices[myIndex * 2 + 0] = myXCoordinates;
```

```
            myTempVertices[myIndex * 2 + 1] = myVertices[myIndex * 2 + 1] = myYCoordinates;
            myIndex++;
        }}
        myBitmapCopy = Bitmap.createBitmap(getWindowManager().
                getDefaultDisplay().getWidth(),getWindowManager().
                getDefaultDisplay().getHeight(), Bitmap.Config.ARGB_8888);
        myCanvas = new Canvas(myBitmapCopy);
        myCanvas.translate(1, 55);
        for (int i = 0; i < myHeight + 1; i++) {
         for (int j = 0; j < myWidth + 1; j++) {
            //通过正弦函数获取偏移量,以实现波浪起伏效果
            float myOffset = (float) Math.sin((float) j / myWidth * 4 * Math.PI + myRatio);
            myVertices[(i * (myWidth + 1) + j) * 2 + 1] =
                    myTempVertices[(i * (myWidth + 1) + j) * 2 + 1] + myOffset * 50;
        }}
        myCanvas.drawColor(Color.TRANSPARENT, PorterDuff.Mode.CLEAR);
        myCanvas.drawBitmapMesh(myBitmap, myWidth, myHeight,
                    myVertices, 0, null, 0, null);         //绘制波浪起伏的图像
        myImageView.setImageBitmap(myBitmapCopy);
    }
    public void onClickBtn2(View v) {                      //响应单击"显示原始图像"按钮
        myBitmap = BitmapFactory.decodeResource(getResources(), R.mipmap.myimage);
        myImageView.setImageBitmap(myBitmap);
    }}
```

上面这段代码在 MyCode\MySample806\app\src\main\java\com\bin\luo\mysample\MainActivity.java 文件中。在这段代码中,float myOffset =（float）Math.sin((float) j / myWidth * 4 * Math.PI + myRatio)表示使用正弦函数计算图像坐标的偏移量,然后使用 drawBitmapMesh()方法重新绘制图像。drawBitmapMesh()方法的语法声明如下:

```
public void drawBitmapMesh(@NonNull Bitmap bitmap, int meshWidth, int meshHeight, @NonNull float[]
verts, int vertOffset, @Nullable int[] colors, int colorOffset,@Nullable Paint paint)
```

其中,参数 Bitmap bitmap 指定将要改变的图像;参数 int meshWidth 和 int meshHeight 分别规定在水平和垂直方向上应该将图像划分为多少个单元格;参数 float[] verts 表示图像改变之后的顶点坐标数组;参数 int vertOffset 表示从参数 verts 中的第几个位置处开始绘制;参数 int[] colors 指定每个顶点位置对应的颜色值;参数 int colorOffset 表示从参数 colors 中的第几个颜色坐标处开始绘制;参数 Paint paint 表示绘制图像所使用的画笔对象。

此实例的完整项目在 MyCode\MySample806 文件夹中。

088 使用 BitmapFactory 控制图像采样比例

此实例主要通过在 BitmapFactory 类的 decodeFile()方法中设置采样像素比例,实现通过控制采样像素比例压缩图像的文件大小和宽高。当实例运行之后,在"图像文件:"输入框中输入图像文件的路径;在"缩放比例:"输入框中输入采样像素比例,"3"表示压缩之后的比例是1/3,"2"表示压缩之后的比例是1/2,以此类推;单击"显示图像"按钮,将在下面显示按照比例压缩之后的图像,效果分别如图 088-1 的左图和右图所示。

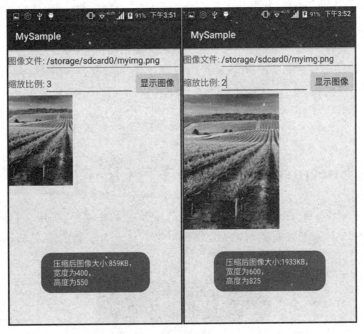

图 088-1

主要代码如下：

```java
public class MainActivity extends Activity {
    EditText myFile;
    EditText mySize;
    ImageView myImage;
    @Override
    protected void onCreate(Bundle savedInstanceState) {
        super.onCreate(savedInstanceState);
        setContentView(R.layout.activity_main);
        myFile = (EditText) findViewById(R.id.myFile);
        mySize = (EditText) findViewById(R.id.mySize);
        myImage = (ImageView) findViewById(R.id.myImage);
        //SD卡的图像文件myimg.png
        myFile.setText(Environment.getExternalStorageDirectory() + "/myimg.png");
        mySize.setText("4");
    }
    public void onClickmyBtn1(View v) {                    //响应单击"显示图像"按钮
        try {
            BitmapFactory.Options myOptions = new BitmapFactory.Options();
            Number myScale = NumberFormat.getNumberInstance(
                    Locale.CHINA).parse(mySize.getText().toString());
            myOptions.inSampleSize = myScale.intValue();
            Bitmap myBmp = BitmapFactory.decodeFile(
                    myFile.getText().toString(), myOptions);
            myImage.setImageBitmap(myBmp);
            String myInfo = "压缩后图像大小:" + (myBmp.getByteCount() / 1024) +
                    "KB,\n宽度为" + myBmp.getWidth() + ",\n高度为" + myBmp.getHeight();
            Toast.makeText(MainActivity.this, myInfo, Toast.LENGTH_SHORT).show();
        } catch (Exception e) {
            Toast.makeText(MainActivity.this,
                    e.getMessage().toString(), Toast.LENGTH_SHORT).show();
        }}}
```

上面这段代码在 MyCode\MySample233\app\src\main\java\com\bin\luo\mysample\MainActivity.java 文件中。在这段代码中，myBmp = BitmapFactory.decodeFile(myFile.getText().toString(),myOptions)用于将指定路径文件根据参数 myOptions 解码为一个位图对象，在参数 myOptions 中，可以控制采样像素比例；这主要通过在该参数的 inSampleSize 属性中设置采样比例。此外，在 SD 卡操作文件需要在 AndroidManifest.xml 文件中添加< uses-permission android:name="android.permission.READ_EXTERNAL_STORAGE"/>权限。此实例的完整项目在 MyCode\MySample233 文件夹中。

089　使用 SweepGradient 创建多色扫描图

此实例主要通过在 SweepGradient 类的构造函数中指定多种颜色，实现在自定义 View 中绘制多色的扫描渐变图形的功能。当实例运行之后，在自定义 View 中绘制的多色扫描渐变图形的效果如图 089-1 所示。

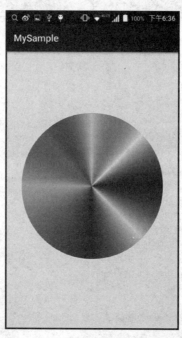

图　089-1

主要代码如下：

```
public class MainActivity extends Activity {
@Override
protected void onCreate(Bundle savedInstanceState) {
 setContentView(new MyView(this));
 super.onCreate(savedInstanceState);
}
class MyView extends View{
 public MyView(Context context) { super(context); }
 @Override
 protected void onDraw(Canvas myCanvas) {
  Display myDisplay = getWindowManager().getDefaultDisplay();
  int myWidth = myDisplay.getWidth() - 200;
```

```
        int myHeight = myDisplay.getHeight() - 100;
        SweepGradient mySweepGradient = new SweepGradient(myWidth/2,
                myHeight/2, new int[] {Color.BLUE,Color.CYAN,
                Color.DKGRAY,Color.GRAY,Color.LTGRAY,Color.MAGENTA,
                Color.GREEN,Color.TRANSPARENT, Color.BLUE }, null);
        Paint myPaint = new Paint();
        myPaint.setShader(mySweepGradient);
        myCanvas.translate(100, -100);              //平移画布到指定位置
        myCanvas.drawCircle(myWidth/2, myHeight/2, myWidth/2, myPaint);
}}}
```

上面这段代码在 MyCode\MySample094\app\src\main\java\com\bin\luo\mysample\MainActivity.java 文件中。在这段代码中，SweepGradient mySweepGradient ＝ new SweepGradient(myWidth/2，myHeight/2，new int［］{Color.BLUE，Color.CYAN，Color.DKGRAY，Color.GRAY，Color.LTGRAY，Color.MAGENTA，Color.GREEN，Color.TRANSPARENT，Color.BLUE}，null)用于创建一个多色扫描渐变的SweepGradient，该构造函数的语法声明如下：

```
public SweepGradient(float cx, float cy, int[] colors, float[] positions)
```

其中，参数 cx 表示渲染中心点水平坐标；参数 cy 表示渲染中心点垂直坐标；参数 colors 表示围绕中心点渲染的颜色数组，至少有两种颜色值；参数 positions 表示颜色相对位置的数组，可为 null，若为 null，颜色沿渐变线均匀分布。

此实例的完整项目在 MyCode\MySample094 文件夹中。

090　使用 RadialGradient 绘制电波扩散图

此实例主要通过使用透明色 TRANSPARENT 创建 RadialGradient 类的实例，实现在图像上显示电波波纹扩散的效果。当实例运行之后，单击图像的任意一点，将以该点为中心生成电波波纹由里向外扩散的效果，效果分别如图 090-1 的左图和右图所示。

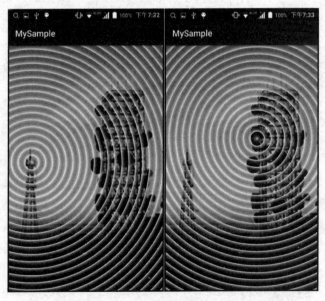

图　090-1

主要代码如下：

```java
public class MainActivity extends Activity {
@Override
protected void onCreate(Bundle savedInstanceState) {
setContentView(new MyView(this));
super.onCreate(savedInstanceState);
}
public class MyView extends View {
Shader myShader = null;
Bitmap myBitmap = null;
Paint myPaint = null;
Shader myRadialGradient = null;
ShapeDrawable myShapeDrawable = null;
public MyView(Context context) {
 super(context);
Bitmap myBitmap = ((BitmapDrawable) getResources().getDrawable(
                   R.mipmap.myimage)).getBitmap();        //获取图像资源
DisplayMetrics myMetrics = getResources().getDisplayMetrics();
this.myBitmap = Bitmap.createScaledBitmap(myBitmap, myMetrics.widthPixels,
 myMetrics.heightPixels, true);                   //创建与当前使用的设备窗口大小一致的图像
myShader = new BitmapShader(myBitmap, Shader.TileMode.REPEAT,
           Shader.TileMode.MIRROR);                //根据新图像创建 BitmapShader
myPaint = new Paint();
}
@Override
protected void onDraw(Canvas myCanvas) {
 super.onDraw(myCanvas);
myShapeDrawable = new ShapeDrawable(null);
myShapeDrawable.getPaint().setShader(myShader);
myShapeDrawable.setBounds(0, 0,
           myBitmap.getWidth(),myBitmap.getHeight()); //设置显示区域
myShapeDrawable.draw(myCanvas);
if (myRadialGradient != null) {                     //绘制图像
myPaint.setShader(myRadialGradient);
myCanvas.drawCircle(0, 0, 2000, myPaint);
}
}
public boolean onTouchEvent(MotionEvent event) {    //覆写触摸屏事件
myRadialGradient = new RadialGradient(event.getX(), event.getY(), 48,
 new int[]{Color.WHITE, Color.TRANSPARENT}, null, Shader.TileMode.REPEAT);
postInvalidate();                                   //刷新屏幕
return true;
}}}
```

上面这段代码在 MyCode\MySample096\app\src\main\java\com\bin\luo\mysample\MainActivity.java 文件中。在这段代码中，myRadialGradient = new RadialGradient（event.getX()，event.getY()，48，new int[]{Color.WHITE，Color.TRANSPARENT}，null，Shader.TileMode.REPEAT）用于创建一个由不透明色（白色）向透明色径向渐变的 RadialGradient，event.getX()表示圆心 X 坐标，event.getY()表示圆心 Y 坐标，48 表示径向渐变的半径（径向范围），new int[] {Color.WHITE，Color.TRANSPARENT}表示渲染径向渐变的颜色是由白色向透明色渐变，null 表示白色和透明色的渐变相对位置沿渐变方向均匀分布，Shader.TileMode.REPEAT 表示重复渲染。此实例

的完整项目在 MyCode\MySample096 文件夹中。

091 使用 BlurMaskFilter 为图像添加轮廓阴影

此实例主要通过使用 BlurMaskFilter 定制阴影画笔,实现为 PNG 格式的图像添加轮廓阴影。当实例运行之后,单击"添加阴影"按钮,则图像的轮廓在添加阴影之后的效果如图 091-1 的左图所示。单击"移除阴影"按钮,将显示原始图像,如图 091-1 的右图所示。

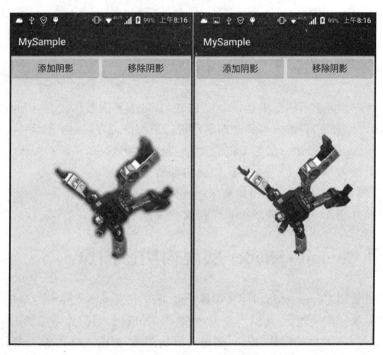

图 091-1

主要代码如下:

```
public void onClickBtn1(View v) {                              //响应单击"添加阴影"按钮
    Bitmap myBitmap =
            BitmapFactory.decodeResource(getResources(), R.mipmap.myimage1);
    BlurMaskFilter myBlurMaskFilter = new BlurMaskFilter(20,
                    BlurMaskFilter.Blur.OUTER);                //初始化外阴影过滤器
    Paint myPaint = new Paint();
    myPaint.setMaskFilter(myBlurMaskFilter);                   //在画笔中应用外阴影过滤器
    int[] myOffset = new int[2];
    Bitmap myImageShadow =
                myBitmap.extractAlpha(myPaint, myOffset);      //创建底图阴影
    Bitmap myShadowImageCopy =
    myImageShadow.copy(Bitmap.Config.ARGB_8888, true);         //创建底图阴影的副本
    Canvas myCanvas =
                new Canvas(myShadowImageCopy);                 //根据底图阴影副本初始化画布
    myCanvas.drawBitmap(myBitmap,
                - myOffset[0], - myOffset[1], null);           //在画布上绘制原图
    myImageView.setImageBitmap(myShadowImageCopy);             //显示阴影图像
}
```

```
public void onClickBtn2(View v) {                    //响应单击"移除阴影"按钮
    Bitmap myBitmap =
            BitmapFactory.decodeResource(getResources(), R.mipmap.myimage1);
    myImageView.setImageBitmap(myBitmap);
}
```

上面这段代码在 MyCode\MySample787\app\src\main\java\com\bin\luo\mysample\MainActivity.java 文件中。在这段代码中，myBlurMaskFilter = new BlurMaskFilter(20, BlurMaskFilter.Blur.OUTER)用于创建半径为 20 的外阴影过滤器。BlurMaskFilter()构造函数的语法声明如下：

```
public BlurMaskFilter(float radius, Blur style)
```

其中，参数 float radius 表示阴影半径。参数 Blur style 表示阴影样式，常用的选项如下所示。
（1）NORMAL，对图像边界的内部和外部都将进行模糊处理（添加阴影）。
（2）SOLID，图像边界外产生一层与 Paint 颜色一致的阴影效果，不影响图像本身。
（3）OUTER，图像边界外产生一层阴影，图像内部镂空。
（4）INNER，在图像边界内部产生模糊效果，外部不绘制。
此实例的完整项目在 MyCode\MySample787 文件夹中。

092　使用 ComposeShader 实现内阴影图像

此实例主要通过使用 ComposeShader 在图像上叠加由中心向四周径向变暗的渐变层，实现在图像上添加内阴影的效果。当实例运行之后，单击"特效图像"按钮，则图像在添加内阴影之后的效果如图 092-1 的左图所示；单击"原始图像"按钮，则显示原始图像，如图 092-1 的右图所示。

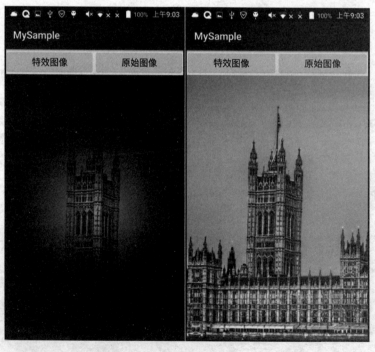

图　092-1

主要代码如下：

```java
public void onClickBtn1(View v) {                        //响应单击"特效图像"按钮
    int myWidth = myBitmap.getWidth();
    int myHeight = myBitmap.getHeight();
    Paint myPaint = new Paint();
    BitmapShader myBitmapShader = new BitmapShader(myBitmap,
            Shader.TileMode.REPEAT, Shader.TileMode.REPEAT); //使用图像创建着色器
    RadialGradient myRadialGradient = new RadialGradient(myWidth / 2,
                myHeight / 2, myWidth / 2, Color.TRANSPARENT,  Color.BLACK,
                Shader.TileMode.CLAMP);                  //创建径向渐变对象(中心透明,周围黑色)
    //在图像着色器上叠加径向渐变对象
    ComposeShader myComposeShader = new ComposeShader(myBitmapShader,
            myRadialGradient, new PorterDuffXfermode(PorterDuff.Mode.SRC_OVER));
    myPaint.setShader(myComposeShader);
    Bitmap myBitmapCopy = Bitmap.createBitmap(myWidth,
                                            myHeight, Bitmap.Config.ARGB_8888);
    Canvas myCanvas = new Canvas(myBitmapCopy);
    //在与图像相同大小的画布上绘制矩形,因为Paint包含图像着色器和径向渐变对象
    //所以实际上呈现的效果是在图像上叠加径向渐变透明层
    myCanvas.drawRect(0, 0, myWidth, myHeight, myPaint);
    myImageView.setImageBitmap(myBitmapCopy);
}
public void onClickBtn2(View v) {                        //响应单击"原始图像"按钮
    myImageView.setImageBitmap(myBitmap);
}
```

上面这段代码在 MyCode\MySample791\app\src\main\java\com\bin\luo\mysample\MainActivity.java 文件中。在这段代码中，ComposeShader myComposeShader = new ComposeShader(myBitmapShader,myRadialGradient,new PorterDuffXfermode(PorterDuff.Mode.SRC_OVER))用于将图像和渐变层(内阴影)叠加在一起。ComposeShader()构造函数的语法声明如下：

```
public ComposeShader(Shader shaderA, Shader shaderB, Xfermode mode)
```

其中，参数 Shader shaderA 和参数 Shader shaderB 表示要叠加的两种着色器对象；参数 Xfermode mode 表示要应用的叠加模式，该参数支持下列类型。

（1）CLEAR，图像所有像素点的 Alpha 通道值和 RGB 颜色值均为 0。
（2）SRC，只保留源图像的 Alpha 通道值和 RGB 颜色值。
（3）DST，只保留目标图像的 Alpha 通道值和 RGB 颜色值。
（4）SRC_OVER，在目标图像上层绘制源图像。
（5）DST_OVER，在源图像上层绘制目标图像。
（6）SRC_IN，在两者交集部分绘制源图像，并且会受到目标图像 Alpha 通道值的影响。
（7）DST_IN，在两者交集部分绘制目标图像，并且会受到源图像 Alpha 通道值的影响。
（8）SRC_OUT，在两者不相交处绘制源图像，相交处根据目标图像的 Alpha 通道值进行过滤，完全不透明处则完全过滤，完全透明则不过滤。
（9）DST_OUT，在两者不相交处绘制目标图像，相交处根据源图像的 Alpha 通道值进行过滤，完全不透明处则完全过滤，完全透明则不过滤。

（10）SRC_ATOP，在源图像和目标图像相交处绘制源图像，不相交处绘制目标图像，并且相交处会受到源图像和目标图像的 Alpha 通道值的影响。

（11）DST_ATOP，在源图像和目标图像相交处绘制目标图像，不相交处绘制源图像，并且相交处会受到源图像和目标图像的 Alpha 通道值的影响。

（12）XOR，在两者不相交处绘制源图像和目标图像，相交处受到对应 Alpha 通道值和颜色值影响，若都不透明则相交处不绘制。

（13）DARKEN，进行对应像素值比较，取较暗值，如果色值相同则进行混合。

（14）LIGHTEN，进行对应像素值比较，取较亮值，如果色值相同则进行混合。

（15）MULTIPLY，查看每个通道中的颜色信息，并将基色与混合色复合，结果色总是较暗的颜色。任何颜色与黑色复合产生黑色，任何颜色与白色复合保持不变。

（16）SCREEN，保留两个图层中较亮的部分，较暗的部分被遮盖；当一层使用了 SCREEN 模式时，图层中纯黑部分变成完全透明，纯白部分变成完全不透明，其他的颜色根据颜色级别产生半透明的效果。

（17）ADD，对目标图像和源图像进行饱和度叠加操作。

（18）OVERLAY，根据底层图像颜色值判断当前像素是以 MULTIPLY 模式混合，还是 SCREEN 模式混合，但底层颜色的高光与阴影部分的亮度细节会被保留。

此实例的完整项目在 MyCode\MySample791 文件夹中。

093　使用 EmbossMaskFilter 强化图像轮廓

此实例主要通过使用不同的光源方向和光照强度创建 EmbossMaskFilter，实现强化 PNG 格式图像的轮廓效果。当实例运行之后，单击"添加特效"按钮，图像的轮廓将更加突出，效果如图 093-1 的左图所示。单击"移除特效"按钮，将显示原始图像，如图 093-1 的右图所示。

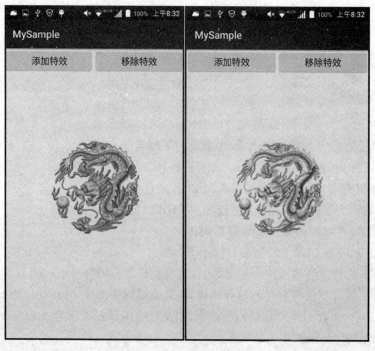

图 093-1

主要代码如下：

```java
public void onClickBtn1(View v) {              //响应单击"添加特效"按钮
    Bitmap myBitmap =
            BitmapFactory.decodeResource(getResources(), R.mipmap.myimage1);
    float[] myDirection = new float[]{10, 10, 10};
    float myAmbient = 0.9f;
    float mySpecular = 1;
    float myBlurRadius = 1;
    EmbossMaskFilter myEmbossMaskFilter = new EmbossMaskFilter(myDirection,
                                myAmbient, mySpecular, myBlurRadius);
    Paint myPaint = new Paint();
    myPaint.setMaskFilter(myEmbossMaskFilter);
    int[] myOffset = new int[5];
    Bitmap myImageFilter = myBitmap.extractAlpha(myPaint, myOffset);
    Bitmap myImageFilterCopy = myImageFilter.copy(Bitmap.Config.ARGB_8888, true);
    Canvas myCanvas = new Canvas(myImageFilterCopy);
    myCanvas.drawBitmap(myBitmap, - myOffset[0], - myOffset[1], null);
    myImageView.setImageBitmap(myImageFilterCopy);
}
public void onClickBtn2(View v) {              //响应单击"移除特效"按钮
    Bitmap myBitmap =
            BitmapFactory.decodeResource(getResources(), R.mipmap.myimage1);
    myImageView.setImageBitmap(myBitmap);
}
```

上面这段代码在 MyCode\MySample788\app\src\main\java\com\bin\luo\mysample\MainActivity.java 文件中。在这段代码中，myEmbossMaskFilter = new EmbossMaskFilter(myDirection，myAmbient，mySpecular，myBlurRadius)用于根据指定的参数创建光照特效过滤器 EmbossMaskFilter，在使用光照特效过滤器 EmbossMaskFilter 之后，将会在目标（图像）上产生轮廓强化的效果。EmbossMaskFilter 构造函数的语法声明如下：

```java
public EmbossMaskFilter(float[] direction,
                float ambient, float specular, float blurRadius)
```

其中，参数 float[] direction 包含 3 个值 X、Y、Z，用来指示光源照射的方向；参数 float ambient 表示光的强度系数，范围为 0～1；参数 float specular 表示亮度系数；参数 float blurRadius 表示模糊半径。

此实例的完整项目在 MyCode\MySample788 文件夹中。

094　使用 GradientDrawable 创建渐变色边框

此实例主要通过使用 GradientDrawable 创建渐变图形作为 ImageView 控件的背景，实现为图像添加渐变色边框的特效。当实例运行之后，单击"渐变色边框"按钮，则图像在添加渐变色边框之后的效果如图 094-1 的左图所示；单击"纯色边框"按钮，则图像在添加纯色边框之后的效果如图 094-1 的右图所示。

图　094-1

主要代码如下：

```java
public void onClickBtn1(View v) {                    //响应单击"渐变色边框"按钮
    GradientDrawable myGradientDrawable = new GradientDrawable();
    //使用扫描渐变,以实现渐变色边框特效
    myGradientDrawable.setGradientType(GradientDrawable.SWEEP_GRADIENT);
    myGradientDrawable.setColors(new int[]{Color.RED,
                    Color.GREEN, Color.BLUE, Color.YELLOW, Color.RED});
    myImageView.setBackground(myGradientDrawable);
}
public void onClickBtn2(View v) {                    //响应单击"纯色边框"按钮
    GradientDrawable myGradientDrawable = new GradientDrawable();
    //使用径向渐变,以实现纯色边框特效
    myGradientDrawable.setGradientType(GradientDrawable.RADIAL_GRADIENT);
    myGradientDrawable.setColors(new int[]{Color.TRANSPARENT, Color.GREEN});
    myImageView.setBackground(myGradientDrawable);
}
```

上面这段代码在 MyCode \ MySample838 \ app \ src \ main \ java \ com \ bin \ luo \ mysample \ MainActivity.java 文件中。在这段代码中，myGradientDrawable.setGradientType（GradientDrawable.SWEEP_GRADIENT）用于指定 GradientDrawable 的渐变类型为 SWEEP_GRADIENT，该方法支持的其他两种类型分别是 RADIAL_GRADIENT 和 LINEAR_GRADIENT。此实例的完整项目在 MyCode\MySample838 文件夹中。

095　使用 VectorDrawable 调整矢量图形亮度

此实例主要通过使用 VectorDrawable 的 setColorFilter()方法，实现对 SVG 矢量图像（形）使用自定义颜色过滤器调节亮度。当实例运行之后，单击"图像变亮"按钮，则矢量图形变亮之后的效果如

图 095-1 的左图所示；单击"图像变暗"按钮，则矢量图形变暗之后的效果如图 095-1 的右图所示。

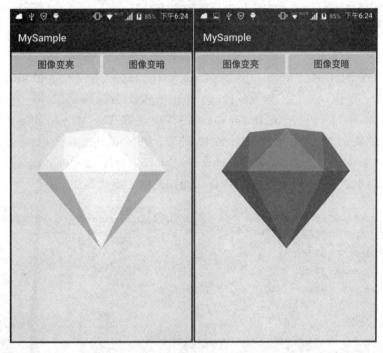

图　095-1

主要代码如下：

```
public class MainActivity extends Activity {
 ImageView myImageView;
 VectorDrawable myVectorDrawable;
 @Override
 protected void onCreate(Bundle savedInstanceState) {
  super.onCreate(savedInstanceState);
  setContentView(R.layout.activity_main);
  myImageView = (ImageView) findViewById(R.id.myImageView);
  myVectorDrawable = (VectorDrawable) myImageView.getDrawable();
 }
 public void onClickBtn1(View v) {                    //响应单击"图像变亮"按钮
  ColorMatrix myColorMatrix = new ColorMatrix();
  float myBrightness = (65 - 50) * 2 * 255 * 0.01f;
  myColorMatrix.set(new float[]{1, 0, 0, 0, myBrightness,
                    0, 1, 0, 0, myBrightness,
                    0, 0, 1, 0, myBrightness,
                    0, 0, 0, 1, 0});        //自定义变亮矩阵
  //将该矩阵对象添加至颜色矩阵过滤器中,并应用于 SVG 图像
  myVectorDrawable.setColorFilter(new ColorMatrixColorFilter(myColorMatrix));
 }
 public void onClickBtn2(View v) {                    //响应单击"图像变暗"按钮
  ColorMatrix myColorMatrix = new ColorMatrix();
  float myBrightness = (35 - 50) * 2 * 255 * 0.01f;
  myColorMatrix.set(new float[]{1, 0, 0, 0, myBrightness,
                    0, 1, 0, 0, myBrightness,
                    0, 0, 1, 0, myBrightness,
```

```
                            0,0,0,1,0});                        //自定义变暗矩阵
    //将该矩阵对象添加至颜色矩阵过滤器中,并应用于SVG图像
    myVectorDrawable.setColorFilter(new ColorMatrixColorFilter(myColorMatrix));
}}
```

上面这段代码在 MyCode\MySample774\app\src\main\java\com\bin\luo\mysample\MainActivity.java 文件中。在这段代码中,myVectorDrawable =(VectorDrawable)myImageView.getDrawable()用于从 ImageView 控件中获取 VectorDrawable 对象,在此实例中,ImageView 控件加载的即是一个 SVG 矢量图形,如 android:src = "@drawable/mysvg"。myVectorDrawable.setColorFilter(new ColorMatrixColorFilter(myColorMatrix))通过颜色矩阵调节亮的结果即作用于该 SVG 矢量图形。SVG 矢量图形的主要代码如下:

```xml
<vector xmlns:android = "http://schemas.android.com/apk/res/android"
    android:width = "256dp"
    android:height = "232dp"
    android:viewportHeight = "232.0"
    android:viewportWidth = "256.0">
<path android:fillColor = "#FDB300"
    android:pathData =
    "M128,0l-72.15,7.63l-55.85,74.98l128,149.05l128,-149.05l-55.85,-74.98z"/>
<path android:fillColor = "#EB6C00"
    android:pathData = "M0,82.61l128,149.05l-76.15,-149.05z"/>
<path android:fillColor = "#EB6C00"
    android:pathData = "M204.15,82.61l-76.15,149.05l128,-149.05z"/>
<path android:fillColor = "#FDAD00"
    android:pathData = "M51.85,82.61l76.15,149.05l76.15,-149.05z"/>
<path android:fillColor = "#FDD231"
    android:pathData = "M55.85,7.63l-4.01,74.98l76.15,-82.61z"/>
<path android:fillColor = "#FDD231"
    android:pathData = "M204.15,82.61l-4.01,-74.98l-72.14,-7.63z"/>
<path android:fillColor = "#FDAD00"
    android:pathData = "M204.15,82.61l51.85,0l-55.85,-74.98z"/>
<path android:fillColor = "#FDAD00"
    android:pathData = "M0,82.61l51.85,0l4.01,-74.98z"/>
<path android:fillColor = "#FEEEB7"
    android:pathData = "M128,0l-76.15,82.61l152.31,0z"/>
</vector>
```

上面这段代码在 MyCode\MySample774\app\src\main\res\drawable\mysvg.xml 文件中。此实例的完整项目在 MyCode\MySample774 文件夹中。

096 使用 ClipDrawable 裁剪图像实现拉幕效果

此实例主要通过使用 ClipDrawable 控制裁剪区域,实现拉幕式的图像展开效果。当实例运行之后,单击"水平展开"按钮,图像将从中心向左右两端展开,犹如幕布从中心向两端拉开一样,原图及"水平展开"效果分别如图 096-1 的左图和右图所示。单击"垂直展开"按钮,则图像将从中心向上下两端展开。

图 096-1

主要代码如下：

```java
public class MainActivity extends Activity {
    ImageView myImageView;
    ClipDrawable myClipDrawable;
    int myLevel = 0;
    Bitmap myBitmap;
    Timer myTimer = new Timer();
    Handler myHandler = new Handler() {
        @Override
        public void handleMessage(Message msg) {
            if (myLevel >= 10000) {
                myTimer.cancel();                        //超过最大值时停止定时器
            } else {
                myClipDrawable.setLevel(myLevel += 200);
            }                                            //更新ClipDrawable对象的level属性值
        }
    };
    @Override
    protected void onCreate(Bundle savedInstanceState) {
        super.onCreate(savedInstanceState);
        setContentView(R.layout.activity_main);
        myImageView = (ImageView) findViewById(R.id.myImageView);
        myBitmap = BitmapFactory.decodeResource(getResources(), R.mipmap.myimage1);
    }
    public void onClickBtn1(View v) {                    //响应单击"水平展开"按钮
        myTimer = new Timer();
        myLevel = 0;
        myClipDrawable = new ClipDrawable(new BitmapDrawable(getResources(), myBitmap),
                Gravity.CENTER, ClipDrawable.HORIZONTAL);   //设定为水平展开模式
        myImageView.setImageDrawable(myClipDrawable);
```

```
        myTimer.schedule(new TimerTask() {
            @Override
            public void run() { myHandler.sendEmptyMessage(0); } },
            0, 25); //异步发送消息,并通过 Handler 更新 ClipDrawable 的 level 属性值
    }
    public void onClickBtn2(View v) {                    //响应单击"垂直展开"按钮
        myTimer = new Timer();
        myLevel = 0;
        myClipDrawable = new ClipDrawable(new BitmapDrawable(getResources(), myBitmap),
                Gravity.CENTER, ClipDrawable.VERTICAL);  //设定为垂直展开模式
        myImageView.setImageDrawable(myClipDrawable);
        myTimer.schedule(new TimerTask() {
            @Override
            public void run() { myHandler.sendEmptyMessage(0);} },
            0, 25);                                       //异步发送消息,并通过 Handler 更新 ClipDrawable 的 level 属性值
    } }
```

上面这段代码在 MyCode\MySample780\app\src\main\java\com\bin\luo\mysample\MainActivity.java 文件中。在这段代码中,myClipDrawable.setLevel(myLevel += 200)用于改变裁剪区域,参数取值范围为 0~10000,为 0 时完全不显示,为 10000 时完全显示。myClipDrawable = new ClipDrawable(new BitmapDrawable(getResources(), myBitmap), Gravity.CENTER, ClipDrawable.HORIZONTAL)用于创建 ClipDrawable 对象,该构造函数的语法声明如下:

```
public ClipDrawable(Drawable drawable, int gravity, int orientation)
```

其中,参数 Drawable drawable 表示该 ClipDrawable 引用的 drawable 资源(通常为图像)。参数 int orientation 表示裁剪的方向,ClipDrawable.HORIZONTAL 表示水平方向,ClipDrawable.VERTICAL 表示垂直方向。参数 int gravity 指定从哪个位置开始裁剪,必须是下面一个或多个选项(多个选项之间用"|"分隔)。

(1) TOP,该选项表示将对象放在容器的顶部,不改变其大小,当参数 orientation 是 VERTICAL 时,则从底部开始裁剪。

(2) BOTTOM,该选项表示将对象放在容器的底部,不改变其大小,当参数 orientation 是 VERTICAL,则从顶部开始裁剪。

(3) LEFT,该选项表示将对象放在容器的左边,不改变其大小。当参数 orientation 是 HORIZONTAL,则从右边开始裁剪,这也是默认值。

(4) RIGHT,该选项表示将对象放在容器的右边,不改变其大小。当参数 orientation 是 HORIZONTAL,则从左边开始裁剪。

(5) CENTER_VERTICAL,该选项表示将对象放在垂直方向的中间,不改变其大小。裁剪的情况和 CENTER 一样。

(6) FILL_VERTICAL,该选项表示在垂直方向上不发生裁剪,除非 drawable 的 level 是 0,才会不可见,表示全部裁剪完。

(7) CENTER_HORIZONTAL,该选项表示将对象放在水平方向的中间,不改变其大小。裁剪的情况和 CENTER 一样。

(8) FILL_HORIZONTAL,该选项表示在水平方向上不发生裁剪。除非 drawable 的 level 是 0,才会不可见,表示全部裁剪完。

(9) CENTER,该选项表示将对象放在水平垂直方向的中间,不改变其大小。当参数 orientation

是 HORIZONTAL 时,裁剪发生在左右。当参数 orientation 是 VERTICAL 时,裁剪发生在上下。

(10) FILL,该选项表示填充整个容器,不会发生裁剪。除非 drawable 的 level 是 0,才会不可见,表示全部裁剪完。

(11) CLIP_VERTICAL,该选项是额外的选项,它能够把容器的上下边界,设置为子对象的上下边缘的裁剪边界。裁剪基于对象垂直重心设置。如果重心设置为 TOP,则裁剪下边;如果设置为 BOTTOM,则裁剪上边;否则上下两边都要裁剪。

(12) CLIP_HORIZONTAL,该选项是额外的选项,它能够把容器的左右边界,设置为子对象的左右图像的裁剪边界。裁剪基于对象水平重心设置。如果重心设置为 RIGHT,则裁剪左边;如果设置为 LEFT,则裁剪右边;否则左右两边都要裁剪。

需要注意的是,参数 int gravity 和参数 int orientation 应该配合使用,不同的组合,裁剪的效果也不同。此实例的完整项目在 MyCode\MySample780 文件夹中。

097　使用 ShapeDrawable 裁剪五角星图像

此实例主要通过以路径围成的五角星 PathShape 创建 ShapeDrawable,并使用图像填充五角星形状的 ShapeDrawable,实现裁剪五角星图像的效果。当实例运行之后,单击"显示原始图像"按钮,将在屏幕上显示原始图像,如图 097-1 的左图所示;单击"显示五角星图像"按钮,将以五角星形状在屏幕上显示裁剪的图像,如图 097-1 的右图所示。

图　097-1

主要代码如下:

```
public class MainActivity extends Activity {
    ImageView myImageView;
    Drawable myImageDrawable;
    @Override
```

```
protected void onCreate(Bundle savedInstanceState) {
    super.onCreate(savedInstanceState);
    setContentView(R.layout.activity_main);
    myImageView = (ImageView) findViewById(R.id.myImageView);
    myImageDrawable = getDrawable(R.mipmap.myimage1);
}
public void onClickBtn1(View v) {                          //响应单击"显示原始图像"按钮
    myImageView.setBackground(myImageDrawable);
}
public void onClickBtn2(View v) {                          //响应单击"显示五角星图像"按钮
    Bitmap myBitmap =
            BitmapFactory.decodeResource(getResources(), R.mipmap.myimage1);
    int myWidth = myBitmap.getWidth();
    Path myPath = new Path();
    myPath.moveTo(myWidth * 0.5f, myWidth * 0);
    myPath.lineTo(myWidth * 0.63f, myWidth * 0.38f);
    myPath.lineTo(myWidth, myWidth * 0.38f);
    myPath.lineTo(myWidth * 0.69f, myWidth * 0.59f);
    myPath.lineTo(myWidth * 0.82f, myWidth);
    myPath.lineTo(myWidth * 0.5f, myWidth * 0.75f);
    myPath.lineTo(myWidth * 0.18f, myWidth);
    myPath.lineTo(myWidth * 0.31f, myWidth * 0.59f);
    myPath.lineTo(0, myWidth * 0.38f);
    myPath.lineTo(myWidth * 0.37f, myWidth * 0.38f);
    myPath.close();                                        //根据五角星的10个顶点创建封闭路径
    ShapeDrawable myShapeDrawable = new ShapeDrawable(
        new PathShape(myPath, myImageDrawable.getIntrinsicWidth(),
        myImageDrawable.getIntrinsicHeight()));            //根据五角星创建 ShapeDrawable
    BitmapShader myBitmapShader = new BitmapShader(myBitmap,
        Shader.TileMode.CLAMP, Shader.TileMode.CLAMP);     //根据图像创建 BitmapShader
    //使用图像 BitmapShader 填充五角星样式的 ShapeDrawable
    myShapeDrawable.getPaint().setShader(myBitmapShader);
    myImageView.setBackground(myShapeDrawable);
}}
```

上面这段代码在 MyCode\MySample822\app\src\main\java\com\bin\luo\mysample\MainActivity.java 文件中。在这段代码中，myPath 主要用于创建封闭的五角星路径。myShapeDrawable 主要根据 myPath 创建五角星样式的 ShapeDrawable。myBitmapShader 则根据图像创建着色器。myShapeDrawable.getPaint().setShader(myBitmapShader) 则使用该图像着色器填充（绘制）五角星的内部，从而实现裁剪五角星图像的效果。此实例的完整项目在 MyCode\MySample822 文件夹中。

098 使用 NinePatchDrawable 设置背景

此实例主要通过使用 NinePatchDrawable 获取 drawable 目录下 .9.png 格式的能够实现部分缩放的图像文件，设置 TextView 控件的背景图像，实现在 TextView 控件的内容增加或减少时，背景图像的 4 个顶点围成的边框（线）根据规则变化，中心部分的大小随着内容的多少同比例缩放。当实例运行之后，单击"使用 png 图设置背景"按钮，则背景图像根据内容按比例缩放，如图 098-1 的左图所示；单击"使用 .9 图设置背景"按钮，则背景图像的 4 个顶点围成的边框（线）变化较小，中心部分按照比例缩放，如图 098-1 的右图所示。

图 098-1

主要代码如下：

```
public void onClickBtn1(View v) {                    //响应单击"使用png图设置背景"按钮
    BitmapDrawable myBitmapDrawable = new BitmapDrawable(
            BitmapFactory.decodeResource(getResources(),R.mipmap.myimage));
    myTextView.setBackground(myBitmapDrawable);
}
public void onClickBtn2(View v) {                    //响应单击"使用.9图设置背景"按钮
    NinePatchDrawable myNinePatchDrawable =
            (NinePatchDrawable)getDrawable(R.drawable.myimage);
    myTextView.setBackground(myNinePatchDrawable);
}
```

上面这段代码在 MyCode \ MySample830 \ app \ src \ main \ java \ com \ bin \ luo \ mysample \ MainActivity.java 文件中。在这段代码中，NinePatchDrawable 用于管理.9图..9图是一种可以拉伸的图像格式，当把它用作背景图时，Android 系统会根据实际情况来拉伸图像资源。NinePatch 是额外包含了一个像素边界的 PNG 图像，用.9.png 来标识，并且存放在应用的 res/drawable 目录下。上边界和左边界定义了.9图的拉伸规则和静态不变的区域，两条线的交集为一个矩形，这个矩形内的像素可以自由拉伸；右边界和下边界定义了内容的位置，可以理解为 padding。可以利用 sdk/tools/中的 Draw 9-patch 工具根据具体需求在 PNG 图四周加上特定的像素描边生成.9图。此实例的完整项目在 MyCode\MySample830 文件夹中。

099 使用 DashPathEffect 创建虚线边框

此实例主要通过使用 DashPathEffect 创建虚线 Paint，实现为图像添加虚线边框的特效。当实例运行之后，单击"虚线边框"按钮，则图像在添加虚线边框之后的效果如图 099-1 的左图所示；单击"实线边框"按钮，则图像在添加实线边框之后的效果如图 099-1 的右图所示。

图　099-1

主要代码如下：

```java
public void onClickBtn1(View v) {                                    //响应单击"虚线边框"按钮
    ShapeDrawable myShapeDrawable = new ShapeDrawable(new RectShape());
    myShapeDrawable.getPaint().setStrokeWidth(40);                   //设置描边宽度
    myShapeDrawable.getPaint().setStyle(Paint.Style.STROKE);         //设置边框绘制模式
    SweepGradient mySweepGradient = new SweepGradient(
            myImageView.getWidth() / 2, myImageView.getHeight() / 2,
            new int[]{Color.RED, Color.BLUE, Color.GREEN, Color.RED}, null);
    myShapeDrawable.getPaint().setShader(mySweepGradient);           //设置边框为渐变色样式
    myShapeDrawable.getPaint().setPathEffect(new DashPathEffect(
            new float[]{30, 30}, 30));                               //设置边框线为虚线样式
    myImageView.setBackground(myShapeDrawable);                      //应用该 ShapeDrawable 对象
}
public void onClickBtn2(View v) {                                    //响应单击"实线边框"按钮
    ShapeDrawable myShapeDrawable = new ShapeDrawable(new RectShape());
    myShapcDrawable.getPaint().setStrokeWidth(40);
    myShapeDrawable.getPaint().setStyle(Paint.Style.STROKE);
    myShapeDrawable.getPaint().setColor(Color.RED);
    myImageView.setBackground(myShapeDrawable);
}
```

上面这段代码在 MyCode \ MySample843 \ app \ src \ main \ java \ com \ bin \ luo \ mysample \ MainActivity.java 文件中。在这段代码中，myShapeDrawable.getPaint().setPathEffect(new DashPathEffect(new float[]{30，30}，30))用于创建虚线，float 数组里面的元素不限，但至少为 2 个，可以为单数，也可以为双数。如果为单数，多出来的最后一个无效果；如果为双数，则一个代表实线长度，一个代表空白长度。例如，new DashPathEffect(new float[]{30,10,20,10},10)，效果是每一

段都分为 4 部分,先是 30px 的实心线,再是 10px 的空白,再是 20px 的实心线,再是 10px 的空白。

DashPathEffect 是 PathEffect 类的一个子类,可以使 paint 画出类似虚线的样子,并且可以任意指定虚部和实部的排列方式。在 Android 中,PathEffect 类常用的子类如下。

(1) CornerPathEffect,该类可以使用圆角来代替尖锐的角,对基本图形的形状尖锐的边角进行平滑。

(2) DashPathEffect,该类可以创建一个虚线的轮廓(短横线/小圆点),而不是使用实线。还可以指定任意的虚/实线段的重复模式。

(3) DiscretePathEffect,该类与 DashPathEffect 相似,但是添加了随机性;当绘制它的时候,需要指定每一段的长度和与原始路径的偏离度。

(4) PathDashPathEffect,该类可以定义一个新的形状(路径)并将其用作原始路径的轮廓标记。

此实例的完整项目在 MyCode\MySample843 文件夹中。

100 使用 ComposePathEffect 组合路径特效

此实例主要通过使用 ComposePathEffect 组合两种路径特效,使图像的四周产生不规则的图案。当实例运行之后,图像的四周产生的不规则图案效果如图 100-1 所示。

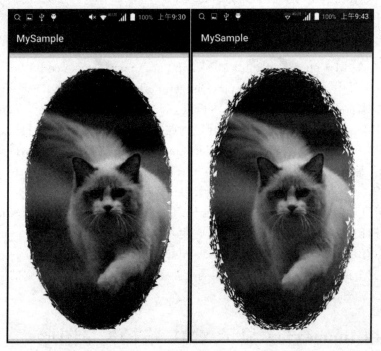

图 100-1

主要代码如下:

```
public class MainActivity extends Activity {
    @Override
    protected void onCreate(Bundle savedInstanceState) {
        setContentView(new MyView(this));
        super.onCreate(savedInstanceState);
    }
```

```
class MyView extends View {
    public MyView(Context context) { super(context); }
    @Override
    protected void onDraw(Canvas myCanvas) {
        Display myDisplay = getWindowManager().getDefaultDisplay();
        int myWidth = myDisplay.getWidth();
        int myHeight = myDisplay.getHeight();
        Bitmap myBitmap =
                BitmapFactory.decodeResource(getResources(),R.mipmap.myimage);
        DiscretePathEffect myDiscretePathEffect1 =
                                new DiscretePathEffect(5.0f, 30.0f);
        DiscretePathEffect myDiscretePathEffect2 =
                                new DiscretePathEffect(1.0f, 7.0f);
        ComposePathEffect myComposePathEffect = new ComposePathEffect
            (myDiscretePathEffect1,myDiscretePathEffect2);       //组合两种路径特效
        Paint myPaint = new Paint();
        myPaint.setPathEffect(myComposePathEffect);              //在画笔上应用组合的路径特效
        myCanvas.drawOval(myWidth/10,myHeight/20,myWidth - myWidth/10,
                                myHeight - myHeight/6,myPaint);  //绘制椭圆
        myPaint.setXfermode(new
                PorterDuffXfermode(PorterDuff.Mode.ADD));        //以椭圆裁剪图像
        myCanvas.drawBitmap(myBitmap, 0, 30, myPaint);           //绘制图像
} } }
```

上面这段代码在 MyCode\MySample105\app\src\main\java\com\bin\luo\mysample\MainActivity.java 文件中。在这段代码中，myDiscretePathEffect1 = new DiscretePathEffect(5.0f, 30.0f)用于创建散列路径特效，5.0f 表示碎片长度，30.0f 表示偏移量。myDiscretePathEffect2 = new DiscretePathEffect(1.0f, 7.0f)也用于创建散列路径特效，1.0f 表示碎片长度，7.0f 表示偏移量。myComposePathEffect = new ComposePathEffect(myDiscretePathEffect1, myDiscretePathEffect2)用于组合上述两种散列路径特效，myDiscretePathEffect 1 表示第一种散列路径特效，myDiscretePathEffect 2 表示第二种散列路径特效。myPaint.setPathEffect(myComposePathEffect)表示在画笔中应用组合路径特效 myComposePathEffect。实际上，DiscretePathEffect 产生的散列路径特效是随机的，即使是相同的特效进行组合，它也不会完全覆盖，例如，myComposePathEffect = new ComposePathEffect(myDiscretePathEffect1, myDiscretePathEffect1)产生的效果如图 100-1 的右图所示，如果是完全覆盖，它应该在图像的四周产生毛刺效果，而不是不规则的图案。此实例的完整项目在 MyCode\MySample105 文件夹中。

101　使用 ImageView 显示 XML 路径矢量图形

此实例主要通过设置 ImageView 控件的 src 属性为 XML 格式的矢量图形文件，实现在 ImageView 控件中显示使用路径描述的矢量图形。当实例运行之后，矢量图形的显示效果如图 101-1 所示。

主要代码如下：

```
< ImageView android:layout_width = "wrap_content"
            android:layout_height = "wrap_content"
            android:src = "@drawable/myvectordata"
            android:layout_centerInParent = "true"/>
```

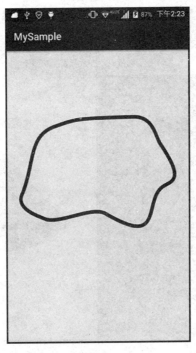

图 101-1

上面这段代码在 MyCode\MySample237\app\src\main\res\layout\activity_main.xml 文件中。在这段代码中,android:src="@drawable/myvectordata" 的 myvectordata 是一个 XML 格式的文件,该文件包含了创建矢量图形的数据。myvectordata.xml 文件的主要内容如下:

```
< vector xmlns:android = "http://schemas.android.com/apk/res/android"
        android:width = "800dp"
        android:height = "800dp"
        android:viewportWidth = "800.0"
        android:viewportHeight = "800.0">
  < path   android:strokeColor = " # 000"
        android:strokeWidth = "15"
        android:pathData = "M215,100.3c97.8 - 32.6,90.5 - 31.9,336 - 37.6c92.4 - 2.1, 98.1,81.6,121.8,
116.4c101.7,149.9,53.5,155.9,14.7,178c - 96.4,54.9,5.4,269 - 257,115.1c - 57 - 33.5 - 203,46.3 - 263.
7,20.1c - 33.5 - 14.5 - 132.5 - 45.5 - 95 - 111.1C125.9,246.6,98.6,139.1,215,100.3z"/>
</vector>
```

上面这段代码在 MyCode\MySample237\app\src\main\res\drawable\myvectordata.xml 文件中。在这段代码中,矢量图形数据主要在 < path >标签中定义,矢量图形的显示与 < vector >标签的 width、height、viewportWidth、viewportHeight 属性有较大关系,否则图形不可见。此实例的完整项目在 MyCode\MySample237 文件夹中。

102 使用 Region 的 INTERSECT 裁剪扇形图像

此实例主要通过使用 Region.Op.INTERSECT 裁剪模式,实现剪裁扇形风格的图像。当实例运行之后,单击"显示原始图像"按钮,将在屏幕上显示原始图像,如图 102-1 的左图所示;单击"显示裁剪图像"按钮,则显示裁剪之后的扇形图像,如图 102-1 的右图所示。

图 102-1

主要代码如下：

```java
public void onClickBtn1(View v) {              //响应单击"显示原始图像"按钮
    myImageView.setImageBitmap(myBitmap);
}
public void onClickBtn2(View v) {              //响应单击"显示裁剪图像"按钮
    Bitmap myBitmap = BitmapFactory.decodeResource(getResources(),
            R.mipmap.myimage1).copy(Bitmap.Config.ARGB_8888,true);
    int myWidth = myBitmap.getWidth();
    Path myPath = new Path();
    myPath.moveTo(myWidth/2,myWidth/2);
    myPath.lineTo(myWidth,myWidth/2);
    myPath.addArc(new RectF(0,0,myWidth,myWidth),0,135);
    myPath.lineTo(myWidth/2,myWidth/2);
    myPath.close();                            //绘制135°扇形
    Region myRegion = new Region();
    myRegion.setPath(myPath,new Region(new Rect(0,0,myBitmap.getWidth(),
            myBitmap.getHeight())));           //根据封闭的扇形创建Region对象
    Canvas myCanvas = new Canvas(myBitmap);
    myCanvas.drawColor(Color.WHITE);
    //Region.Op.INTERSECT 表示获取图像与 Region 的交集
    myCanvas.clipRegion(myRegion,Region.Op.INTERSECT);
    myCanvas.drawBitmap(BitmapFactory.decodeResource(getResources(),
            R.mipmap.myimage1),0,0,null);
    myImageView.setImageBitmap(myBitmap);      //在ImageView上显示裁剪结果
}
```

上面这段代码在 MyCode\MySample827\app\src\main\java\com\bin\luo\mysample\MainActivity.java 文件中。在这段代码中，myCanvas.clipRegion(myRegion, Region.Op. INTERSECT)

表示使用 Region.Op.INTERSECT 模式裁剪图像，Region.Op.INTERSECT 模式表示裁剪结果是两者（扇形和图像）的交集。此实例的完整项目在 MyCode\MySample827 文件夹中。

103　使用裁剪路径将图像从矩形裁剪成椭圆

此实例主要通过使用 Path 的 addOval()方法和 Canvas 的 clipPath()方法，实现将矩形图像裁剪成椭圆图像。当实例运行之后，单击"显示原始图像"按钮，将在屏幕上显示原始的矩形图像，如图 103-1 的左图所示；单击"显示裁剪图像"按钮，矩形图像将被裁剪成椭圆图像，如图 103-1 的右图所示。

图　103-1

主要代码如下：

```java
public void onClickBtn1(View v) {                          //响应单击"显示原始图像"按钮
  myImageView.setImageBitmap(myBitmap);
}
public void onClickBtn2(View v) {                          //响应单击"显示裁剪图像"按钮
  int myWidth = myBitmap.getWidth();
  int myHeight = myBitmap.getHeight();
  Bitmap myNewBitmap = Bitmap.createBitmap(myWidth,
              myHeight, Bitmap.Config.ARGB_8888);           //创建空白位图
  Canvas myCanvas = new Canvas(myNewBitmap);
  myCanvas.drawColor(Color.WHITE);
  Path myPath = new Path();                                 //初始化路径对象
  //根据图像的宽度和高度增加椭圆路径
  myPath.addOval(0,0,myWidth,myHeight,Path.Direction.CW);
  myCanvas.clipPath(myPath);                                //根据指定路径进行裁剪操作
  //在椭圆区域内重新绘制原图
  myCanvas.drawBitmap(BitmapFactory.decodeResource(getResources(),
```

```
               R.mipmap.myimage1),0,0,new Paint(Paint.ANTI_ALIAS_FLAG));
    myImageView.setImageBitmap(myNewBitmap);           //在 ImageView 上显示裁剪结果
}
```

上面这段代码在 MyCode\MySample824\app\src\main\java\com\bin\luo\mysample\MainActivity.java 文件中。此实例的完整项目在 MyCode\MySample824 文件夹中。

104　在自定义 View 中使用扇形裁剪图像

此实例主要通过使用 ArcShape 创建扇形，并使用 BitmapShader 创建图像着色器，实现使用扇形裁剪图像的功能。当实例运行之后，在自定义 View 中使用扇形裁剪图像的效果如图 104-1 所示。

图　104-1

主要代码如下：

```
public class MainActivity extends Activity {
  @Override
  protected void onCreate(Bundle savedInstanceState) {
    setContentView(new MyView(this));
    super.onCreate(savedInstanceState);
  }
  class MyView extends View {
    public MyView(Context context) {
      super(context);
      ArcShape myArcShape =
          new ArcShape(30, 300);    //创建以 30°为开始角度，扫描角度为 300°的扇形
      ShapeDrawable myShapeDrawable = new ShapeDrawable(myArcShape);
      Bitmap myBitmap = ((BitmapDrawable)getResources().getDrawable
                      (R.mipmap.myimage)).getBitmap();         //获取图像资源
```

```
Shader myShader = new BitmapShader(myBitmap,Shader.TileMode.REPEAT,
                Shader.TileMode.REPEAT);            //创建 BitmapShader 图像着色器
myShapeDrawable.getPaint().setShader(myShader);    //使用图像作为画笔填充扇形
this.setBackgroundDrawable(myShapeDrawable);
}
protected void onDraw(Canvas myCanvas) { super.onDraw(myCanvas); }
} }
```

上面这段代码在 MyCode\MySample099\app\src\main\java\com\bin\luo\mysample\MainActivity.java 文件中。在这段代码中，myArcShape = new ArcShape(30，300)表示创建一个以右下角 30°为开始角度，顺时针旋转 300°然后停止的扇形。myShapeDrawable = new ShapeDrawable(myArcShape)表示使用此扇形创建一个 ShapeDrawable。myShader = new BitmapShader(myBitmap，Shader.TileMode.REPEAT，Shader.TileMode.REPEAT)表示使用 myBitmap 创建图像着色器，myBitmap 表示图像，Shader.TileMode.REPEAT（前一个）表示图像在水平方向上的平铺模式，Shader.TileMode.REPEAT（后一个）表示图像在垂直方向上的平铺模式。myShapeDrawable.getPaint().setShader(myShader)表示使用图像着色器 myShader 设置 myShapeDrawable 的画笔。this.setBackgroundDrawable(myShapeDrawable)表示使用刚才裁剪的扇形图像设置自定义 View 的背景。此实例的完整项目在 MyCode\MySample099 文件夹中。

105　根据行列数量将图像切割成碎片并拼图

此实例主要实现了根据指定的行列数量将整幅图像切割成碎片（子图像），然后又将碎片拼成整幅图像的效果。当实例运行之后，单击"正常显示"按钮，将显示整幅图像，如图 105-1 的左图所示；单击"碎片显示"按钮，先将整幅图像切割成 4 行 4 列的 16 个碎片图像，然后将这 16 个碎片图像按照顺序拼成整幅图像，如图 105-1 的右图所示。

图　105-1

主要代码如下:

```java
public void onClickBtn1(View v) {                    //响应单击"正常显示"按钮
    myImageView.setImageResource(R.mipmap.myimage1);
}
public void onClickBtn2(View v) {                    //响应单击"碎片显示"按钮
    //将图像切割成 4 行 4 列, 间距为 8
    List<ImagePart> myImagePartList = SplitImage(BitmapFactory.decodeResource
                              (getResources(), R.mipmap.myimage1), 4, 4, 8);
    int myImageWidth = myImagePartList.get(myImagePartList.size() - 1).myPartX
        + myImagePartList.get(myImagePartList.size() - 1).myPartBitmap.getWidth();
    int myImageHeight = myImagePartList.get(myImagePartList.size() - 1).myPartY
        + myImagePartList.get(myImagePartList.size() - 1).myPartBitmap.getHeight();
    Bitmap myBitmap = Bitmap.createBitmap(myImageWidth,
                              myImageHeight, Bitmap.Config.ARGB_8888);
    Canvas myCanvas = new Canvas(myBitmap);
    //逐个绘制碎片图像
    for (int i = 0; i < myImagePartList.size(); i++) {
        myCanvas.drawBitmap(myImagePartList.get(i).myPartBitmap,
        myImagePartList.get(i).myPartX, myImagePartList.get(i).myPartY, new Paint());
    }
    myImageView.setImageBitmap(myBitmap);
}
class ImagePart {
    public int myPartIndex = 0;
    public int myPartX;
    public int myPartY;
    public Bitmap myPartBitmap = null;
}
//将图像切割成碎片, 并返回碎片数组(即多个子图像)
public List<ImagePart> SplitImage(Bitmap bitmap,
                              int xPiece, int yPiece, int dividerWidth) {
    List<ImagePart> myImagePartList = new ArrayList<ImagePart>(xPiece * yPiece);
    int myBitmapWidth = bitmap.getWidth();
    int myBitmapHeight = bitmap.getHeight();
    int myPartWidth = myBitmapWidth / xPiece;
    int myPartHeight = myBitmapHeight / yPiece;
    for (int i = 0; i < yPiece; i++) {
        for (int j = 0; j < xPiece; j++) {
            ImagePart myPart = new ImagePart();
            myPart.myPartIndex = j + i * xPiece;
            myPart.myPartX = j * myPartWidth;
            myPart.myPartY = i * myPartHeight;
            myPart.myPartBitmap = Bitmap.createBitmap(bitmap, myPart.myPartX +
                dividerWidth, myPart.myPartY + dividerWidth,
                myPartWidth - dividerWidth, myPartHeight - dividerWidth);
            myImagePartList.add(myPart);
        }
    }
    return myImagePartList;
}
```

上面这段代码在 MyCode\MySample749\app\src\main\java\com\bin\luo\mysample\MainActivity.java 文件中。此实例的完整项目在 MyCode\MySample749 文件夹中。

106 使用 BitmapRegionDecoder 加载大图

此实例主要通过在自定义类 LargeImageView 中使用 BitmapRegionDecoder，实现对大尺寸图像通过手势移动进行部分加载和显示。当实例运行之后，手指按住屏幕（大图像）移动，即可查看大图像的任何部分，效果分别如图 106-1 的左图和右图所示。

图 106-1

主要代码如下：

```java
public class MainActivity extends Activity {
 @Override
 protected void onCreate(Bundle savedInstanceState) {
  super.onCreate(savedInstanceState);
  setContentView(R.layout.activity_main);
  try {
   //LargeImageView 是自定义控件
   LargeImageView myLargeImageView =
                   (LargeImageView) findViewById(R.id.myLargeImage);
   //读取 Assets 文件夹中的指定图像并转换为输入流
   InputStream myInputStream = getAssets().open("myimage1.jpg");
   //通过输入流加载该图像
   myLargeImageView.setInputStream(myInputStream);
  } catch (Exception e) { }
 }
}
```

上面这段代码在 MyCode\MySample882\app\src\main\java\com\bin\luo\mysample\MainActivity.java 文件中。在这段代码中，LargeImageView 是一个自定义控件（类），该控件主要用于实现大尺寸图像的加载和拖曳查看。关于 LargeImageView 类的详细内容请参考 MyCode\MySample882\app\src\main\java\com\bin\luo\mysample\LargeImageView.java 文件。此实例的完整项目在 MyCode\MySample882 文件夹中。

第 5 章

动画

107 使用 ObjectAnimator 创建坐标平移动画

此实例主要通过在 ObjectAnimator 的 ofFloat() 方法中指定 X 和 Y 坐标值,实现平移控件。当实例运行之后,单击"播放分段动画"按钮,将首先改变图像的 X 坐标执行水平平移,然后改变 Y 坐标执行垂直平移,即小狗将首先从左平移到右,再从上平移到下,如图 107-1 的左图所示。单击"播放组合动画"按钮,将同时改变图像的 X 坐标和 Y 坐标执行平移运动,即小狗将直接从左上角沿着对角线方向平移到右下角,如图 107-1 的右图所示。

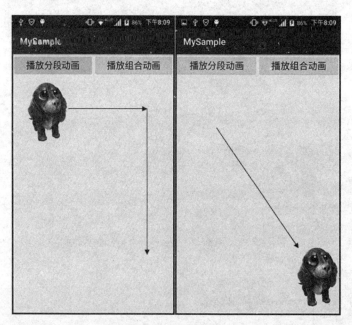

图 107-1

主要代码如下:

```
public void onClickBtn1(View v) {                    //响应单击"播放分段动画"按钮
    ImageView myImageView = (ImageView) findViewById(R.id.myImageView);
    ObjectAnimator myObjectAnimator1 =
                    ObjectAnimator.ofFloat(myImageView, "x", 0, 700);
    ObjectAnimator myObjectAnimator2 =
```

```
                        ObjectAnimator.ofFloat(myImageView, "y", 150, 1200);
    AnimatorSet myAnimatorSet = new AnimatorSet();
    myAnimatorSet.play(myObjectAnimator2).after(myObjectAnimator1);
    myAnimatorSet.setDuration(2000);
    myAnimatorSet.start();
}
public void onClickBtn2(View v) {            //响应单击"播放组合动画"按钮
    ImageView myImageView = (ImageView) findViewById(R.id.myImageView);
    ObjectAnimator myObjectAnimator1 =
                        ObjectAnimator.ofFloat(myImageView, "x", 0, 700);
    ObjectAnimator myObjectAnimator2 =
                        ObjectAnimator.ofFloat(myImageView, "y", 150, 1200);
    AnimatorSet myAnimatorSet = new AnimatorSet();
    myAnimatorSet.play(myObjectAnimator2).with(myObjectAnimator1);
    myAnimatorSet.setDuration(2000);
    myAnimatorSet.start();
}
```

上面这段代码在 MyCode\MySample169\app\src\main\java\com\bin\luo\mysample\MainActivity.java 文件中。在这段代码中，myObjectAnimator1 = ObjectAnimator.ofFloat(myImageView, "x", 0, 700)用于创建一个平移动画，该动画在指定的时间内将 myImageView 控件的 X 坐标值从 0 变到 700。myObjectAnimator2 = ObjectAnimator.ofFloat(myImageView，"y"，150，1200)用于创建一个平移动画，该动画在指定的时间内将 myImageView 控件的 Y 坐标值从 150 变到 1200。myAnimatorSet.play（myObjectAnimator2）.after（myObjectAnimator1）表示将 myObjectAnimator1 动画和 myObjectAnimator2 动画添加到 myAnimatorSet 动画集合中，并且先执行 myObjectAnimator1 动画，然后再执行 myObjectAnimator2 动画。myAnimatorSet.play（myObjectAnimator2）.with（myObjectAnimator1）表示将 myObjectAnimator1 动画和 myObjectAnimator2 动画添加到 myAnimatorSet 动画集合中，并且两个动画同时执行。此实例的完整项目在 MyCode\MySample169 文件夹中。

108 使用 ObjectAnimator 创建波纹扩散动画

此实例主要通过在自定义 RelativeLayout 中，通过迭代多个 ObjectAnimator 动画，实现波纹扩散的动画特效。当实例运行之后，单击"开始动画"按钮，将在图标的周围产生波纹扩散动画；单击"停止动画"按钮，则波纹扩散动画自动消失，效果分别如图 108-1 的左图和右图所示。

主要代码如下：

```
public void startRipple(View v) {            //响应单击"开始动画"按钮
    myRippleLayout.startRippleAnimation();
}
public void stopRipple(View v) {             //响应单击"停止动画"按钮
    myRippleLayout.stopRippleAnimation();
}
```

上面这段代码在 MyCode\MySample885\app\src\main\java\com\bin\luo\mysample\MainActivity.java 文件中。在这段代码中，RippleLayout 是以 RelativeLayout 为基类创建的自定义控件（类），用于实现波纹扩散的动画。关于 RippleLayout 类的详细内容请参考 MyCode\MySample885\app\src\main\java\com\bin\luo\mysample\RippleLayout.java 文件。此实例的完整

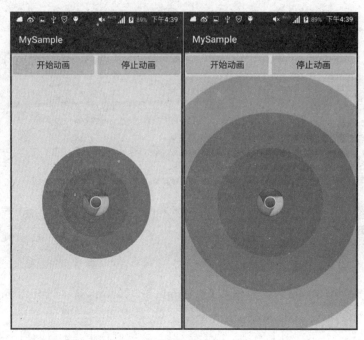

图 108-1

项目在 MyCode\MySample885 文件夹中。

109 使用 ValueAnimator 动态绘制桃心图形

此实例主要通过在自定义 View 中使用 ValueAnimator 动画控制桃心图形的路径（PathMeasure），实现以动画的形式动态绘制桃心图形的功能。当实例运行之后，将从零开始逐点动态绘制桃心图形，效果分别如图 109-1 的左图和右图所示。

图 109-1

主要代码如下：

```xml
<com.bin.luo.mysample.MyView android:id = "@+id/myView"
                              android:layout_width = "match_parent"
                              android:layout_height = "match_parent"/>
```

上面这段代码在 MyCode\MySample846\app\src\main\res\layout\activity_main.xml 文件中。在这段代码中，com.bin.luo.mysample.MyView 即是用于以动画的形式动态绘制桃心图形的自定义控件，MyView 控件（类）的主要代码如下：

```java
public class MyView extends View {
  public MyView(Context context) { super(context); }
  private Paint myPaint;
  private ValueAnimator myValueAnimator;
  private float myLength;
  private PathMeasure myPathMeasure;              //Path路径追踪
  private Path myPath;
  public MyView(Context context, AttributeSet attrs) {
    super(context, attrs);
    init();
    myValueAnimator.start();
  }
  private void init() {
    myPaint = new Paint(Paint.ANTI_ALIAS_FLAG);
    myPaint.setColor(Color.BLUE);
    myPaint.setStyle(Paint.Style.STROKE);
    myPaint.setStrokeWidth(15);
    myPath = new Path();
    myPath.moveTo(396 - 50, 313);
    myPath.cubicTo(207 - 50, 114, 339 - 50, 46, 396 - 50, 111);
    myPath.cubicTo(453 - 50, 46, 585 - 50, 114, 396 - 50, 313);
    myPathMeasure = new PathMeasure(myPath, false);
    myValueAnimator = ValueAnimator.ofFloat(0, myPathMeasure.getLength());
    myValueAnimator.setDuration(5000);
    myValueAnimator.addUpdateListener(new ValueAnimator.AnimatorUpdateListener() {
      @Override
      public void onAnimationUpdate(ValueAnimator animation) {
        myLength = (float) animation.getAnimatedValue();
        invalidate();
      } }); }
  @Override
  protected void onMeasure(int widthMeasureSpec, int heightMeasureSpec) {
    super.onMeasure(widthMeasureSpec, heightMeasureSpec);
  }
  public MyView(Context context, AttributeSet attrs, int defStyleAttr) {
    super(context, attrs, defStyleAttr);
  }
  @Override
  protected void onDraw(Canvas canvas) {
    super.onDraw(canvas);
    canvas.save();
    canvas.translate(200,600);                    //平移画布至(200,600)
    Path path = new Path();
```

```
    myPathMeasure.getSegment(0, myLength,
            path, true);    //取路径片段,0 指从头截取,myLength 指截取的长度
    canvas.drawPath(path, myPaint);
    canvas.restore();
}
public void start() {   myValueAnimator.start(); }
}
```

上面这段代码在 MyCode \ MySample846 \ app \ src \ main \ java \ com \ bin \ luo \ mysample \ MyView.java 文件中。在这段代码中，myValueAnimator = ValueAnimator.ofFloat（0，myPathMeasure.getLength()）表示在指定的时间内（5000ms）将目标值从 0 变为 myPathMeasure. getLength()；myLength = （float）animation.getAnimatedValue()则监听该值的改变，并根据该值（myLength）在 onDraw（Canvas canvas）方法中绘制桃心图形。此实例的完整项目在 MyCode \ MySample846 文件夹中。

110　使用 AnimationSet 组合多个不同的动画

此实例主要通过使用 AnimationSet 组合 ScaleAnimation 和 AlphaAnimation，实现在 ImageView 控件上同时产生两种动画效果。当实例运行之后，单击"开始播放动画"按钮，图像将以自身为中心执行缩小动画直到消失，在图像缩小的过程中，图像的透明度也将逐渐变小，直到为零，效果分别如图 110-1 的左图和右图所示。

图　110-1

主要代码如下：

```
public class MainActivity extends Activity {
    ImageView myImageView;
```

```
AnimationSet myAnimationSet;
@Override
protected void onCreate(Bundle savedInstanceState) {
  super.onCreate(savedInstanceState);
  setContentView(R.layout.activity_main);
  myImageView = (ImageView) findViewById(R.id.myImageView);
  ScaleAnimation myScaleAnimation = new ScaleAnimation(
        1.0f, 0.0f, 1.0f, 0.0f, Animation.RELATIVE_TO_SELF, 0.5f,
        Animation.RELATIVE_TO_SELF, 0.5f);        //创建缩放动画
  myScaleAnimation.setDuration(3000);             //设置动画持续时间3s
  AlphaAnimation myAlphaAnimation =
                        new AlphaAnimation(1.0f,0.0f);  //创建透明度动画
  myAlphaAnimation.setDuration(3000);             //设置动画持续时间3s
  myAnimationSet = new AnimationSet(true);        //创建动画集合对象
  //在动画集合对象中添加缩放动画
  myAnimationSet.addAnimation(myScaleAnimation);
  //在动画集合对象中添加透明度动画
  myAnimationSet.addAnimation(myAlphaAnimation);
  myAnimationSet.setFillAfter(true);              //保留动画最后的状态
  myAnimationSet.setFillEnabled(true);
}
public void onClickBtn1(View v) {                 //响应单击"开始播放动画"按钮
  myImageView.clearAnimation();                   //清除ImageView控件上的动画
  myImageView.setAnimation(myAnimationSet);       //在ImageView控件上添加动画
  myAnimationSet.start();
}
public void onClickBtn2(View v) {                 //响应单击"停止播放动画"按钮
  myImageView.clearAnimation();
} }
```

上面这段代码在 MyCode\MySample160\app\src\main\java\com\bin\luo\mysample\MainActivity.java 文件中。在这段代码中，AnimationSet 通过调用其 addAnimation() 方法将多个不同的动画组织到一起，然后调用 start() 方法执行这些动画。addAnimation() 支持的动画类型包括：AlphaAnimation、RotateAnimation、ScaleAnimation 和 TranslateAnimation，其他动画有可能不支持。此实例的完整项目在 MyCode\MySample160 文件夹中。

111 自定义 TypeEvaluator 合成多方向的位移

此实例主要通过创建自定义估值器 MyTypeEvaluator，配置动画开始值和结束值之间的关系，从而使 ValueAnimator 动画同时实现在水平方向和垂直方向做平移运动。当实例运行之后，单击"开始播放动画"按钮，则图像（足球）将在 3s 内沿着对角线方向从左下角移动到右上角，效果分别如图 111-1 的左图和右图所示。

主要代码如下：

```
public class MainActivity extends Activity {
  ImageView myImageView;
  ValueAnimator myValueAnimator;
  @Override
  protected void onCreate(Bundle savedInstanceState) {
    super.onCreate(savedInstanceState);
```

图 111-1

```
setContentView(R.layout.activity_main);
myImageView = (ImageView)findViewById(R.id.myImageView);
myImageView.setVisibility(View.GONE);
myValueAnimator = ValueAnimator.ofObject(new MyTypeEvaluator(),
                        new Point(10,1300),new Point(830,10));
myValueAnimator.setDuration(3000);
//设置加速减速插值器(动画两头慢,中间快)
myValueAnimator.setInterpolator(new AccelerateDecelerateInterpolator());
myValueAnimator.addUpdateListener(new ValueAnimator.AnimatorUpdateListener(){
  @Override
  public void onAnimationUpdate(ValueAnimator animation) {
    Point myPoint = (Point)animation.getAnimatedValue();
    myImageView.setX(myPoint.x);
    myImageView.setY(myPoint.y);
}});}
public void onClickBtn1(View v) {            //响应单击"开始播放动画"按钮
  myImageView.setVisibility(View.VISIBLE);
  myValueAnimator.start();
}
public void onClickBtn2(View v) {            //响应单击"停止播放动画"按钮
  myValueAnimator.cancel();
}
//自定义估值算法计算开始值与结束值之间的关系
public class MyTypeEvaluator implements TypeEvaluator<Point>{
  @Override
  public Point evaluate(float fraction,Point startValue,Point endValue){
    Point  point = new Point();
    //fraction 代表时间流逝的百分比
    point.x = startValue.x + (int)(fraction * (endValue.x - startValue.x));
    point.y = startValue.y + (int)(fraction * (endValue.y - startValue.y));
    return point;
}}}
```

上面这段代码在 MyCode\MySample163\app\src\main\java\com\bin\luo\mysample\MainActivity.java 文件中。在这段代码中，myValueAnimator = ValueAnimator.ofObject(new MyTypeEvaluator(),new Point(10,1300),new Point(830,10))用于创建一个值动画，该值动画的变化范围在左下角 Point(10,1300)和右上角 Point(830,10)之间。MyTypeEvaluator 自定义估值器用于计算在动画的持续时间内，动画对象的 Point 如何确定。evaluate(float fraction,Point startValue, Point endValue)用于计算动画对象的 Point，fraction 表示动画持续时间的百分比，startValue 表示动画开始值，endValue 表示动画结束值。此实例的完整项目在 MyCode\MySample163 文件夹中。

112　使用 PropertyValuesHolder 实现弹簧动画

此实例主要通过使用 PropertyValuesHolder 的 ofKeyframe()方法加载多个不同的关键帧 Keyframe，从而使 ObjectAnimator 在播放这些关键帧时呈现类似弹簧伸缩的动画效果。当实例运行之后，单击"启动动画"按钮，图像（实际上是按钮的背景图像）的宽度（在水平方向上伸缩）将由大变小，或由小变大，效果分别如图 112-1 的左图和右图所示。

图　112-1

主要代码如下：

```
public class MainActivity extends Activity {
 private Button myButton;
 @Override
 protected void onCreate(Bundle savedInstanceState) {
  super.onCreate(savedInstanceState);
  setContentView(R.layout.activity_main);
  myButton = (Button) findViewById(R.id.myButton);
  myButton.setBackground(getDrawable(R.mipmap.myimage));
 }
 public void onClickmyBtn1(View v) {        //响应单击"启动动画"按钮
```

```
        Keyframe kf0 = Keyframe.ofInt(0, 200);
        Keyframe kf1 = Keyframe.ofInt(0.25f, 1080);
        Keyframe kf2 = Keyframe.ofInt(0.5f, 400);
        Keyframe kf3 = Keyframe.ofInt(0.75f, 880);
        Keyframe kf4 = Keyframe.ofInt(1f, 650);
        PropertyValuesHolder myPropertyValuesHolder =
             PropertyValuesHolder.ofKeyframe("width", kf0, kf1, kf2, kf3, kf4);
        ObjectAnimator myObjectAnimator =
          ObjectAnimator.ofPropertyValuesHolder(myButton, myPropertyValuesHolder);
        myObjectAnimator.setDuration(2000);
        myObjectAnimator.start();
    }}
```

上面这段代码在 MyCode\MySample254\app\src\main\java\com\bin\luo\mysample\MainActivity.java 文件中。在这段代码中，Keyframe kf1 = Keyframe.ofInt(0.25f，1080)用于创建一个关键帧，它表示在动画总时间的 1/4 时，图像的宽度是 1080 像素。KeyFrame 是抽象类，它通过 ofInt()、ofFloat()、ofObject()获得适当的 KeyFrame，然后通过 PropertyValuesHolder.ofKeyframe()方法获得 PropertyValuesHolder 对象，最后通过 ObjectAnimator.ofPropertyValuesHolder()方法获得 Animator。此功能类似于 myObjectAnimator = ObjectAnimator.ofInt(myButton,"width",200, 1080,400,880,650)。此实例的完整项目在 MyCode\MySample254 文件夹中。

113　自定义 selector 实现以动画形式改变透明度

此实例主要通过设置 ImageView 控件的 stateListAnimator 属性为自定义 selector，并在自定义 selector 中使用 objectAnimator 改变 alpha 属性值，实现以动画的形式改变控件的透明度（即淡入淡出效果）。当实例运行之后，4 个 ImageView 控件（4 幅电影海报图像）将以半透明状态显示，因此有点暗淡；单击其中任意一个 ImageView 控件，该 ImageView 控件的 alpha 属性值将由 0.4 增大至 1，即完全不透明，因此有突出显示的效果；在离开该控件时将减少（恢复）至半透明状态，即暗淡状态。效果分别如图 113-1 的左图和右图所示。

图　113-1

主要代码如下：

```xml
<LinearLayout android:layout_width = "match_parent"
              android:layout_height = "wrap_content"
              android:orientation = "vertical">
  <LinearLayout android:layout_weight = "1"
                android:layout_width = "match_parent"
                android:layout_height = "wrap_content"
                android:orientation = "horizontal">
    <ImageView android:alpha = "0.4"
               android:clickable = "true"
               android:stateListAnimator = "@animator/myselector"
               android:layout_width = "match_parent"
               android:layout_height = "match_parent"
               android:layout_weight = "1"
               android:scaleType = "centerCrop"
               android:layout_margin = "2dp"
               android:src = "@mipmap/myimage1"/>
    <ImageView android:alpha = "0.4"
               android:clickable = "true"
               android:stateListAnimator = "@animator/myselector"
               android:layout_width = "match_parent"
               android:layout_height = "match_parent"
               android:layout_weight = "1"
               android:scaleType = "centerCrop"
               android:layout_margin = "2dp"
               android:src = "@mipmap/myimage2"/>
  </LinearLayout>
  <LinearLayout android:layout_weight = "1"
                android:layout_width = "match_parent"
                android:layout_height = "wrap_content"
                android:orientation = "horizontal">
    <ImageView android:alpha = "0.4"
               android:clickable = "true"
               android:stateListAnimator = "@animator/myselector"
               android:layout_width = "match_parent"
               android:layout_height = "match_parent"
               android:layout_weight = "1"
               android:scaleType = "centerCrop"
               android:layout_margin = "2dp"
               android:src = "@mipmap/myimage3"/>
    <ImageView android:alpha = "0.4"
               android:clickable = "true"
               android:stateListAnimator = "@animator/myselector"
               android:layout_width = "match_parent"
               android:layout_height = "match_parent"
               android:layout_weight = "1"
               android:scaleType = "centerCrop"
               android:layout_margin = "2dp"
               android:src = "@mipmap/myimage4"/>
  </LinearLayout>
</LinearLayout>
```

上面这段代码在 MyCode\MySample576\app\src\main\res\layout\activity_main.xml 文件中。

在这段代码中,android:stateListAnimator="@animator/myselector"用于设置动态改变透明度大小的自定义动画 selector。自定义动画 selector 的主要内容如下：

```xml
<?xml version = "1.0" encoding = "utf - 8"?>
<selector xmlns:android = "http://schemas.android.com/apk/res/android">
 <item android:state_pressed = "true">
  <set>
   <objectAnimator android:duration = "@android:integer/config_shortAnimTime"
            android:propertyName = "alpha"
            android:valueTo = "1"
            android:valueType = "floatType" /></set></item>
 <item>
  <set>
   <objectAnimator android:duration = "@android:integer/config_shortAnimTime"
            android:propertyName = "alpha"
            android:valueTo = "0.4"
            android:valueType = "floatType" /></set></item>
</selector>
```

上面这段代码在 MyCode\MySample576\app\src\main\res\animator\myselector.xml 文件中。在这段代码中,<item android:state_pressed="true">表示在控件被按下时执行此标签中的动画。android:propertyName="alpha"表示动画类型是改变 alpha 属性值,android:valueTo="1"表示将 alpha 属性值改变到 1（1 表示完全不透明,0 表示完全透明）。此实例的完整项目在 MyCode\MySample576 文件夹中。

114 使用 StateListAnimator 实现状态切换动画

此实例主要通过使用 AnimatorInflater 的 loadStateListAnimator() 方法根据自定义选择器文件创建 StateListAnimator,并设置该 StateListAnimator 为 ImageView 控件的 stateListAnimator 属性值,实现在触摸 ImageView 控件时使该控件旋转 30°的动画效果。当实例运行之后,如果手指触摸到 ImageView 控件（怪兽图像）,则该控件将顺时针旋转 30°,如图 114-1 的右图所示；在离开该控件时将逆时针旋转（恢复）至正常状态,如图 114-1 的左图所示。

主要代码如下：

```java
public class MainActivity extends Activity {
 ImageView myImageView;
 @Override
 protected void onCreate(Bundle savedInstanceState) {
  super.onCreate(savedInstanceState);
  setContentView(R.layout.activity_main);
  myImageView = (ImageView)findViewById(R.id.myImageView);
  StateListAnimator myStateListAnimator = AnimatorInflater.
       loadStateListAnimator(this, R.animator.myselector);
  myImageView.setStateListAnimator(myStateListAnimator);
  myImageView.setClickable(true);
 } }
```

上面这段代码在 MyCode\MySample577\app\src\main\java\com\bin\luo\mysample\MainActivity.java 文件中。在这段代码中,myImageView.setStateListAnimator(myStateListAnimator)用

图 114-1

于设置 ImageView 控件的 stateListAnimator 属性为自定义的 myStateListAnimator。myStateListAnimator = AnimatorInflater.loadStateListAnimator（this，R.animator.myselector）用于从资源文件 myselector 中加载自定义 selector。myselector 资源文件的主要内容如下：

```xml
<?xml version = "1.0" encoding = "utf-8"?>
<selector xmlns:android = "http://schemas.android.com/apk/res/android">
  <item android:state_pressed = "true">
   <set>
    <objectAnimator
        android:duration = "@android:integer/config_shortAnimTime"
        android:propertyName = "rotation"
        android:valueTo = "30"
        android:valueType = "floatType" /></set></item>
  <item>
   <set>
    <objectAnimator
        android:duration = "@android:integer/config_shortAnimTime"
        android:propertyName = "rotation"
        android:valueTo = "0"
        android:valueType = "floatType" /></set></item>
</selector>
```

上面这段代码在 MyCode\MySample577\app\src\main\res\animator\myselector.xml 文件中。在这段代码中，<item android:state_pressed = "true">表示在单击控件时执行此动画。android:propertyName = "rotation"表示动画类型是旋转，android:valueTo = "30"表示旋转 30°。此实例的完整项目在 MyCode\MySample577 文件夹中。

115　自定义 TypeEvaluator 以加速动画显示字母

此实例主要通过创建自定义估值器 MyTypeEvaluator 配置字母的显示规则，并在 ValueAnimator 动画中使用 AccelerateInterpolator 插值器，实现以由慢变快的加速动画显示所有英文字母。当实例运行之后，单击"启动动画"按钮，将按照字母表的顺序由慢变快显示所有字母，效果如图 115-1 所示。

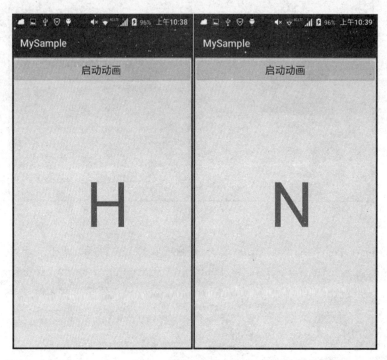

图 115-1

主要代码如下：

```
public class MainActivity extends Activity {
  TextView myTextView;
  @Override
  protected void onCreate(Bundle savedInstanceState) {
    super.onCreate(savedInstanceState);
    setContentView(R.layout.activity_main);
    myTextView = (TextView)findViewById(R.id.myTextView);
  }
  public void onClickmyBtn1(View v) {            //响应单击"启动动画"按钮
    //设置 myValueAnimator 动画显示的开始字符是 A,结束字符是 Z
    ValueAnimator myValueAnimator = ValueAnimator.ofObject(
            new MyTypeEvaluator(),new Character('A'),new Character('Z'));
    myValueAnimator.addUpdateListener(new ValueAnimator.AnimatorUpdateListener(){
      @Override
      public void onAnimationUpdate(ValueAnimator animation) {
        char myChar = (char)animation.getAnimatedValue();
        myTextView.setText(String.valueOf(myChar));//根据时间线显示字符
      } });
```

```
    myValueAnimator.setDuration(2000);              //动画持续时间是2000ms
    //设置由慢变快的加速插值器
    myValueAnimator.setInterpolator(new AccelerateInterpolator());
    myValueAnimator.start();                         //启动动画
}
//自定义 TypeEvaluator 实现根据字母的 ASCII 值进行递增
public class MyTypeEvaluator implements TypeEvaluator<Character>{
    @Override
    public Character evaluate(float fraction,
                        Character startValue, Character endValue){
        int startInt  = (int)startValue;
        int endInt    = (int)endValue;
        int curInt    = (int)(startInt + fraction * (endInt - startInt));
        char result   = (char)curInt;                //根据字母的 ASCII 值转换成字母
        return result;
}}}
```

上面这段代码在 MyCode\ MySample859\ app\ src\ main\ java\ com\ bin\ luo\ mysample\ MainActivity.java 文件中。在这段代码中，myValueAnimator = ValueAnimator.ofObject（new MyTypeEvaluator()，new Character('A')，new Character('Z')）表示根据自定义类 MyTypeEvaluator 的字母显示规则显示 A 至 Z 的字母。65～90 是 A～Z 大写英文字母的 ASCII 值，97～122 是 a～z 小写英文字母的 ASCII 值。此实例的完整项目在 MyCode\ MySample859 文件夹中。

116 使用 BounceInterpolator 实现弹跳动画

此实例主要通过使用 BounceInterpolator 作为 ValueAnimator 的插值器，实现在动画结束之前呈现弹跳动画。当实例运行之后，单击"启动动画"按钮，图像（ImageView 控件）将在 1000ms 内从顶部下落到底部，之后反复弹跳几次，效果分别如图 116-1 的左图和右图所示。

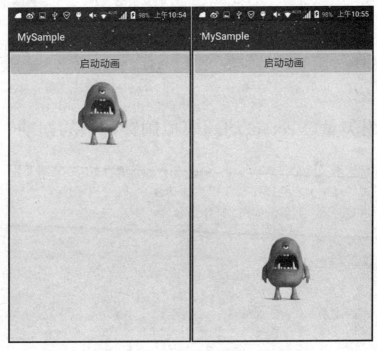

图 116-1

主要代码如下:

```java
public class MainActivity extends Activity {
    ImageView myImageView;
    @Override
    protected void onCreate(Bundle savedInstanceState) {
        super.onCreate(savedInstanceState);
        setContentView(R.layout.activity_main);
        myImageView = (ImageView) findViewById(R.id.myImageView);
    }
    public void onClickmyBtn1(View v) {                    //响应单击"启动动画"按钮
        //设置垂直下落范围为 50～1150
        ValueAnimator myValueAnimator = ValueAnimator.ofInt(50, 1150);
        myValueAnimator.addUpdateListener(new ValueAnimator.AnimatorUpdateListener(){
            @Override
            public void onAnimationUpdate(ValueAnimator animation) {
                //根据时间线在范围内获取当前在垂直方向上的变化值
                int myAnimatedValue = (int) animation.getAnimatedValue();
                //根据变化值重置 myImageView 控件的垂直坐标(相对父容器)
                myImageView.layout(myImageView.getLeft(), myAnimatedValue,
                    myImageView.getRight(), myAnimatedValue + myImageView.getHeight());
            }});
        myValueAnimator.setDuration(1000);                 //设置动画持续时间是 1000ms
        //设置 BounceInterpolator 插值器实现弹跳动画
        myValueAnimator.setInterpolator(new BounceInterpolator());
        myValueAnimator.setRepeatCount(-1);                //设置无限重复
        myValueAnimator.start();
    }}
```

上面这段代码在 MyCode\MySample861\app\src\main\java\com\bin\luo\mysample\MainActivity.java 文件中。在这段代码中,myValueAnimator.setInterpolator(new BounceInterpolator())用于设置 BounceInterpolator 作为 myValueAnimator 的插值器,以实现在动画结束前出现弹跳效果。myImageView.layout(myImageView.getLeft(),myAnimatedValue,myImageView.getRight(),myAnimatedValue+ myImageView.getHeight())用于重置 myImageView 在下落动画执行过程中的大小和位置,4 个参数分别是 myImageView 控件的左上角和右下角坐标,该坐标不是相对于屏幕原点,而是相对于父布局。此实例的完整项目在 MyCode\MySample861 文件夹中。

117 使用矢量(Vector)动画模拟闹钟耳朵的摆动

此实例主要通过使用 AnimatedVectorDrawableCompat 解析 XML 动画文件,实现在 XML 文件中使用矢量绘制的闹钟耳朵呈现绕弧线左右摆动的动画效果。当实例运行之后,单击"执行动画"按钮,闹钟的两个耳朵将左右摆动,效果如图 117-1 所示。

主要代码如下:

```java
public class MainActivity extends Activity {
    ImageView myImageView;
    @Override
    protected void onCreate(Bundle savedInstanceState) {
```

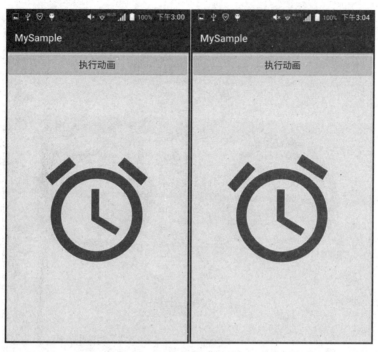

图 117-1

```
super.onCreate(savedInstanceState);
setContentView(R.layout.activity_main);
myImageView = (ImageView)findViewById(R.id.myImageView);
AnimatedVectorDrawableCompat myAnimatedVectorDrawableCompat =
    AnimatedVectorDrawableCompat.create(this, R.drawable.myanimation);
myImageView.setImageDrawable(myAnimatedVectorDrawableCompat);
}
public void onClickmyBtn1(View v) {                //响应单击"执行动画"按钮
    ((Animatable) myImageView.getDrawable()).start();
} }
```

上面这段代码在 MyCode\MySample612\app\src\main\java\com\bin\luo\mysample\MainActivity.java 文件中。在这段代码中，myAnimatedVectorDrawableCompat = AnimatedVectorDrawableCompat.create(this, R.drawable.myanimation)用于加载动画文件 myanimation 中的矢量动画。关于 myanimation 动画文件的详细内容请参考 MyCode\MySample612\app\src\main\res\drawable\myanimation.xml 文件。需要注意的是，使用此实例的相关类需要在 gradle 中引入 compile 'com.android.support:appcompat-v7:24.0.0'依赖项。此实例的完整项目在 MyCode\MySample612 文件夹中。

118 控制 trimPathEnd 动态生成非连续矢量图

此实例主要通过在多个 objectAnimator 动画中将 trimPathEnd 属性值从 0 变到 1(0%～100%)，并且通过 android:startOffset 属性控制间隔时间，实现动态依次生成非连续矢量图的动画效果。当实例运行之后，单击屏幕，先绘制外圆，然后绘制里面的上半圆，最后绘制里面的下半圆，效果分别如图 118-1 的左图和右图所示。

图 118-1

主要代码如下：

```java
public class MainActivity extends Activity {
  private ImageView myImageView;
  private boolean bChecked = false;
  @Override
  protected void onCreate(Bundle savedInstanceState) {
    super.onCreate(savedInstanceState);
    setContentView(R.layout.activity_main);
    myImageView = (ImageView) findViewById(R.id.myImageView);
    myImageView.setOnClickListener(new View.OnClickListener() {
      @Override
      public void onClick(View v) {
        AnimatedVectorDrawable myAnimatedVectorDrawable =
                  (AnimatedVectorDrawable) myImageView.getDrawable();
        if(!bChecked) myAnimatedVectorDrawable.start();
        else myAnimatedVectorDrawable.stop();
        bChecked = !bChecked;
} }); } }
```

上面这段代码在 MyCode\MySample622\app\src\main\java\com\bin\luo\mysample\MainActivity.java 文件中。在这段代码中，myAnimatedVectorDrawable.start()表示启动 ImageView 控件的矢量图绘制动画，myAnimatedVectorDrawable.stop()表示停止 ImageView 控件的矢量图绘制动画，因此 ImageView 控件的动画不能直接为纯粹的图像资源，而应是包括有关动画内容的 XML 资源。如 android:src="@drawable/mytransition"，表示在 ImageView 控件上加载的 XML 动画 mytransition。mytransition 文件的主要内容如下：

```
<animated-vector xmlns:android="http://schemas.android.com/apk/res/android"
    android:drawable="@drawable/myvector">
    <target android:name="mypath1"
        android:animation="@animator/myanimation1"/>
    <target android:name="mypath2"
        android:animation="@animator/myanimation2"/>
    <target android:name="mypath3"
        android:animation="@animator/myanimation3"/>
</animated-vector>
```

上面这段代码在 MyCode\MySample622\app\src\main\res\drawable\mytransition.xml 文件中。在这段代码中，android:name="mypath1"表示自定义的路径(矢量图)名称(该名称必须与 myvector 文件中的名称一一对应)，android:animation="@animator/myanimation1"中的 myanimation1 文件是 mypath1 矢量图对应的动画文件，因此 3 段矢量图就有 3 个动画文件。关于这 3 个动画文件的详细内容请参考 MyCode\MySample622\app\src\main\res\animator 文件夹下的 myanimation1.xml、myanimation2.xml 和 myanimation3.xml。此实例的完整项目在 MyCode\MySample622 文件夹中。

119　改变矢量数据实现不同图形数字的平滑过渡

此实例主要通过在 objectAnimator 动画中改变矢量图形数字的 pathData 属性值，实现矢量图形数字 1、2、3 在相互之间切换时发生轮廓形变的平滑过渡效果。当实例运行之后，将显示矢量图形数字 1(非字符，直接用线条绘制的数字)；单击矢量图形数字 1，则矢量图形数字 1 的形状发生形变，并慢慢过渡到矢量图形数字 2，如图 119-1 的左图所示；单击矢量图形数字 2，则矢量图形数字 2 的形状发生形变，并慢慢过渡到矢量图形数字 3，如图 119-1 的右图所示。

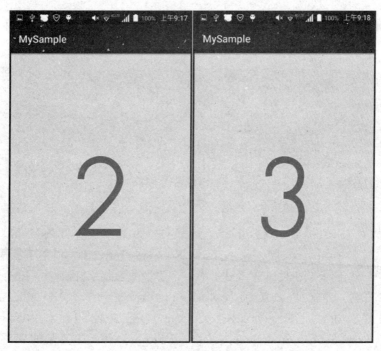

图　119-1

主要代码如下：

```java
public class MainActivity extends Activity {
 private ImageView myImageView;
 private int myCurrentNum = -1;
 private static final int[] myStates = {
                        R.attr.state1, R.attr.state2, R.attr.state3};
@Override
protected void onCreate(Bundle savedInstanceState) {
 super.onCreate(savedInstanceState);
 setContentView(R.layout.activity_main);
 myImageView = (ImageView) findViewById(R.id.myImageView);
 myImageView.setOnClickListener(new View.OnClickListener() {
  @Override
  public void onClick(View v) {
   myCurrentNum++;
   if (myCurrentNum > 2) { myCurrentNum = 0; }
   int[] myStates = new int[MainActivity.myStates.length];
   for (int i = 0; i < MainActivity.myStates.length; i++) {
    if (i == myCurrentNum) { myStates[i] = MainActivity.myStates[i]; }
    else { myStates[i] = -MainActivity.myStates[i]; }
   }
   myImageView.setImageState(myStates, true);
  } });
  myImageView.callOnClick();         //自动单击 ImageView 控件
} }
```

上面这段代码在 MyCode\MySample624\app\src\main\java\com\bin\luo\mysample\MainActivity.java 文件中。在这段代码中，myStates = {R.attr.state1, R.attr.state2, R.attr.state3}是自定义属性，用于确定当前矢量图形数字的状态，该属性定义在 attrs 文件中，attrs 文件的主要内容如下：

```xml
<?xml version = "1.0" encoding = "utf-8"?>
<resources>
 <declare-styleable name = "DigitState">
  <attr name = "state1" format = "boolean"/>
  <attr name = "state2" format = "boolean"/>
  <attr name = "state3" format = "boolean"/>
 </declare-styleable>
</resources>
```

上面这段代码在 MyCode\MySample624\app\src\main\res\values\attrs.xml 文件中。在 MainActivity.java 文件中，myImageView.setImageState(myStates，true)用于根据当前 myStates 的值，确定在单击 myImageView 控件时应该显示哪一个矢量图形数字，因此 myImageView 的 src 属性不能直接为图像资源，而应是 XML 资源。如 android:src = "@drawable/mystage"，表示在 myImageView 控件上加载动画 mystage。关于 mystage 文件的详细内容请参考 MyCode\MySample624\app\src\main\res\drawable\mystage.xml 文件。此实例的完整项目在 MyCode\MySample624 文件夹中。

120 自定义 TimeInterpolator 控制转圈进度动画

此实例主要通过使用 TimeInterpolator 自定义时间插值器,然后将(圆点)运动位置动态转换为指定贝赛尔曲线或直线的坐标值,实现自定义转圈动画的过渡效果。当实例运行之后,单击"启动动画"按钮,转圈进度动画将按照自定义规则进行旋转,效果分别如图 120-1 的左图和右图所示。单击"停止动画"按钮,则转圈动画立即停止。

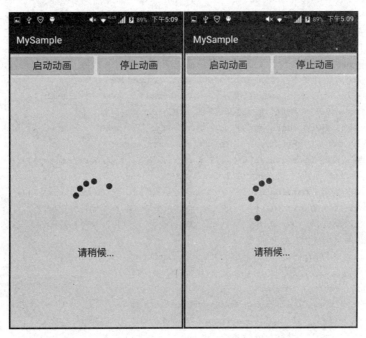

图 120-1

主要代码如下:

```
public class MainActivity extends Activity {
 RelativeLayout myLoadingIcon, myLoadingFrame;
 float myDotDegree;
 @Override
 protected void onCreate(Bundle savedInstanceState) {
  super.onCreate(savedInstanceState);
  setContentView(R.layout.activity_main);
  myLoadingFrame = (RelativeLayout) findViewById(R.id.myLoadingFrame);
  myLoadingIcon = (RelativeLayout) findViewById(R.id.myLoadingIcon);
  myDotDegree = MyUtils.InitLoadingDot(myLoadingIcon, this, myDotDegree);
 }
 public void onClickBtn1(View v) {                    //响应单击"启动动画"按钮
  InitLoadingAnim();
  myLoadingFrame.setVisibility(View.VISIBLE);
 }
 public void onClickBtn2(View v) {                    //响应单击"停止动画"按钮
  myLoadingFrame.setVisibility(View.GONE);
 }
 public void InitLoadingAnim() {
```

```java
        AnimatorSet myAnimatorSet = new AnimatorSet();
        List<Animator> myAnimatorList = new ArrayList(myLoadingIcon.getChildCount());
        for (int i = 0; i < myLoadingIcon.getChildCount(); i++) {
            long myDuration = 3000;                              //动画持续时间
            final View myDotView = myLoadingIcon.getChildAt(i);
            //最小执行单位时间所占总时间的比例
            double myRatio = (myDuration / 100f) / myDuration;
            final float myOffset = (float)(i * (100 - 86) * myRatio /
                    (myLoadingIcon.getChildCount() - 1));  //计算每个圆点的偏移量
            //获取 X、Y 坐标值集合
            final double[] myXCoordinates = MyUtils.getXCoordinates(myOffset, myRatio);
            final double[] myYCoordinates = MyUtils.getYCoordinates();
            //根据上述坐标,计算在不同情况下每种贝塞尔曲线的控制点 Y 坐标值
            final float myPointY_1 = MyUtils.calcPointY(myXCoordinates[2],
                    myYCoordinates[2], myXCoordinates[3], myYCoordinates[3], myXCoordinates[1]);
            final float myPointY_2 = MyUtils.calcPointY(myXCoordinates[2],
                    myYCoordinates[2], myXCoordinates[3], myYCoordinates[3], myXCoordinates[4]);
            final float myPointY_3 = MyUtils.calcPointY(myXCoordinates[5],
                    myYCoordinates[5], myXCoordinates[6], myYCoordinates[6], myXCoordinates[4]);
            final float myPointY_4 = MyUtils.calcPointY(myXCoordinates[5],
                    myYCoordinates[5], myXCoordinates[6], myYCoordinates[6], myXCoordinates[7]);
            ObjectAnimator myObjectAnimator = ObjectAnimator.ofFloat(myLoadingIcon.
                    getChildAt(i), "rotation", -myDotDegree * i, 720 - myDotDegree * i);
            //设置动画时长和重复次数(永不停歇)
            myObjectAnimator.setDuration(myDuration).setRepeatCount(-1);
            //应用自定义插值器
            myObjectAnimator.setInterpolator(
                    new CustomInterpolator(myDotView, myXCoordinates,
                    myYCoordinates, myPointY_1, myPointY_2, myPointY_3, myPointY_4));
            myAnimatorList.add(myObjectAnimator);
        }
        myAnimatorSet.playTogether(myAnimatorList);
        myAnimatorSet.start();
    } }
```

上面这段代码在 MyCode\MySample773\app\src\main\java\com\bin\luo\mysample\MainActivity.java 文件中。在这段代码中,CustomInterpolator 是以 TimeInterpolator 为基类创建的自定义插值器,关于 CustomInterpolator 类的详细内容请参考 MyCode\MySample773\app\src\main\java\com\bin\luo\mysample\CustomInterpolator.java 文件。此实例的完整项目在 MyCode\MySample773 文件夹中。

121 使用 animated-selector 实现轮播多幅图像

此实例主要通过在 animated-selector 中设置自定义属性,实现轮播多幅图像。当实例运行之后,单击 ImageView 控件(电影海报图像),则切换到下一幅电影海报图像;单击 9 次,则显示 9 幅不同的电影海报图像;若全部显示完毕,则重新开始。效果如图 121-1 所示。

主要代码如下:

```java
public class MainActivity extends Activity {
    private ImageView myImageView;
```

图 121-1

```
private int myCurrentNum = 0;
private static final int[] myImages = {R.attr.state1,
        R.attr.state2, R.attr.state3, R.attr.state4, R.attr.state5,
        R.attr.state6, R.attr.state7, R.attr.state8, R.attr.state9};
@Override
protected void onCreate(Bundle savedInstanceState) {
  super.onCreate(savedInstanceState);
  setContentView(R.layout.activity_main);
  myImageView = (ImageView) findViewById(R.id.myImageView);
  myImageView.setOnClickListener(new View.OnClickListener() {
    @Override
    public void onClick(View v) {
      myCurrentNum++;
      if (myCurrentNum > 8) { myCurrentNum = 0; }
      int[] myStates = new int[myImages.length];
      for (int i = 0; i < myImages.length; i++) {
        if (i == myCurrentNum) { myStates[i] = myImages[i]; }
        else { myStates[i] = - myImages[i]; }
      }
      myImageView.setImageState(myStates, true);
    } });
  myImageView.callOnClick(); //自动单击 ImageView 控件
} }
```

上面这段代码在 MyCode \ MySample623 \ app \ src \ main \ java \ com \ bin \ luo \ mysample \ MainActivity.java 文件中。在这段代码中，myImageView.setImageState(myStates, true)用于根据当前 myStates 的值，确定在单击 myImageView 控件时究竟显示哪一幅图像；因此 myImageView 的 src 属性就不能直接为图像资源，而应是 XML 资源。如 android:src = "@drawable/myselector"，表示 myImageView 显示的电影海报图像在 myselector 选择器资源中。myselector 资源的主要内容如下：

```xml
<?xml version="1.0" encoding="utf-8"?>
<animated-selector xmlns:android="http://schemas.android.com/apk/res/android"
                   xmlns:app="http://schemas.android.com/apk/res-auto">
    <item android:id="@+id/myimage1"
          android:drawable="@drawable/myimage1"
          app:state1="true"/>
    <item android:id="@+id/myimage2"
          android:drawable="@drawable/myimage2"
          app:state2="true"/>
    <item android:id="@+id/myimage3"
          android:drawable="@drawable/myimage3"
          app:state3="true"/>
    <item android:id="@+id/myimage4"
          android:drawable="@drawable/myimage4"
          app:state4="true"/>
    <item android:id="@+id/myimage5"
          android:drawable="@drawable/myimage5"
          app:state5="true"/>
    <item android:id="@+id/myimage6"
          android:drawable="@drawable/myimage6"
          app:state6="true"/>
    <item android:id="@+id/myimage7"
          android:drawable="@drawable/myimage7"
          app:state7="true"/>
    <item android:id="@+id/myimage8"
          android:drawable="@drawable/myimage8"
          app:state8="true"/>
    <item android:id="@+id/myimage9"
          android:drawable="@drawable/myimage9"
          app:state9="true"/>
</animated-selector>
```

上面这段代码在 MyCode\MySample623\app\src\main\res\drawable\myselector.xml 文件中。在这段代码中,app:state9、app:state8、app:state7、app:state6 等是自定义属性,用于设置 item 的两种状态。该自定义属性在 attrs.xml 文件中,attrs 文件的主要内容如下:

```xml
<?xml version="1.0" encoding="utf-8"?>
<resources>
    <declare-styleable name="DigitState">
        <attr name="state1" format="boolean"/>
        <attr name="state2" format="boolean"/>
        <attr name="state3" format="boolean"/>
        <attr name="state4" format="boolean"/>
        <attr name="state5" format="boolean"/>
        <attr name="state6" format="boolean"/>
        <attr name="state7" format="boolean"/>
        <attr name="state8" format="boolean"/>
        <attr name="state9" format="boolean"/>
    </declare-styleable>
</resources>
```

上面这段代码在 MyCode\MySample623\app\src\main\res\values\attrs.xml 文件中。此实例

的完整项目在 MyCode\MySample623 文件夹中。

122 使用 animation-list 实现两幅图像的切换

此实例主要通过在 animation-list 中设置两幅图像,并使用 AnimationDrawable 轮播这两幅图像,实现单击 ImageView 控件时切换显示这两幅图像。当实例运行之后,单击图像(ImageView 控件),每幅图像在停留 500ms 后自动显示下一幅图像,效果分别如图 122-1 的左图和右图所示。

图 122-1

主要代码如下:

```java
public class MainActivity extends Activity {
  AnimationDrawable myAnimationDrawable;
  ImageView myImageView;
  @Override
  protected void onCreate(Bundle savedInstanceState) {
    super.onCreate(savedInstanceState);
    setContentView(R.layout.activity_main);
    myImageView = (ImageView) findViewById(R.id.myImageView);
    myImageView.setBackgroundResource(R.drawable.myimages);
    myAnimationDrawable = (AnimationDrawable) myImageView.getBackground();
  }
  public boolean onTouchEvent(MotionEvent event) {        //响应单击应用屏幕
    if (event.getAction() == MotionEvent.ACTION_DOWN) {
      myAnimationDrawable.stop();                          //停止播放多幅图像
      myAnimationDrawable.start();                         //开始播放多幅图像
      return true;
    }
    return super.onTouchEvent(event);
  } }
```

上面这段代码在 MyCode\MySample252\app\src\main\java\com\bin\luo\mysample\MainActivity.java 文件中。在这段代码中，myImageView.setBackgroundResource（R.drawable.myimages）表示使用 myimages 文件中的多幅图像设置 myImageView 控件的背景，myimages 文件的主要内容如下：

```xml
<animation-list xmlns:android="http://schemas.android.com/apk/res/android"
    android:oneshot="true">
<item android:drawable="@drawable/myimage1" android:duration="500" />
<item android:drawable="@drawable/myimage2" android:duration="500" />
</animation-list>
```

上面这段代码在 MyCode\MySample252\app\src\main\res\drawable\myimages.xml 文件中。在这段代码中，<animation-list>元素为根节点，<item>节点定义了显示每一幅图像的参数，即图像和间隔时间。设置 android:oneshot 属性为 true，表示此次动画只执行一次，并且停留在最后一帧。设置 android:oneshot 为 false，则表示动画循环播放。此实例的完整项目在 MyCode\MySample252 文件夹中。

123　使用 AnimationDrawable 逐帧播放图像

此实例主要通过使用 AnimationDrawable，实现以动画的形式连续逐帧播放多幅图像文件。当实例运行之后，单击"开始播放"按钮，则开始逐帧以动画形式播放多幅图像，如图 123-1 的左图所示。单击"停止播放"按钮，则停止播放动画图像，如图 123-1 的右图所示。可以停止和播放任意一幅图像。

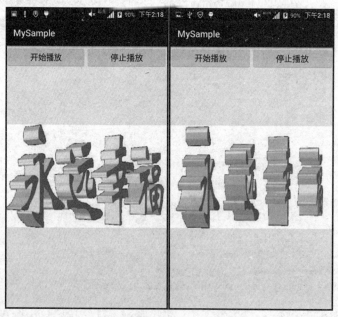

图 123-1

主要代码如下：

```java
public class MainActivity extends Activity {
    AnimationDrawable myAnimationDrawable;
    @Override
```

```java
protected void onCreate(Bundle savedInstanceState) {
  super.onCreate(savedInstanceState);
  setContentView(R.layout.activity_main);
  ImageView myImageView = (ImageView)findViewById(R.id.myImageView);
  //获取 AnimationDrawable 动画对象
  myAnimationDrawable = (AnimationDrawable)myImageView.getBackground();
}
public void onClickBtn1(View v) {      //响应单击"开始播放"按钮
  myAnimationDrawable.start();
}
public void onClickBtn2(View v) {      //响应单击"停止播放"按钮
  myAnimationDrawable.stop();
  }
}
```

上面这段代码在 MyCode\MySample145\app\src\main\java\com\bin\luo\mysample\MainActivity.java 文件中。在这段代码中，AnimationDrawable 用于管理动画图像资源，此处的动画图像资源在 ImageView 的 background 属性中指定，主要代码如下：

```xml
<ImageView android:id = "@+id/myImageView"
        android:layout_marginTop = "120dp"
        android:layout_width = "wrap_content"
        android:layout_height = "wrap_content"
        android:background = "@drawable/mybmps"
        android:scaleType = "center"/>
```

上面这段代码在 MyCode\MySample145\app\src\main\res\layout\activity_main.xml 文件中。在这段代码中，android:background="@drawable/mybmps" 中的 mybmps 是负责配置动画的 XML 文件，该文件的主要内容如下：

```xml
<?xml version = "1.0" encoding = "utf-8"?>
<animation-list xmlns:android = "http://schemas.android.com/apk/res/android"
                android:oneshot = "false">
<item android:drawable = "@drawable/mybmp001" android:duration = "60" />
<item android:drawable = "@drawable/mybmp002" android:duration = "60" />
<item android:drawable = "@drawable/mybmp003" android:duration = "60" />
<item android:drawable = "@drawable/mybmp004" android:duration = "60" />
<item android:drawable = "@drawable/mybmp005" android:duration = "60" />
<item android:drawable = "@drawable/mybmp006" android:duration = "60" />
<item android:drawable = "@drawable/mybmp007" android:duration = "60" />
<item android:drawable = "@drawable/mybmp008" android:duration = "60" />
</animation-list>
```

上面这段代码在 MyCode\MySample145\app\src\main\res\drawable\mybmps.xml 文件中。在这段代码中，<animation-list> 元素为根节点，<item> 节点定义了每一帧的参数，表示一个 drawable 资源的帧（图像）和帧间隔。设置 android:oneshot 属性为 true，表示此次动画只执行一次，最后停留在最后一帧。设置 android:oneshot 为 false，则表示动画循环播放。此实例的完整项目在 MyCode\MySample145 文件夹中。

124 使用 AnimatedVectorDrawable 旋转图形

此实例主要通过使用 AnimatedVectorDrawable 将 SVG 矢量图形和自定义动画关联起来，从而实现以动画的形式旋转 SVG 矢量图形。当实例运行之后，将显示按照 SVG 标准绘制的矢量图形（太极图案），单击"开始旋转"按钮，太极图案将沿着顺时针方向一直旋转；单击"停止旋转"按钮，旋转动画立即停止。效果分别如图 124-1 的左图和右图所示。

图 124-1

主要代码如下：

```java
public class MainActivity extends Activity {
    AnimatedVectorDrawable myAnimatedVectorDrawable;
    @Override
    protected void onCreate(Bundle savedInstanceState) {
        super.onCreate(savedInstanceState);
        setContentView(R.layout.activity_main);
        ImageView myImageView = (ImageView) findViewById(R.id.myImageView);
        //读取 SVG 图像与动画关联文件并转换为 AnimatedVectorDrawable 对象
        myAnimatedVectorDrawable = (AnimatedVectorDrawable)
                    getResources().getDrawable(R.drawable.myanimvector);
        //应用 AnimatedVectorDrawable 对象
        myImageView.setImageDrawable(myAnimatedVectorDrawable);
    }
    public void onClickBtn1(View v) {         //响应单击"开始旋转"按钮
        myAnimatedVectorDrawable.start();
    }
    public void onClickBtn2(View v) {         //响应单击"停止旋转"按钮
        myAnimatedVectorDrawable.stop();
    }
}
```

上面这段代码在 MyCode\MySample772\app\src\main\java\com\bin\luo\mysample\MainActivity.java 文件中。在这段代码中，myAnimatedVectorDrawable =（AnimatedVectorDrawable）getResources().getDrawable(R.drawable.myanimvector) 的 myanimvector 用于组合旋转动画 myanim 和矢量图形 myvector，此处即是将 myanimvector 转换为 AnimatedVectorDrawable 对象。myanimvector 是一个 XML 格式的文件，该文件的主要内容如下：

```xml
<animated-vector xmlns:android="http://schemas.android.com/apk/res/android"
        android:drawable="@drawable/myvector">
    <target android:name="myTaiji"
            android:animation="@animator/myanim" />
</animated-vector>
```

上面这段代码在 MyCode\MySample772\app\src\main\res\drawable\myanimvector.xml 文件中。android:drawable="@drawable/myvector" 的 myvector 是按照 SVG 标准绘制的矢量图形，即太极图案，myvector.xml 文件的主要内容如下：

```xml
<vector xmlns:android="http://schemas.android.com/apk/res/android"
        android:width="400dp"
        android:height="400dp"
        android:viewportHeight="400"
        android:viewportWidth="400">
    <group android:name="myTaiji"
            android:pivotX="200"
            android:pivotY="200">
        <path android:fillColor="#ffffff"
            android:pathData="M200 200,m0 -100,a100 100 0 1 1 0 200z"/>
        <path android:fillColor="#000000"
            android:pathData="M200 200,m0 -100,a100 100 0 1 0 0 200z"/>
        <path android:fillColor="#ffffff"
            android:pathData="M200 150,m0 50,a50 50 0 1 1 0 -100,a50 50 0 1 1 0 100z"/>
        <path android:fillColor="#000000"
            android:pathData="M200 250,m0 50,a50 50 0 1 0 0 -100z"/>
        <path android:fillColor="#000000"
            android:pathData="M200 150,m0 10,a10 10 0 1 1 0 -20,a10 10 0 1 1 0 20z"/>
        <path android:fillColor="#ffffff"
            android:pathData="M200 250,m0 10,a10 10 0 1 0 0 -20,a10 10 0 1 0 0 20z"/>
    </group>
</vector>
```

上面这段代码在 MyCode\MySample772\app\src\main\res\drawable\myvector.xml 文件中。在这段代码中，android:name="myTaiji" 表示矢量图形名称，在动画操作中，需要访问此名称。旋转动画文件 myanim.xml 的主要内容如下：

```xml
<objectAnimator xmlns:android="http://schemas.android.com/apk/res/android"
        android:duration="2000"
        android:propertyName="rotation"
        android:valueFrom="0"
        android:repeatCount="infinite"
        android:valueType="floatType"
        android:valueTo="360" />
```

上面这段代码在 MyCode\MySample772\app\src\main\res\animator\myanim.xml 文件中。此实例的完整项目在 MyCode\MySample772 文件夹中。

125 以旋转淡出的动画效果切换两个 Activity

此实例主要通过使用 overridePendingTransition()方法，实现以旋转淡出的动画效果切换两个 Activity。当实例运行之后，在图 125-1 左图所示的图像中，孔雀所代表的 MainActivity 将淡出，同时天鹅所代表的 SecondActivity 将旋转放大，直到完全铺满屏幕，如图 125-1 的右图所示。

图　125-1

主要代码如下：

```java
public class MainActivity extends Activity {
    @Override
    protected void onCreate(Bundle savedInstanceState) {
        super.onCreate(savedInstanceState);
        setContentView(R.layout.activity_main);
        new Handler().postDelayed(new Runnable() {
            @Override
            public void run() {
                Intent myIntent = new Intent(MainActivity.this, SecondActivity.class);
                MainActivity.this.startActivity(myIntent);
                MainActivity.this.finish();
                //以旋转淡出的动画切换两个 Activity
                overridePendingTransition(R.anim.myscalerotate, R.anim.myalpha);
            }
        }, 500);
    }
}
```

上面这段代码在 MyCode\MySample302\app\src\main\java\com\bin\luo\mysample\MainActivity.java 文件中。在这段代码中，overridePendingTransition()方法用于在参数中加载旋转淡出的两个动画，该方法通常在 startActivity()方法或者 finish()方法执行之后调用。旋转放大动画 myscalerotate 的主要内容如下：

```xml
<?xml version = "1.0" encoding = "utf-8"?>
<set xmlns:android = "http://schemas.android.com/apk/res/android">
<scale android:duration = "2000"
       android:fromXScale = "0.0"
       android:fromYScale = "0.0"
       android:pivotX = "50%"
       android:pivotY = "50%"
       android:repeatCount = "0"
       android:startOffset = "20"
       android:toXScale = "1.0"
       android:toYScale = "1.0"></scale>
<rotate android:duration = "2000"
        android:fromDegrees = "0"
        android:pivotX = "50%"
        android:pivotY = "50%"
        android:toDegrees = " + 720"/>
</set>
```

上面这段代码在 MyCode\MySample302\app\src\main\res\anim\myscalerotate.xml 文件中。淡出动画 myalpha 的主要内容如下：

```xml
<?xml version = "1.0" encoding = "utf-8"?>
<set xmlns:android = "http://schemas.android.com/apk/res/android">
<alpha android:duration = "2000"
       android:fromAlpha = "1.0"
       android:toAlpha = "0"/>
</set>
```

上面这段代码在 MyCode\MySample302\app\src\main\res\anim\myalpha.xml 文件中。此外，当在项目中新增了 SecondActivity，则需要在 AndroidManifest.xml 文件中注册该 Activity，如 <activity android:name=".SecondActivity"/>。此实例的完整项目在 MyCode\MySample302 文件夹中。

126　在切换 Activity 的转场动画中共享不同元素

此实例主要通过在 ActivityOptions 的 makeSceneTransitionAnimation()方法中设置共享元素，实现在两个不同 Activity 的共享元素（Button 控件和 ImageView 控件）间产生转场动画效果。当实例运行之后，由于 MainActivity 的"进入 SecondActivity"按钮和 SecondActivity 的 ImageView 控件（台北 101 大楼图像）被设置为共享元素（myShare），因此单击"进入 SecondActivity"按钮，则 ImageView 控件（台北 101 大楼图像）从上向下拉伸；在 SecondActivity 中单击"返回 MainActivity"按钮，则 ImageView 控件（台北 101 大楼图像）从下向上折叠。效果分别如图 126-1 的左图和右图所示。

图 126-1

主要代码如下：

```java
public class MainActivity extends Activity {
  Button myButton;
  @Override
  protected void onCreate(Bundle savedInstanceState) {
    super.onCreate(savedInstanceState);
    setContentView(R.layout.activity_main);
    myButton = (Button)findViewById(R.id.myButton);
  }
  public void onClickmyBtnEnter(View v) {     //响应单击"进入 SecondActivity"按钮
    Intent myIntent = new Intent(MainActivity.this,SecondActivity.class);
    startActivity(myIntent, ActivityOptions.makeSceneTransitionAnimation(this, myButton," myShare").toBundle());
  } }
```

上面这段代码在 MyCode\MySample533\app\src\main\java\com\bin\luo\mysample\MainActivity.java 文件中。在这段代码中，startActivity（myIntent，ActivityOptions.makeSceneTransitionAnimation(this，myButton,"myShare").toBundle())中的 myShare 表示在两个 Activity 中指定的共享元素（Button 控件和 ImageView 控件）的共享名称，如 android：transitionName＝"myShare"，主要代码如下：

```xml
<LinearLayout android:layout_width = "match_parent"
              android:layout_height = "match_parent"
              android:orientation = "vertical">
  <Button android:id = "@+id/myButton"
          android:layout_width = "match_parent"
          android:layout_height = "match_parent"
```

```
            android:layout_weight = "10"
            android:onClick = "onClickmyBtnEnter"
            android:text = "进入 SecondActivity"
            android:textAllCaps = "false"
            android:textSize = "20dp"
            android:transitionName = "myShare"/>
    <ImageView android:layout_width = "match_parent"
            android:layout_height = "match_parent"
            android:layout_weight = "1"
            android:scaleType = "fitXY"
            android:src = "@mipmap/myimage1"/>
</LinearLayout>
```

上面这段代码在 MyCode\MySample533\app\src\main\res\layout\activity_main.xml 文件中。在这段代码中，android:transitionName = "myShare"用于指定 android:id = "@+id/myButton"控件为共享元素。当在 activity_main.xml 文件中指定了共享元素之后，还必须在 activity_second.xml 文件中指定同名的共享元素，主要代码如下：

```
<LinearLayout android:layout_width = "match_parent"
            android:layout_height = "match_parent"
            android:orientation = "vertical">
    <ImageView android:layout_width = "match_parent"
            android:layout_height = "match_parent"
            android:layout_weight = "1"
            android:transitionName = "myShare"
            android:scaleType = "fitXY"
            android:src = "@mipmap/myimage2"/>
    <Button android:layout_width = "match_parent"
            android:layout_height = "match_parent"
            android:layout_weight = "10"
            android:onClick = "onClickmyBtnExit"
            android:text = "返回 MainActivity"
            android:textAllCaps = "false"
            android:textSize = "20dp"/>
</LinearLayout>
```

上面这段代码在 MyCode\MySample533\app\src\main\res\layout\activity_second.xml 文件中。在这段代码中，android:transitionName = "myShare"用于指定 ImageView 控件为共享元素，共享名称 myShare 必须与在 activity_main.xml 文件中指定的共享元素（Button 控件）的名称相同。android:onClick = "onClickmyBtnExit" 表示在单击按钮后执行 SecondActivity 中的 onClickmyBtnExit()方法，主要代码如下：

```
public class SecondActivity extends Activity {
    @Override
    protected void onCreate(Bundle savedInstanceState) {
        super.onCreate(savedInstanceState);
        setContentView(R.layout.activity_second);
    }
    public void onClickmyBtnExit(View v) {      //响应单击"返回 MainActivity"按钮
        this.onBackPressed();
    }}
```

上面这段代码在 MyCode\MySample533\app\src\main\java\com\bin\luo\mysample\SecondActivity.java 文件中。需要注意的是，当在项目中新增 SecondActivity 之后，则应该在 AndroidManifest.xml 文件中注册该 Activity，如< activity android:name= ".SecondActivity"/>。此实例的完整项目在 MyCode\MySample533 文件夹中。

127　在过渡 Activity 时禁止部分控件产生动画

此实例主要通过使用 android:excludeId 属性指定需要禁止的控件，从而实现两个 Activity 在使用过渡动画切换时，禁止指定的控件执行过渡动画。当实例运行之后，单击 MainActivity 的 ImageView 控件（电影海报图像），将以滑动动画切换到 SecondActivity；在此实例中，由于 SecondActivity 上面的 ImageView 控件（《风语者》电影海报图像）被设置了 android:excludeId 属性值，因此在从 MainActivity 切换到 SecondActivity 时，SecondActivity 上面的 ImageView 控件（《风语者》电影海报图像）直接显示，没有滑动动画效果，仅 SecondActivity 下面的 ImageView 控件（电影海报图像）有滑动动画效果，效果分别如图 127-1 的左图和右图所示。

图　127-1

主要代码如下：

```
public class MainActivity extends Activity {
  @Override
  protected void onCreate(Bundle savedInstanceState) {
    super.onCreate(savedInstanceState);
    setContentView(R.layout.activity_main);
  }
  public void onClickmyImageEnter(View v) {      //响应单击图像进入 SecondActivity
    Intent myIntent = new Intent(MainActivity.this,SecondActivity.class);
    startActivity(myIntent,
        ActivityOptions.makeSceneTransitionAnimation(this).toBundle());
} }
```

上面这段代码在 MyCode\MySample578\app\src\main\java\com\bin\luo\mysample\MainActivity.java 文件中。在这段代码中，startActivity(myIntent, ActivityOptions.makeSceneTransitionAnimation(this).toBundle())表示从 MainActivity 切换到 SecondActivity 时使用转场动画。转场动画则在 styles.xml 文件中设置，主要代码如下：

```xml
<resources>
<!-- Base application theme. -->
<style name = "AppTheme" parent = "android:Theme.Material.Light.NoActionBar">
<!-- Customize your theme here. -->
<item name = "android:windowExitTransition">@transition/myexit</item>
<item name = "android:windowEnterTransition">@transition/myenter</item>
</style>
</resources>
```

上面这段代码在 MyCode\MySample578\app\src\main\res\values\styles.xml 文件中。在这段代码中，<item name="android:windowExitTransition">@transition/myexit</item>表示退场动画使用 transition 目录下的 myexit.xml 文件定义的动画。<item name = "android:windowEnterTransition">@transition/myenter</item>表示进场动画使用 transition 目录下的 myenter.xml 文件定义的动画。myexit.xml 文件的主要内容如下：

```xml
<?xml version = "1.0" encoding = "utf-8"?>
<transitionSet xmlns:android = "http://schemas.android.com/apk/res/android"
        android:duration = "5000">
 <slide>
  <targets>
   <target android:excludeId = "@android:id/statusBarBackground"/>
   <target android:excludeId = "@android:id/navigationBarBackground"/>
   <target android:excludeId = "@id/myImageView2"/></targets></slide>
</transitionSet>
```

上面这段代码在 MyCode\MySample578\app\src\main\res\transition\myexit.xml 文件中。myenter.xml 文件的主要内容如下：

```xml
<?xml version = "1.0" encoding = "utf-8"?>
<transitionSet xmlns:android = "http://schemas.android.com/apk/res/android"
        android:duration = "5000">
 <slide>
  <targets>
   <target android:excludeId = "@android:id/statusBarBackground"/>
   <target android:excludeId = "@android:id/navigationBarBackground"/>
   <target android:excludeId = "@id/myImageView2"/></targets></slide>
</transitionSet>
```

上面这段代码在 MyCode\MySample578\app\src\main\res\transition\myenter.xml 文件中。在这段代码中，<target android:excludeId="@android:id/statusBarBackground"/>表示禁止状态栏在过渡过程中执行动画。<target android:excludeId="@android:id/navigationBarBackground"/>表示禁止导航栏在过渡过程中执行动画。<target android:excludeId="@id/myImageView2"/>表示禁止 myImageView2 控件（即 SecondActivity 上面的《风语者》电影海报图像控件）在过渡过程中执行动画。此外，当在项目中新增了 SecondActivity 之后，则需要在 AndroidManifest.xml 文件中注册

该 Activity，如< activity android：name＝".SecondActivity"/>。此实例的完整项目在 MyCode\MySample578 文件夹中。

128　使用指定的裁剪区域动态切换两个 Activity

此实例主要通过使用 ActivityOptionsCompat 的 makeClipRevealAnimation（）方法，实现以指定的区域过渡动画来切换两个 Activity。当实例运行之后，在 MainActivity 中单击图像，该图像将以指定的右下角裁剪区域过渡到 SecondActivity，效果分别如图 128-1 的左图和右图所示。

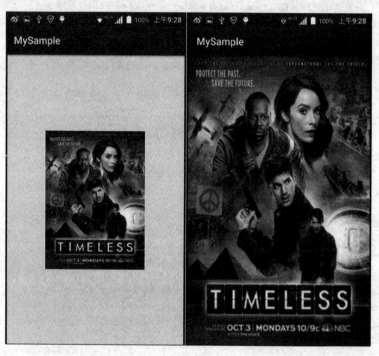

图　128-1

主要代码如下：

```
public class MainActivity extends Activity {
 @Override
 protected void onCreate(Bundle savedInstanceState) {
  super.onCreate(savedInstanceState);
  setContentView(R.layout.activity_main);
 }
 //单击图像之后切换到 SecondActivity
 public void onClickToSecondActivity(View myView) {
  Intent myIntent = new Intent(this, SecondActivity.class);
  ActivityOptionsCompat myActivityOptionsCompat =
        ActivityOptionsCompat.makeClipRevealAnimation(myView,
            myView.getWidth()/2,myView.getHeight()/2,
            myView.getWidth()/2,myView.getHeight()/2);
  ActivityCompat.startActivity(MainActivity.this,
      myIntent, myActivityOptionsCompat.toBundle());
  //this.startActivity(myIntent);
 }}
```

上面这段代码在 MyCode\MySample589\app\src\main\java\com\bin\luo\mysample\MainActivity.java 文件中。在这段代码中,makeClipRevealAnimation()方法用于指定过渡区域,该方法的语法声明如下:

```
makeClipRevealAnimation(View source,
                int startX, int startY, int width, int height)
```

其中,参数 View source 是一个 View 对象,用于确定动画启动初始坐标的控件;参数 int startX 表示相对于 source 的 X 坐标;参数 int startY 表示相对于 source 的 Y 坐标;参数 int width 表示新 Activity 的初始宽度;参数 int height 表示新 Activity 的初始高度。

此外,使用此实例的相关类需要在 gradle 中引入 compile 'com.android.support:design:25.0.1' 依赖项,并且应在 Android 6.0 中测试。需要注意的是,当在项目中新增 SecondActivity 之后,则应在 AndroidManifest.xml 文件中注册该 Activity,如< activity android:name=".SecondActivity"/>。此实例的完整项目在 MyCode\MySample589 文件夹中。

129 在关闭应用(Activity)时显示退场动画

此实例主要通过在 overridePendingTransition()方法的参数中设置进场和退场动画,实现在调用 onBackPressed()方法关闭应用时显示退场(桌面是进场动画)动画。当实例运行之后,单击手机的后退键,则应用的窗口逐渐缩小,直到消失;桌面的窗口逐渐扩大,直到铺满。效果分别如图 129-1 的左图和右图所示。

图 129-1

主要代码如下:

```
public class MainActivity extends Activity {
    @Override
```

```
protected void onCreate(Bundle savedInstanceState) {
  super.onCreate(savedInstanceState);
  setContentView(R.layout.activity_main);
}
@Override
public void onBackPressed() {
  super.onBackPressed();
  overridePendingTransition(R.anim.myenter,R.anim.myexit);
} }
```

上面这段代码在 MyCode\MySample508\app\src\main\java\com\bin\luo\mysample\MainActivity.java 文件中。在这段代码中，overridePendingTransition（R.anim.myenter，R.anim.myexit）表示在单击后退键时执行转场动画 myenter 和 myexit，myenter 动画的作用是使手机桌面（进场动画）由小变大，myexit 动画的作用是使当前应用（退场动画）的窗口由大变小。myenter 动画的主要内容如下：

```xml
<?xml version = "1.0" encoding = "utf-8"?>
<set xmlns:android = "http://schemas.android.com/apk/res/android">
<scale android:duration = "5000"
       android:fromXScale = "0.0"
       android:fromYScale = "0.0"
       android:pivotX = "50%"
       android:pivotY = "50%"
       android:toXScale = "1.0"
       android:toYScale = "1.0"/>
</set>
```

上面这段代码在 MyCode\MySample508\app\src\main\res\anim\myenter.xml 文件中。myexit 动画的主要内容如下：

```xml
<?xml version = "1.0" encoding = "utf-8"?>
<set xmlns:android = "http://schemas.android.com/apk/res/android">
<scale android:duration = "5000"
       android:fromXScale = "1.0"
       android:fromYScale = "1.0"
       android:pivotX = "50%"
       android:pivotY = "50%"
       android:toXScale = "0.0"
       android:toYScale = "0.0"/>
</set>
```

上面这段代码在 MyCode\MySample508\app\src\main\res\anim\myexit.xml 文件中。此实例的完整项目在 MyCode\MySample508 文件夹中。

130 使用转场动画 Slide 切换两个 Activity

此实例主要通过在 Window 的 setEnterTransition（）方法中使用 Slide 动画作为参数，实现以滑动的方式切换两个 Activity。当实例运行之后，单击"以 Slide 动画进入第二个 Activity"按钮，第二个 Activity（SecondActivity）将从屏幕底部滑向顶部；在 SecondActivity 中单击"以 Slide 动画返回到第一个 Activity"按钮,第二个 Activity(SecondActivity)将从顶部向底部滑出,从而显示第一个
Activity。效果分别如图 130-1 的左图和右图所示。

图 130-1

主要代码如下:

```
public class MainActivity extends Activity {
  @Override
  protected void onCreate(Bundle savedInstanceState) {
    super.onCreate(savedInstanceState);
    setContentView(R.layout.activity_main);
  }
  //响应单击"以 Slide 动画进入第二个 Activity"按钮
  public void onClickmyBtnEnter(View v) {
    Intent myIntent = new Intent(MainActivity.this,SecondActivity.class);
    startActivity(myIntent,
        ActivityOptions.makeSceneTransitionAnimation(this).toBundle());
  } }
```

上面这段代码在 MyCode \ MySample504 \ app \ src \ main \ java \ com \ bin \ luo \ mysample \
MainActivity. java 文件中。在这段代码中,startActivity(myIntent, ActivityOptions.
makeSceneTransitionAnimation(this).toBundle())表示从 MainActivity 跳转到 SecondActivity 时使
用转场动画。SecondActivity 的主要代码如下:

```
public class SecondActivity extends Activity {
  @Override
  protected void onCreate(Bundle savedInstanceState) {
    super.onCreate(savedInstanceState);
    setContentView(R.layout.activity_second);
```

```
    Window myWindow = getWindow();
    //使用 Slide 设置进场动画
    myWindow.setEnterTransition(new Slide().setDuration(2000));
}
//响应单击"以 Slide 动画返回到第一个 Activity"按钮
public void onClickmyBtnExit(View v) {
    this.onBackPressed();
} }
```

上面这段代码在 MyCode\MySample504\app\src\main\java\com\bin\luo\mysample\SecondActivity.java 文件中。在这段代码中，myWindow.setEnterTransition(new Slide().setDuration(2000))表示在进入 SecondActivity 时，使用转场动画 Slide，持续时间是 2s。此外，当在项目中新增了 SecondActivity 之后，则需要在 AndroidManifest.xml 文件中注册该 Activity，如<activity android:name=".SecondActivity"/>。此实例的完整项目在 MyCode\MySample504 文件夹中。

131 使用 TransitionSet 组合 Explode 和 Fade 动画

此实例主要通过使用 TransitionSet 的 addTransition()方法增加动画类型，实现在两个布局过渡的过程中同时呈现多种动画效果。当实例运行之后，《泰坦尼克》电影海报位于屏幕的正中（即 myscene1）。单击屏幕，则《泰坦尼克》电影海报将以 Explode 和 Fade 动画风格逐渐过渡到 myscene2，即《妈妈》电影海报；在 myscene2 中单击屏幕，则《妈妈》电影海报将以 Explode 和 Fade 动画风格逐渐过渡到 myscene1，即《泰坦尼克》电影海报。效果分别如图 131-1 的左图和右图所示。

图 131-1

主要代码如下：

```
public class MainActivity extends Activity {
    @Override
```

```java
protected void onCreate(Bundle savedInstanceState) {
    super.onCreate(savedInstanceState);
    setContentView(R.layout.activity_main);
}
//单击 myscene1 过渡到 myscene2(由《泰坦尼克》电影海报过渡到《妈妈》电影海报)
public void onClickScene1(View v) {
    ViewGroup myRootView = (ViewGroup)findViewById(R.id.myRootView);
    Scene myScene2 = Scene.getSceneForLayout(myRootView, R.layout.myscene2, this);
    Explode myExplode = new Explode();
    Fade myFade = new Fade();
    TransitionSet myTransitionSet = new TransitionSet();
    myTransitionSet.addTransition(myExplode).addTransition(myFade);
    myTransitionSet.setDuration(5000);
    TransitionManager.go(myScene2, myTransitionSet);
}
//单击 myscene2 过渡到 myscene1(由《妈妈》电影海报过渡到《泰坦尼克》电影海报)
public void onClickScene2(View v) {
    ViewGroup myRootView = (ViewGroup)findViewById(R.id.myRootView);
    Scene myScene1 = Scene.getSceneForLayout(myRootView, R.layout.myscene1, this);
    Explode myExplode = new Explode();
    Fade myFade = new Fade();
    TransitionSet myTransitionSet = new TransitionSet();
    myTransitionSet.addTransition(myExplode).addTransition(myFade);
    myTransitionSet.setDuration(5000);
    TransitionManager.go(myScene1, myTransitionSet);
} }
```

上面这段代码在 MyCode\MySample550\app\src\main\java\com\bin\luo\mysample\MainActivity.java 文件中。在这段代码中，myTransitionSet.addTransition(myExplode).addTransition(myFade)用于在 myTransitionSet 过渡动画集合中增加 myExplode 和 myFade 动画。myRootView＝(ViewGroup)findViewById(R.id.myRootView)用于根据 activity_main 布局中的 RelativeLayout 创建动画容器。activity_main 布局的主要内容如下：

```xml
<!-- 这个 RelativeLayout 用来做动画的容器 -->
<RelativeLayout android:id = "@+id/myRootView"
                android:layout_width = "match_parent"
                android:layout_height = "match_parent"
                android:layout_centerInParent = "true">
    <include layout = "@layout/myscene1"></include>
</RelativeLayout>
```

上面这段代码在 MyCode\MySample550\app\src\main\res\layout\activity_main.xml 文件中。在这段代码中，<include layout＝"@layout/myscene1"></include>表示容器的子布局是 myscene1，myscene1 布局的主要内容如下：

```xml
<?xml version = "1.0" encoding = "utf-8"?>
<RelativeLayout xmlns:android = "http://schemas.android.com/apk/res/android"
                android:background = "#000"
                android:onClick = "onClickScene1"
                android:layout_width = "match_parent"
                android:layout_height = "match_parent">
```

```xml
<ImageView android:layout_width="match_parent"
    android:layout_height="match_parent"
    android:src="@mipmap/myimage1"
    android:layout_centerInParent="true"/>
</RelativeLayout>
```

上面这段代码在 MyCode\MySample550\app\src\main\res\layout\myscene1.xml 文件中。在使用 TransitionManager.go() 方法实现过渡动画时，通常需要两个布局（场景），即此实例的 myscene1 和 myscene2，该方法的第一个参数代表目标布局，第二个参数代表过渡动画类型。myscene2 布局的主要代码如下：

```xml
<RelativeLayout xmlns:android="http://schemas.android.com/apk/res/android"
    android:background="#000"
    android:onClick="onClickScene2"
    android:layout_width="match_parent"
    android:layout_height="match_parent">
<ImageView android:layout_width="match_parent"
    android:layout_height="match_parent"
    android:src="@mipmap/myimage2"
    android:layout_centerInParent="true"/>
</RelativeLayout>
```

上面这段代码在 MyCode\MySample550\app\src\main\res\layout\myscene2.xml 文件中。此外，使用此实例的相关类需要在 gradle 中引入 compile 'com.android.support:design:25.0.1' 依赖项。此实例的完整项目在 MyCode\MySample550 文件夹中。

132 使用 TransitionManager 实现缩放过渡动画

此实例主要通过指定 TransitionManager 的 beginDelayedTransition() 方法参数为容器（控件的父节点），然后在修改布局参数后调用 setLayoutParams() 方法改变控件大小，实现在修改尺寸参数后的子节点（控件）产生缩放效果的过渡动画。当实例运行之后，小狗图像（ImageView 控件）位于屏幕中心，单击屏幕，则小狗图像放大 3 倍；再次单击屏幕，则小狗图像将缩小至原始状态。效果分别如图 132-1 的左图和右图所示。

主要代码如下：

```java
public class MainActivity extends Activity {
    RelativeLayout myLayout;
    boolean bChanged = false;
    @Override
    protected void onCreate(Bundle savedInstanceState) {
        super.onCreate(savedInstanceState);
        setContentView(R.layout.activity_main);
        myLayout = (RelativeLayout)findViewById(R.id.myLayout);
    }
    public void onClickScreen(View v) {          //单击屏幕实现图像缩放动画切换
        RelativeLayout.LayoutParams myLayoutParams;
        ViewGroup myRootView = (ViewGroup)findViewById(R.id.myRootView);
        TransitionManager.beginDelayedTransition(myRootView);
```

图 132-1

```
if (bChanged) {
  myLayoutParams = new RelativeLayout.LayoutParams(
          myLayout.getMeasuredWidth()/3,myLayout.getMeasuredHeight()/3);
  myLayout.setLayoutParams(myLayoutParams);
} else {
  myLayoutParams = new RelativeLayout.LayoutParams(
          myLayout.getMeasuredWidth() * 3,myLayout.getMeasuredHeight() * 3);
  myLayout.setLayoutParams(myLayoutParams);
}
bChanged = !bChanged;
}}
```

上面这段代码在 MyCode\MySample593\app\src\main\java\com\bin\luo\mysample\MainActivity.java 文件中。在这段代码中，TransitionManager.beginDelayedTransition（myRootView）表示在父容器 myRootView 中进行缩放过渡动画。myRootView＝(ViewGroup) findViewById(R.id.myRootView)用于根据 activity_main 布局中的 RelativeLayout 创建动画容器。activity_main 布局的主要内容如下：

```xml
<!-- 这个 RelativeLayout 用来做动画的父布局 -->
<RelativeLayout android:id = "@ + id/myRootView"
        android:layout_width = "match_parent"
        android:layout_height = "match_parent"
        android:gravity = "center_horizontal|center_vertical">
    <RelativeLayout android:id = "@ + id/myLayout"
            android:layout_width = "150dp"
            android:layout_height = "150dp">
        <ImageView android:layout_width = "match_parent"
            android:layout_height = "match_parent"
            android:src = "@mipmap/myimage1"/>
    </RelativeLayout>
</RelativeLayout>
```

上面这段代码在 MyCode\MySample593\app\src\main\res\layout\activity_main.xml 文件中。此外,使用此实例的相关类需要在 gradle 中引入 compile 'com.android.support:design:25.0.1'依赖项。此实例的完整项目在 MyCode\MySample593 文件夹中。

133 使用 TransitionManager 实现绕 Y 轴旋转动画

此实例主要通过在 TransitionManager 的 go()方法中设置 ChangeTransform 参数,实现在两个布局过渡的过程中围绕 Y 轴旋转相同 ID 的子控件。当实例运行之后,ImageView 控件(电影海报图像)位于屏幕的正中,如图 133-1 的左图所示(即 myscene1)。单击屏幕,ImageView 控件(电影海报图像)将围绕 Y 轴正向旋转 45°,如图 133-1 的右图所示(即 myscene2)。在 myscene2 中单击屏幕,ImageView 控件(电影海报图像)将反向旋转 45°,如图 133-1 的左图所示(即 myscene1)。

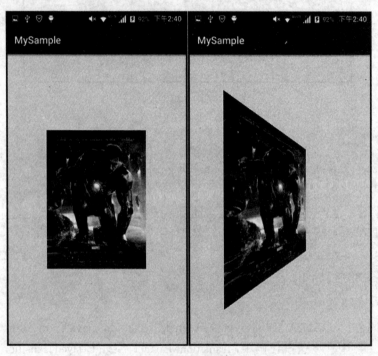

图 133-1

主要代码如下:

```java
public class MainActivity extends Activity {
    @Override
    protected void onCreate(Bundle savedInstanceState) {
        super.onCreate(savedInstanceState);
        setContentView(R.layout.activity_main);
    }
    //单击 myscene1 过渡到 myscene2(图像围绕 Y 轴正向旋转 45°)
    public void onClickScene1(View v) {
        ViewGroup myRootView = (ViewGroup)findViewById(R.id.myRootView);
        Scene myScene2 = Scene.getSceneForLayout(myRootView, R.layout.myscene2, this);
        TransitionManager.go(myScene2, new ChangeTransform().setDuration(5000));
    }
    //单击 myscene2 过渡到 myscene1(图像围绕 Y 轴反向旋转 45°)
```

```
public void onClickScene2(View v) {
    ViewGroup myRootView = (ViewGroup)findViewById(R.id.myRootView);
    Scene myScene1 = Scene.getSceneForLayout(myRootView, R.layout.myscene1, this);
    TransitionManager.go(myScene1, new ChangeTransform().setDuration(5000));
} }
```

上面这段代码在 MyCode\MySample556\app\src\main\java\com\bin\luo\mysample\MainActivity.java 文件中。在这段代码中,TransitionManager.go(myScene2,new ChangeTransform().setDuration(5000))表示以 ChangeTransform 动画风格(此例为围绕 Y 轴旋转 45°)在 5s 内从 myScene1 过渡到 myScene2。ViewGroup myRootView=(ViewGroup) findViewById(R.id.myRootView)用于根据 activity_main 布局中的 RelativeLayout 创建动画容器。activity_main 布局的主要内容如下:

```
<!-- 这个 RelativeLayout 用来做动画的父布局 -->
<RelativeLayout android:id = "@ + id/myRootView"
                android:layout_width = "match_parent"
                android:layout_height = "wrap_content"
                android:layout_centerInParent = "true">
    <include layout = "@layout/myscene1"></include>
</RelativeLayout>
```

上面这段代码在 MyCode\MySample556\app\src\main\res\layout\activity_main.xml 文件中。在这段代码中,<include layout="@layout/myscene1"></include>表示容器的子布局是 myscene1,myscene1 布局的主要内容如下:

```
<?xml version = "1.0" encoding = "utf-8"?>
<RelativeLayout xmlns:android = "http://schemas.android.com/apk/res/android"
                android:onClick = "onClickScene1"
                android:padding = "20dp"
                android:layout_width = "match_parent"
                android:layout_height = "match_parent">
    <ImageView android:layout_width = "200dp"
               android:layout_height = "300dp"
               android:src = "@mipmap/myimage1"
               android:layout_centerInParent = "true"
               android:id = "@ + id/myImageView"/>
</RelativeLayout>
```

上面这段代码在 MyCode\MySample556\app\src\main\res\layout\myscene1.xml 文件中。在使用 TransitionManager.go()方法实现过渡动画时,通常需要两个布局(场景),即此实例的 myscene1 和 myscene2,当 myscene1 的某个控件的 ID 值(android:id="@+id/myImageView")与 myscene2 的某个控件的 ID 值(android:id="@+id/myImageView")相同,且 android:rotationY 属性不同,在从 myscene1 过渡到 myscene2 时,这对控件就会执行围绕 Y 轴的旋转动画。myscene2 布局的主要代码如下:

```
<RelativeLayout xmlns:android = "http://schemas.android.com/apk/res/android"
                android:padding = "20dp"
                android:onClick = "onClickScene2"
```

```
                    android:layout_width = "match_parent"
                    android:layout_height = "match_parent">
    <ImageView android:layout_width = "200dp"
            android:layout_height = "300dp"
            android:rotationY = "45"
            android:src = "@mipmap/myimage1"
            android:layout_centerInParent = "true"
            android:id = "@ + id/myImageView" />
</RelativeLayout>
```

上面这段代码在 MyCode\MySample556\app\src\main\res\layout\myscene2.xml 文件中。此外，使用此实例的相关类需要在 gradle 中引入 compile 'com.android.support:design:25.0.1' 依赖项。此实例的完整项目在 MyCode\MySample556 文件夹中。

134 使用多个 TranslateAnimation 实现抖动窗口

此实例主要通过使用 AnimationSet 将多个 TranslateAnimation 动画组合在一起，从而实现抖动窗口的效果。当实例运行之后，单击"抖动窗口"按钮，应用主窗口将上下左右抖动；单击"振动并抖动窗口"按钮，应用主窗口将上下左右抖动，同时手机振动。效果分别如图 134-1 的左图和右图所示。

图 134-1

主要代码如下：

```
public class MainActivity extends Activity {
    private AnimationSet myAnimationSet;
    private RelativeLayout myRelativeLayout;
    @Override
    protected void onCreate(Bundle savedInstanceState) {
```

```
    super.onCreate(savedInstanceState);
    setContentView(R.layout.activity_main);
    myRelativeLayout = (RelativeLayout) findViewById(R.id.activity_main);
    myAnimationSet = new AnimationSet(true);
    TranslateAnimation myTranslateAnimation1 =
                        new TranslateAnimation(0, -10, 0, -10);
    myTranslateAnimation1.setDuration(300);
    TranslateAnimation myTranslateAnimation2 =
                        new TranslateAnimation(0, 10, 0, -10);
    myTranslateAnimation2.setDuration(300);
    myTranslateAnimation2.setStartOffset(300);
    TranslateAnimation myTranslateAnimation3 =
                        new TranslateAnimation(0, -10, 0, 10);
    myTranslateAnimation3.setDuration(300);
    myTranslateAnimation3.setStartOffset(600);
    TranslateAnimation myTranslateAnimation4 =
                        new TranslateAnimation(0, 10, 0, 10);
    myTranslateAnimation4.setDuration(300);
    myTranslateAnimation4.setStartOffset(900);
    myAnimationSet.addAnimation(myTranslateAnimation1);
    myAnimationSet.addAnimation(myTranslateAnimation2);
    myAnimationSet.addAnimation(myTranslateAnimation3);
    myAnimationSet.addAnimation(myTranslateAnimation4);
    myAnimationSet.setInterpolator(new CycleInterpolator(2));
}
//响应单击"抖动窗口"按钮
public void onClickBtn1(View view) {
    myRelativeLayout.startAnimation(myAnimationSet);
}
//响应单击"振动并抖动窗口"按钮
public void onClickBtn2(View view) {
    Vibrator myVibrator =
            (Vibrator) getSystemService(Context.VIBRATOR_SERVICE);
    long[] myPattern = new long[]{0, 300, 300, 300};
    myVibrator.vibrate(myPattern, -1);                          //振动手机
    myRelativeLayout.startAnimation(myAnimationSet);            //抖动窗口
    }
}
```

上面这段代码在 MyCode\MySample729\app\src\main\java\com\bin\luo\mysample\MainActivity.java 文件中。在这段代码中，TranslateAnimation myTranslateAnimation4 = new TranslateAnimation(0，10，0，10)用于根据指定的参数创建平移动画,该构造函数的语法声明如下：

```
TranslateAnimation(float fromXDelta,
            float toXDelta, float fromYDelta, float toYDelta)
```

其中,参数 float fromXDelta 表示动画开始点离当前 View(控件)的 X 坐标上的差值；参数 float toXDelta 参数表示动画结束点离当前 View 的 X 坐标上的差值；参数 float fromYDelta 参数表示动画开始点离当前 View 的 Y 坐标上的差值；参数 float toYDelta 表示动画结束点离当前 View 的 Y 坐标上的差值。即如果 View 在 A(x,y)点,那么动画就是从 B 点(x+fromXDelta，y+fromYDelta)移动到 C 点(x+toXDelta,y+toYDelta)。

myTranslateAnimation4.setStartOffset(900)用于设置动画启动时间,单位为 ms。系统默认在执行 TranslateAnimation 的 start()方法后立刻执行动画,但是当使用该方法后,将延迟一定的时间再执行动画。在此实例中,由于是将 4 个 TranslateAnimation 动画组合在一起,setStartOffset()方法的作用就相当于设置相邻两个动画之间的时间间隔。

此外,使用手机的振动功能需要在 AndroidManifest.xml 文件中添加< uses-permission android:name= "android.permission.VIBRATE"/>权限。此实例的完整项目在 MyCode\MySample729 文件夹中。

135 使用 LayoutTransition 实现布局改变动画

此实例主要通过在 LinearLayout 的 setLayoutTransition()方法中使用过渡动画 LayoutTransition,实现在布局容器中新增控件时显示自定义动画。当实例运行之后,单击"新增图像"按钮,则新增的图像将首先在水平方向压缩 50%,然后在水平方向上将图像拉伸到 100%,效果分别如图 135-1 的左图和右图所示。

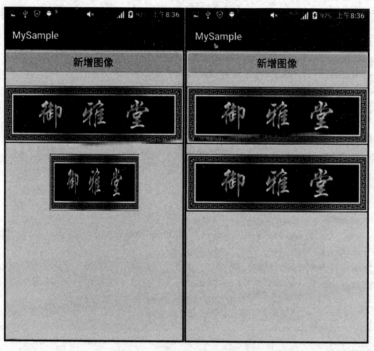

图 135-1

主要代码如下:

```
public class MainActivity extends Activity {
    private LinearLayout myLayout;
    @Override
    protected void onCreate(Bundle savedInstanceState) {
        super.onCreate(savedInstanceState);
        setContentView(R.layout.activity_main);
        myLayout = (LinearLayout) findViewById(R.id.myLayout);
    }
    //响应单击"新增图像"按钮
```

```
public void onClickmyBtn1(View v) {
    ImageView myImageView = new ImageView(this);
    myImageView.setImageResource(R.mipmap.myimage);
    //创建 ObjectAnimator 动画
    ObjectAnimator myObjectAnimator = (ObjectAnimator)
                AnimatorInflater.loadAnimator(this, R.animator.myanimator);
    myObjectAnimator.setTarget(myImageView);
    LayoutTransition myLayoutTransition = new LayoutTransition();
    //在新增 ImageView 时展示动画
    myLayoutTransition.setAnimator(LayoutTransition.APPEARING, myObjectAnimator);
    //为布局添加 LayoutTransition
    myLayout.setLayoutTransition(myLayoutTransition);
    myLayout.addView(myImageView);
    }
}
```

上面这段代码在 MyCode\MySample253\app\src\main\java\com\bin\luo\mysample\MainActivity.java 文件中。在这段代码中，myLayoutTransition.setAnimator（LayoutTransition.APPEARING，myObjectAnimator）表示在控件出现（APPEARING）时执行动画（myObjectAnimator），可以在此方法中设置下列四种动画类型：

（1）APPEARING，View 控件出现动画；

（2）DISAPPEARING，消失动画；

（3）CHANGE_APPEARING，由于新增了其他 View 控件而需要改变位置的动画；

（4）CHANGE_DISAPPEARING，由于移除了其他 View 控件而需要改变位置的动画。

默认情况下，DISAPPEARING 和 CHANGE_APPEARING 动画是立即开始的，其他动画都有一个默认的开始延迟。这是因为，当一个新 View 控件出现的时候，其他 View 控件需要立即执行 CHANGE_APPEARING 动画以腾出位置，而新出现的 View 控件在一定延迟之后再执行 APPEARING。相反地，当一个 View 控件消失的时候，它需要执行 DISAPPEARING 动画，而其他的 View 控件需要等待它消失后再执行 CHANGE_DISAPPEARING。当然这些默认的行为都可以通过 setDuration（int，long）和 setStartDelay（int，long）等方法改变。如果需要使用默认的动画，可以在 XML 布局文件中把 android：animateLayoutchanges 属性设置为 true。myObjectAnimator =（ObjectAnimator）AnimatorInflater.loadAnimator（this，R.animator.myanimator）则用于根据 myanimator 的内容创建 myObjectAnimator 动画，myanimator 的主要内容如下：

```
<?xml version="1.0" encoding="utf-8"?>
<objectAnimator xmlns:android="http://schemas.android.com/apk/res/android"
        android:propertyName="scaleX"
        android:valueFrom="0.5"
        android:valueTo="1.0">
</objectAnimator>
```

上面这段代码在 MyCode\MySample253\app\src\main\res\animator\myanimator.xml 文件中。在这段代码中，android：propertyName="scaleX"表示该动画在水平方向进行缩放，android：valueFrom="0.5"表示动画开始值为 50%，android：valueTo="1.0"表示动画结束值为 100%。注意，animator 是新建的文件夹，默认的项目中没有此文件夹。此实例的完整项目在 MyCode\MySample253 文件夹中。

136 使用 TransitionDrawable 动态改变图像颜色

此实例主要通过使用 TransitionDrawable，实现在指定的时间内从一幅（彩色）图像过渡到另一幅（黑白）图像。当实例运行之后，将显示《妈妈》电影海报的彩色图像，如图 136-1 的左图所示，单击"从彩色过渡到黑白"按钮，则将在 3s 内，以动态的方式过渡到该电影海报的黑白图像，如图 136-1 的右图所示。单击"从黑白过渡到彩色"按钮，仍将在 3s 内，以动态的方式从该电影海报的黑白图像过渡到彩色图像。

图 136-1

主要代码如下：

```java
public class MainActivity extends Activity {
    ImageView myImageView;
    Bitmap myColorBitmap,myGrayBitmap;
    @Override
    protected void onCreate(Bundle savedInstanceState) {
        super.onCreate(savedInstanceState);
        setContentView(R.layout.activity_main);
        myImageView = (ImageView)findViewById(R.id.myImageView);
        myColorBitmap = BitmapFactory.decodeResource(getResources(),
                                R.mipmap.myimage1);    //获取彩色图像位图
        Paint myPaint = new Paint();
        ColorMatrix myColorMatrix = new ColorMatrix();
        myColorMatrix.setSaturation(0);                //设置图像的饱和度(0-1),0 为黑白,1 为彩色
        myPaint.setColorFilter(new ColorMatrixColorFilter(myColorMatrix));
        myGrayBitmap = Bitmap.createBitmap(myColorBitmap.getWidth(),
                        myColorBitmap.getHeight(),Bitmap.Config.ARGB_8888);
        Canvas myCanvas = new Canvas(myGrayBitmap);
        myCanvas.drawBitmap(myColorBitmap,0,0,myPaint);    //生成黑白图像
    }
    public void onClickBtn1(View v) {                      //响应单击"从彩色过渡到黑白"按钮
        TransitionDrawable myTransitionDrawable = new TransitionDrawable(
```

```
                new Drawable[]{new BitmapDrawable(getResources(),myColorBitmap),
                    new BitmapDrawable(getResources(),myGrayBitmap)});
        myImageView.setImageDrawable(myTransitionDrawable);
        //设置过渡时长 3000ms 并开始过渡动画
        myTransitionDrawable.startTransition(3000);
    }
    public void onClickBtn2(View v) {        //响应单击"从黑白过渡到彩色"按钮
        TransitionDrawable myTransitionDrawable = new TransitionDrawable(
                new Drawable[]{new BitmapDrawable(getResources(),myGrayBitmap),
                    new BitmapDrawable(getResources(),myColorBitmap)});
        //应用该 TransitionDrawable 对象
        myImageView.setImageDrawable(myTransitionDrawable);
        //设置过渡时长 3000ms 并开始过渡动画
        myTransitionDrawable.startTransition(3000);
    }}
```

上面这段代码在 MyCode \ MySample767 \ app \ src \ main \ java \ com \ bin \ luo \ mysample \ MainActivity.java 文件中。在这段代码中,myTransitionDrawable = new TransitionDrawable(new Drawable[] {new BitmapDrawable(getResources(), myGrayBitmap), new BitmapDrawable(getResources(), myColorBitmap)}) 表示使用两幅图像创建 TransitionDrawable 的实例,该构造函数的参数是一个 Drawable 数组,当执行 myTransitionDrawable.startTransition(3000)方法时,将在指定的时间内,从 Drawable 数组的第一个元素(黑白图像)依次过渡到最后一个元素(彩色图像)。此实例的完整项目在 MyCode\MySample767 文件夹中。

137　在 GridView 的各个网格中实现 Explode 动画

此实例主要通过在 TransitionManager 的 beginDelayedTransition()方法中设置 GridView 控件为过渡动画 Explode 的容器,从而在显示 GridView 时出现各个网格图像从四周向中心聚拢的动画效果。当实例运行之后,单击"执行 Explode 动画"按钮,GridView 的各个网格图像将向四周散开,直到完全消失;再次单击"执行 Explode 动画"按钮,GridView 的各个网格图像将从四周向中心聚拢,直到完全显示。效果分别如图 137-1 的左图和右图所示。

图　137-1

主要代码如下：

```java
public class MainActivity extends Activity {
  private GridView myGridView = null;
  boolean bChanged = false;
  @Override
  protected void onCreate(Bundle savedInstanceState) {
    super.onCreate(savedInstanceState);
    setContentView(R.layout.activity_main);
    myGridView = (GridView) findViewById(R.id.myGridView);
    myGridView.setAdapter(new ImageAdapter(MainActivity.this));
  }
  public void onClickmyBtn1(View v) {        //响应单击"执行 Explode 动画"按钮
    TransitionManager.beginDelayedTransition(myGridView,
                          new Explode().setDuration(1000));
    if(bChanged){ myGridView.setAdapter(new ImageAdapter(MainActivity.this));}
    else {myGridView.setAdapter(null);}
    bChanged = !bChanged;
  }
  public class ImageAdapter extends BaseAdapter {
  private Context myContext;
  private int[] myImages =
        {R.drawable.myimage1,R.drawable.myimage2,R.drawable.myimage3,
         R.drawable.myimage4,R.drawable.myimage5,R.drawable.myimage6,
         R.drawable.myimage7,R.drawable.myimage8,R.drawable.myimage9};
  public ImageAdapter(Context context) { this.myContext = context; }
  public int getCount() { return myImages.length; }
  public Object getItem(int arg0) { return null; }
  public long getItemId(int position) {  return 0; }
  public View getView(int position, View convertView, ViewGroup parent) {
    ImageView myImageView;
    if (convertView == null) {
      myImageView = new ImageView(myContext);
      myImageView.setLayoutParams(new GridView.LayoutParams(340, 500));
      myImageView.setScaleType(ImageView.ScaleType.CENTER_CROP);
      myImageView.setPadding(2, 0, 0, 0);
    } else { myImageView = (ImageView) convertView; }
    myImageView.setImageResource(myImages[position]);
    return myImageView;                  //设置网格内容
}}}
```

上面这段代码在 MyCode\MySample608\app\src\main\java\com\bin\luo\mysample\MainActivity.java 文件中。在这段代码中，TransitionManager.beginDelayedTransition（myGridView，new Explode().setDuration(1000))表示在 myGridView 中执行 Explode 动画。Explode 动画通常在显示或隐藏控件的时候就执行，无需其他代码。在执行 Explode 动画时通常需要一个容器（一般为布局管理器），但是 GridView 控件天然就是一个容器，它里面可以容纳多个 Item（网格），因此也无需其他代码，从 activity_main 布局中就可以看出实现此动画没有其他的容器。activity_main 布局的主要内容如下：

```xml
<LinearLayout android:layout_width = "match_parent"
              android:layout_height = "match_parent"
```

```xml
        android:orientation = "vertical">
<Button android:layout_width = "match_parent"
        android:layout_height = "wrap_content"
        android:text = "执行 Explode 动画"
        android:textSize = "20dp"
        android:textAllCaps = "false"
        android:onClick = "onClickmyBtn1"/>
<GridView android:id = "@ + id/myGridView"
        android:padding = "5dp"
        android:layout_width = "match_parent"
        android:layout_height = "match_parent"
        android:gravity = "center"
        android:numColumns = "3"
        android:paddingTop = "2dp"
        android:stretchMode = "columnWidth"
        android:verticalSpacing = "5dip"/>
</LinearLayout>
```

上面这段代码在 MyCode\MySample608\app\src\main\res\layout\activity_main.xml 文件中。此实例的完整项目在 MyCode\MySample608 文件夹中。

138　使用 layoutAnimation 平移 RecyclerView 网格

此实例主要通过使用 XML 动画文件设置 RecyclerView 的 layoutAnimation 属性，实现按照指定的顺序和方向在 RecyclerView 的每个网格中实现透明度（从无到有显示）和平移（从下向上）动画。当实例运行之后，将按照从下向上、从左到右的顺序在 RecyclerView 的每个网格中逐个演示透明度和平移动画，效果分别如图 138-1 的左图和右图所示。

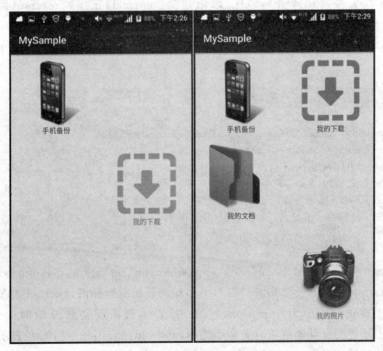

图　138-1

主要代码如下：

```xml
<android.support.v7.widget.RecyclerView
    android:id="@+id/MyRecyclerView"
    android:layout_width="match_parent"
    android:layout_height="match_parent"
    android:layout_alignParentStart="true"
    android:layoutAnimation="@anim/mylayoutanimation"/>
```

上面这段代码在 MyCode\MySample866\app\src\main\res\layout\activity_main.xml 文件中。在这段代码中，android:layoutAnimation="@anim/mylayoutanimation" 用于使用 mylayoutanimation 设置 RecyclerView 的 layoutAnimation 属性，mylayoutanimation 的主要内容如下：

```xml
<?xml version="1.0" encoding="utf-8"?>
<layoutAnimation xmlns:android="http://schemas.android.com/apk/res/android"
    android:delay="1"
    android:animationOrder="normal"
    android:animation="@anim/myanim"/>
```

上面这段代码在 MyCode\MySample866\app\src\main\res\anim\mylayoutanimation.xml 文件中。在这段代码中，android:delay="1"表示每个 Item（风格）动画延时，取值是 android:animation 所指定动画时长的倍数，取值类型可以是 float 类型，也可以是百分数，默认是 0.5，如果动画是 android:animation="@anim/myanim"，则在 myanim 文件中指定 android:duration="1000"，即单次动画的时长是 1000ms，并且指定 android:delay="1"，则一个 Item 动画将在上一个 Item 动画完成后延时单次动画时长的一倍时间后开始，即延时 1000ms 后开始。android:animationOrder="normal" 用于指定 RecyclerView 控件动画开始顺序，取值包括：normal（正序）、reverse（倒序）、random（随机）。android:animation="@anim/myanim"指定每个 Item 入场动画，仅能指定 res/anim 目录下的 Animation 定义的动画，不可使用 Animator 动画，如 res/anim 目录下的 myanim。myanim 的主要内容如下：

```xml
<?xml version="1.0" encoding="utf-8"?>
<set
    xmlns:android="http://schemas.android.com/apk/res/android"
    android:duration="1000">
    <translate android:fromYDelta="100%p"
        android:toYDelta="0"/>
    <alpha android:fromAlpha="0.0"
        android:toAlpha="1.0"/>
</set>
```

上面这段代码在 MyCode\MySample866\app\src\main\res\anim\myanim.xml 文件中。在这段代码中，android:fromYDelta="100%p"表示平移动画开始纵坐标值，android:toYDelta="0"表示平移动画结束纵坐标值，android:fromAlpha="0.0"表示透明度动画开始时完全透明，android:toAlpha="1.0"表示透明度动画结束时完全不透明，android:duration="1000"表示动画持续时间是 1000ms。此实例的完整项目在 MyCode\MySample866 文件夹中。

139　在 ListView 列表项上实现抽屉式滑动动画

此实例主要通过在 ListView 的 setLayoutAnimation（）方法中使用自定义动画控制器 LayoutAnimationController，实现 ListView 在加载列表项时为每个列表项添加抽屉式的滑动动画。当实例运行之后，ListView 的每个列表项将按照添加顺序依次执行滑动动画。图 139-1 左图所示效果是第 3 个列表项从右向左滑动，并且颜色逐渐由模糊变清晰。图 139-1 右图所示效果是第 4 个列表项从右向左滑动，并且颜色逐渐由模糊变清晰。

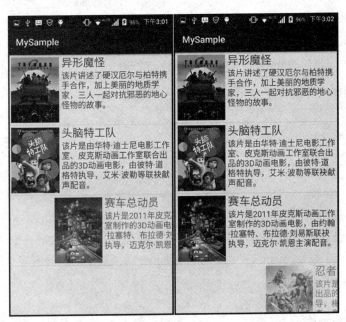

图　139-1

主要代码如下：

```java
public class MainActivity extends ListActivity {
  private  int[] myImages = {R.mipmap.myimage1,
            R.mipmap.myimage2, R.mipmap.myimage3, R.mipmap.myimage4};
  private String[] myTitle = {"异形魔怪", "头脑特工队", "赛车总动员", "忍者神龟"};
  private String[] myContent = {"该片讲述了硬汉厄尔与柏特携手合作,加上美丽的地质学家,三人一起对抗邪恶的地心怪物的故事.","该片是由华特·迪士尼电影工作室、皮克斯动画工作室联合出品的 3D 动画电影,由彼特·道格特执导,艾米·波勒等联袂献声配音.","该片是 2011 年皮克斯动画工作室制作的 3D 动画电影,由约翰·拉塞特、布拉德·刘易斯联袂执导,迈克尔·凯恩主演配音.","该片是由美国派拉蒙影业公司出品的动作片,由戴夫·格林执导,梅根·福克斯、阿伦·瑞奇森、泰勒·派瑞联合主演."};
  ArrayList< Map< String, Object >> myItems = new ArrayList< Map< String, Object >>();
  @Override
  protected void onCreate(Bundle savedInstanceState) {
    int myLength = myTitle.length;
    for (int i = 0; i < myLength; i++) {
      Map< String, Object > myItem = new HashMap< String, Object >();
      myItem.put("image", myImages[i]);
      myItem.put("title", myTitle[i]);
      myItem.put("content", myContent[i]);
      myItems.add(myItem);                             //向每个列表项添加内容
    }
```

```java
SimpleAdapter myAdapter = new SimpleAdapter(this, myItems,
        R.layout.activity_main,new String[]{"image", "title", "content"},
        new int[]{R.id.image, R.id.title, R.id.content});
setListAdapter(myAdapter);                                      //实现列表项数据映射
ListView myListView = getListView();
myListView.setChoiceMode(ListView.CHOICE_MODE_SINGLE);  //设置列表项单选模式
//为列表项添加单击事件响应方法
myListView.setOnItemClickListener(new OnItemClickListener() {
    @Override
    public void onItemClick(AdapterView<?> adapterView,
                              View view, int position, long id) {
        Toast.makeText(MainActivity.this, "您选择的电影是: "
                         + myTitle[position], Toast.LENGTH_LONG).show();
}});
Animation myAnimation = AnimationUtils.loadAnimation(this,
        R.anim.myanim);                                   //通过加载 XML 动画文件创建 Animation 对象
//根据 myAnimation 创建 LayoutAnimationController 对象
LayoutAnimationController myLayoutAnimationController =
                        new LayoutAnimationController(myAnimation);
myLayoutAnimationController.setDelay(1.5f);
myListView.setLayoutAnimation(myLayoutAnimationController);
super.onCreate(savedInstanceState);
}}
```

上面这段代码在 MyCode\MySample466\app\src\main\java\com\bin\luo\mysample\ MainActivity.java 文件中。在这段代码中，AnimationUtils.loadAnimation(this，R.anim.myanim)用于根据 anim 目录下的动画资源 myanim 创建 Animation 动画对象。myanim 动画资源的主要内容如下：

```xml
<?xml version = "1.0" encoding = "utf-8"?>
<set xmlns:android = "http://schemas.android.com/apk/res/android">
<translate android:duration = "1300"
        android:fromXDelta = "100 % p"
        android:startOffset = "100"
        android:toXDelta = "0"/>
<alpha android:fromAlpha = "0.0"
        android:toAlpha = "1.0"
        android:duration = "1300"
        android:startOffset = "100"/>
</set>
```

上面这段代码在 MyCode\MySample466\app\src\main\res\anim\myanim.xml 文件中。在这段代码中，android:fromXDelta="100%p"和 android:toXDelta="0"组合在一起，实现从右向左滑动；如果 android:fromXDelta="－100%p"和 android:toXDelta="0"组合在一起，则实现从左向右滑动。

在 MainActivity.java 文件中，myAdapter = new SimpleAdapter(this，myItems，R.layout.activity_main，new String[]{"image"，"title"，"content"}，new int[]{R.id.image，R.id.title，R.id.content})用于根据 activity_main 布局设置列表项。activity_main 布局的主要内容如下：

```xml
<RelativeLayout android:layout_width = "match_parent"
        android:layout_height = "150dp">
<ImageView android:id = "@ + id/image"
        android:layout_width = "wrap_content"
        android:layout_height = "match_parent"
```

```
                android:layout_alignParentBottom = "true"
                android:layout_alignParentTop = "true"
                android:adjustViewBounds = "true"
                android:padding = "5dip"/>
    <TextView android:id = "@ + id/title"
                android:layout_width = "wrap_content"
                android:layout_height = "wrap_content"
                android:layout_alignParentRight = "true"
                android:layout_alignParentTop = "true"
                android:layout_alignWithParentIfMissing = "true"
                android:layout_toRightOf = "@ + id/image"
                android:textSize = "24dip"/>
    <TextView android:paddingRight = "5dip"
                android:id = "@ + id/content"
                android:layout_width = "match_parent"
                android:layout_height = "wrap_content"
                android:layout_alignParentRight = "true"
                android:layout_below = "@ + id/title"
                android:layout_toRightOf = "@ + id/image"
                android:ellipsize = "marquee"
                android:textSize = "18dip"/>
</RelativeLayout>
```

上面这段代码在 MyCode\MySample466\app\src\main\res\layout\activity_main.xml 文件中。此实例的完整项目在 MyCode\MySample466 文件夹中。

140 在 ViewPager 中实现渐变淡入的转场动画

此实例主要通过使用 ViewPager.PageTransformer 接口创建自定义转场动画类 MyPageTransformer，从而在 ViewPager 中实现以渐变淡入的转场动画切换 View。当实例运行之后，如果向左图滑动图像，即右图的图像露出的部分越大，则亮度（透明度）越大，否则亮度（透明度）越暗。效果分别如图 140-1 的左图和右图所示。

图 140-1

主要代码如下：

```java
public class MainActivity extends AppCompatActivity {
 private ViewPager myViewPager;
 private ViewPagerAdapter myViewPagerAdapter;
 private int[] myImages;
 @Override
 protected void onCreate(Bundle savedInstanceState) {
  super.onCreate(savedInstanceState);
  setContentView(R.layout.activity_main);
  myViewPager = (ViewPager) findViewById(R.id.myViewPager);
  myImages = new int[]{R.mipmap.myimage1, R.mipmap.myimage2, R.mipmap.myimage3,
          R.mipmap.myimage4, R.mipmap.myimage5, R.mipmap.myimage6,
          R.mipmap.myimage7, R.mipmap.myimage8, R.mipmap.myimage9};
  myViewPagerAdapter = new ViewPagerAdapter(MainActivity.this, myImages);
  myViewPager.setAdapter(myViewPagerAdapter);
  myViewPager.setPageTransformer(true, new MyPageTransformer());
 }
 public class ViewPagerAdapter extends PagerAdapter {
  private Context myContext;
  private int[] myImages;
  public ViewPagerAdapter(Context context, int[] datas) {
   myImages = datas;
   myContext = context;
  }
  @Override
  public int getCount() {  return myImages.length; }
  @Override
  public boolean isViewFromObject(View view, Object object){return view == object;}
  @Override
  public Object instantiateItem(ViewGroup container, int position) {
   ImageView myImageView = createImageView(myContext, position);
   container.addView(myImageView);                    //添加 View
   return myImageView;
  }
  @Override
  public void destroyItem(ViewGroup container, int position, Object object) {
   container.removeView((View) object);               //移除 View
  }
  private ImageView createImageView(Context mContext, int position) {
   ImageView myImageView = new ImageView(mContext);
   ViewPager.LayoutParams myLayoutParams = new ViewPager.LayoutParams();
   myImageView.setLayoutParams(myLayoutParams);
   myImageView.setImageResource(myImages[position]);
   myImageView.setScaleType(ImageView.ScaleType.CENTER_CROP);
   return myImageView;                                //创建 ImageView
 } }
 public class MyPageTransformer implements ViewPager.PageTransformer {
  public void transformPage(View view, float position) {
   int pageWidth = view.getWidth();
   if (position < -1) {
    view.setAlpha(0);
   } else if (position <= 0) {
    view.setAlpha(1);
    view.setTranslationX(0);
   } else if (position <= 1) {
    view.setAlpha(1 - position);                      //根据当前位置改变透明度
```

```
            view.setTranslationX(pageWidth * -position);
        } else {
            view.setAlpha(0);
}}}}
```

上面这段代码在 MyCode\MySample519\app\src\main\java\com\bin\luo\mysample\MainActivity.java 文件中。在这段代码中,view.setAlpha(1-position)用于根据当前图像(View)的位置设置其透明度,该参数的变化范围为 0.0~1.0,0 表示完全透明,1 表示完全不透明;在透明的情况下,显示容器的内容,由于实例的容器(FrameLayout)的背景颜色是黑色,因此产生了从暗到亮的效果。此外,使用此实例的相关控件需要在 gradle 中引入 compile 'com.android.support:design:23.3.0'依赖项。此实例的完整项目在 MyCode\MySample519 文件夹中。

141 使用 FragmentTransaction 实现转场动画

此实例主要通过使用 FragmentTransaction 的 setTransition()方法,实现在切换 Fragment 时显示转场动画效果。当实例运行之后,单击第二个标签"头脑特工队",则选项卡的显示效果如图 141-1 的左图所示。单击第三个标签"赛车总动员",则选项卡的显示效果如图 141-1 的右图所示。单击第一个标签"异形魔怪"的显示效果与前两者类似。每个选项卡在显示时均会出现淡入淡出产生的闪烁效果。

图 141-1

主要代码如下:

```
<?xml version = "1.0" encoding = "utf-8"?>
<RelativeLayout xmlns:android = "http://schemas.android.com/apk/res/android"
    xmlns:tools = "http://schemas.android.com/tools"
    android:id = "@ + id/activity_main"
```

```
                android:layout_width = "match_parent"
                android:layout_height = "match_parent"
                tools:context = "com.bin.luo.mysample.MainActivity">
<!-- 使用 FrameLayout 布局为 Fragment 占位 -->
    <FrameLayout android:id = "@ + id/myContent"
                android:layout_width = "match_parent"
                android:layout_height = "match_parent">
    </FrameLayout>
<!-- 选项卡的三个标签页,用 RadioGroup 来实现 -->
    <include layout = "@layout/mybar" />
</RelativeLayout>
```

上面这段代码在 MyCode\MySample512\app\src\main\res\layout\activity_main.xml 文件中。在这段代码中,FrameLayout 控件主要用于为 Fragment 占位,在每次切换标签时,即使用新的 Fragment 代替当前的 Fragment。< include layout = "@layout/mybar" />用于定义选项卡的标签,mybar 是一个布局文件,关于该文件的详细内容请参考 MyCode\ MySample512\app\src\main\res\layout\mybar.xml 文件。此外,使用此实例的相关类需要在 gradle 中引入 compile 'com.android.support:support-v4:24.2.0'依赖项。此实例的完整项目在 MyCode\MySample512 文件夹中。

142　使用 PatternPathMotion 实现路径过渡动画

此实例主要通过在 PatternPathMotion 的 setPatternPath()方法的参数中使用半圆弧路径,实现在两个布局(myscene1 和 myscene2)切换时,ImageView 控件沿着指定的半圆弧路径平移。当实例运行之后,蜘蛛图像(ImageView 控件)位于屏幕顶端,如图 142-1 的左图所示(即 myscene1)。单击屏幕,蜘蛛图像将沿着细实线(不可见)所示的右半圆弧路径向下平移到屏幕底部,如图 142-1 的右图所示(即 myscene2)。在 myscene2 中单击屏幕,蜘蛛图像将沿着细实线(不可见)所示的左半圆弧路径向上平移到屏幕顶部,如图 142-1 的左图所示(即 myscene1)。

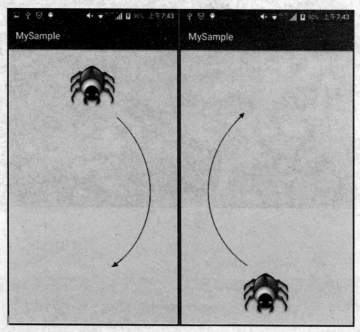

图　142-1

主要代码如下：

```java
public class MainActivity extends Activity {
    @Override
    protected void onCreate(Bundle savedInstanceState) {
        super.onCreate(savedInstanceState);
        setContentView(R.layout.activity_main);
    }
    //单击myscene1过渡到myscene2(蜘蛛图像沿右半圆弧线向下平移)
    public void onClickScene1(View v) {
        ViewGroup myRootView = (ViewGroup)findViewById(R.id.myRootView);
        Scene myScene2 = Scene.getSceneForLayout(myRootView, R.layout.myscene2, this);
        Path myPath = new Path();
        RectF myRect = new RectF(50, 50, 300, 400);
        myPath.arcTo(myRect, 140, 140);              //在路径中添加圆弧
        PatternPathMotion myPatternPathMotion = new PatternPathMotion();
        myPatternPathMotion.setPatternPath(myPath);    //在移动模型中添加路径
        ChangeBounds myChangeBounds = new ChangeBounds();
        myChangeBounds.setPathMotion(myPatternPathMotion);
        myChangeBounds.setDuration(5000);
        TransitionManager.go(myScene2, myChangeBounds);
    }
    //单击myscene2过渡到myscene1(蜘蛛图像沿左半圆弧线向上平移)
    public void onClickScene2(View v) {
        ViewGroup myRootView = (ViewGroup)findViewById(R.id.myRootView);
        Scene myScene1 = Scene.getSceneForLayout(myRootView, R.layout.myscene1, this);
        Path myPath = new Path();
        RectF myRect = new RectF(50, 50, 300, 400);
        myPath.arcTo(myRect, 140, 140);
        PatternPathMotion myPatternPathMotion = new PatternPathMotion();
        myPatternPathMotion.setPatternPath(myPath);
        ChangeBounds myChangeBounds = new ChangeBounds();
        myChangeBounds.setPathMotion(myPatternPathMotion);
        myChangeBounds.setDuration(5000);
        TransitionManager.go(myScene1, myChangeBounds);
    }
}
```

上面这段代码在 MyCode \ MySample585 \ app \ src \ main \ java \ com \ bin \ luo \ mysample \ MainActivity.java 文件中。在这段代码中，myChangeBounds.setPathMotion（myPatternPathMotion）用于设置动画路径。TransitionManager.go(myScene2，myChangeBounds)表示以定制的ChangeBounds动画风格（即半圆弧路径）在 5s 内从 myScene1 过渡到 myScene2。myRootView ＝（ViewGroup）findViewById(R.id.myRootView)用于根据activity_main布局中的RelativeLayout创建动画容器。activity_main布局的主要内容如下：

```xml
<!--这个RelativeLayout用来做动画的父布局-->
<RelativeLayout android:id = "@+id/myRootView"
                android:layout_width = "match_parent"
                android:layout_height = "wrap_content"
                android:layout_centerInParent = "true">
    <include layout = "@layout/myscene1"></include>
</RelativeLayout>
```

上面这段代码在 MyCode\MySample585\app\src\main\res\layout\activity_main.xml 文件中。在这段代码中，<include layout="@layout/myscene1"></include>表示容器的子布局是 myscene1，myscene1 布局的主要内容如下：

```xml
<?xml version="1.0" encoding="utf-8"?>
<RelativeLayout xmlns:android="http://schemas.android.com/apk/res/android"
            android:layout_width="match_parent"
            android:layout_height="match_parent"
            android:gravity="center_horizontal"
            android:onClick="onClickScene1"
            android:padding="20dp">
    <ImageView android:id="@+id/myImageView"
            android:layout_width="100dp"
            android:layout_height="100dp"
            android:layout_alignParentTop="true"
            android:src="@mipmap/myimage1"/>
</RelativeLayout>
```

上面这段代码在 MyCode\MySample585\app\src\main\res\layout\myscene1.xml 文件中。在使用 TransitionManager.go() 方法实现过渡动画时，通常需要两个布局（场景），即此实例的 myscene1 和 myscene2，当 myscene1 的某个控件的 ID 值与 myscene2 的某个控件的 ID 值相同，且位置或大小不同时，在从 myscene1 过渡到 myscene2 时，这对控件就会执行平移或缩放动画。myscene2 布局的主要内容如下：

```xml
<RelativeLayout xmlns:android="http://schemas.android.com/apk/res/android"
            android:layout_width="match_parent"
            android:layout_height="match_parent"
            android:gravity="center_horizontal"
            android:onClick="onClickScene2"
            android:padding="20dp">
    <ImageView android:id="@+id/myImageView"
            android:layout_width="100dp"
            android:layout_height="100dp"
            android:layout_alignParentBottom="true"
            android:src="@mipmap/myimage1"/>
</RelativeLayout>
```

上面这段代码在 MyCode\MySample585\app\src\main\res\layout\myscene2.xml 文件中。此外，使用此实例的相关类需要在 gradle 中引入 compile 'com.android.support:design:25.0.1' 依赖项。此实例的完整项目在 MyCode\MySample585 文件夹中。

143 使用 RippleDrawable 创建波纹扩散动画

此实例主要通过使用 RippleDrawable 创建指定颜色的中心波纹扩散动画并设置为控件的背景，实现在单击 ImageView 控件时产生波纹扩散效果。当实例运行之后，单击 ImageView 控件（苹果图像），将在该控件上产生黄色的中心波纹扩散效果。效果分别如图 143-1 的左图和右图所示。

第5章 动画

图 143-1

主要代码如下：

```java
public class MainActivity extends Activity {
    @Override
    protected void onCreate(Bundle savedInstanceState) {
        super.onCreate(savedInstanceState);
        setContentView(R.layout.activity_main);
    }
    //响应单击图像产生波纹扩散的效果
    public void onClickImageView(View myView) {
        //ColorStateList myColorStateList =
        // getResources().getColorStateList(android.R.color.holo_blue_dark);
        ColorStateList myColorStateList = createColorStateList(0xffffffff,
                                    0xffffff00, 0xff0000ff, 0xffff0000);
        RippleDrawable myRippleDrawable =
                        new RippleDrawable(myColorStateList,null,null);
        myView.setBackground(myRippleDrawable);
    }
    private ColorStateList createColorStateList(int normal,
                                int pressed, int focused, int unable) {
        int[] myColors =
            new int[] { pressed, focused, normal, focused, unable, normal };
        int[][] myStates = new int[6][];
        myStates[0] = new int[] { android.R.attr.state_pressed,
                                    android.R.attr.state_enabled };
        myStates[1] = new int[] { android.R.attr.state_enabled,
                                    android.R.attr.state_focused };
        myStates[2] = new int[] { android.R.attr.state_enabled };
        myStates[3] = new int[] { android.R.attr.state_focused };
```

```
    myStates[4] = new int[] { android.R.attr.state_window_focused };
    myStates[5] = new int[] {};
    ColorStateList myColorStateList = new ColorStateList(myStates, myColors);
    return myColorStateList;
} }
```

上面这段代码在 MyCode \ MySample564 \ app \ src \ main \ java \ com \ bin \ luo \ mysample \ MainActivity.java 文件中。在这段代码中，myRippleDrawable = new RippleDrawable(myColorStateList，null，null)用于根据指定的颜色创建波纹动画。myView.setBackground(myRippleDrawable)表示使用波纹动画 myRippleDrawable 作为控件 myView 的背景。此实例的完整项目在 MyCode\MySample564 文件夹中。

144　自定义 GLSurfaceView 实现波浪起伏的动画

此实例主要通过使用 GLSurfaceView.Renderer 创建自定义控件，动态更改该图像在 X、Y、Z 方向上的顶点坐标，实现为图像添加波浪起伏的动画特效。当实例运行之后，图像就像波浪（或风吹红旗）一样起伏不停。效果分别如图 144-1 的左图和右图所示。

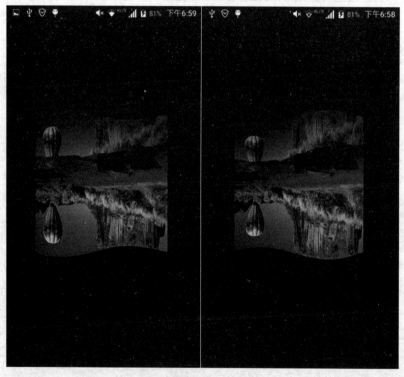

图　144-1

主要代码如下：

```
public class MainActivity extends Activity {
    @Override
    protected void onCreate(Bundle savedInstanceState) {
        super.onCreate(savedInstanceState);
        setContentView(R.layout.activity_main);
```

```
        Bitmap myBitmap = BitmapFactory.decodeResource(getResources(),
                                                        R.mipmap.myimage);
        GLRenderer myGLRenderer = new GLRenderer(myBitmap);
        GLSurfaceView myGLSurfaceView =
                   (GLSurfaceView)findViewById(R.id.myGLSurfaceView);
        myGLSurfaceView.setZOrderOnTop(true);             //将控件置于Z轴顶部
        myGLSurfaceView.setRenderer(myGLRenderer);        //应用渲染器对象
}}
```

上面这段代码在 MyCode\MySample777\app\src\main\java\com\bin\luo\mysample\MainActivity.java 文件中。在这段代码中，myGLRenderer = new GLRenderer(myBitmap)中的 GLRenderer 是实现了 GLSurfaceView.Renderer 接口的自定义类，关于 GLRenderer 类的详细内容请参考 MyCode\MySample777\app\src\main\java\com\bin\luo\mysample\ GLRenderer.java 文件。此实例的完整项目在 MyCode\MySample777 文件夹中。

145 自定义 Animation 实现硬币正反面绕 Y 轴旋转

此实例主要通过在自定义 RotateYAnimation 中使用 Matrix 的 preTranslate()、postTranslate() 方法和 Camera 的 rotateY() 方法，实现在围绕 Y 轴旋转时，硬币的正反面自然切换。当实例运行之后，正反面硬币围绕 Y 轴的旋转效果分别如图 145-1 的左图和右图所示。

图 145-1

主要代码如下：

```
public class MainActivity extends Activity {
    @Override
    protected void onCreate(Bundle savedInstanceState) {
```

```java
super.onCreate(savedInstanceState);
setContentView(R.layout.activity_main);
final ImageView myImageView = (ImageView) findViewById(R.id.myImageView);
final RotateYAnimation myFrontAnimation = new RotateYAnimation();
myFrontAnimation.setMyOriginDegrees(90);                    //设置旋转前角度
myFrontAnimation.setMyRotateDegrees(180);                   //设置需要旋转角度
myFrontAnimation.setAnimationListener(new Animation.AnimationListener() {
  @Override
  public void onAnimationStart(Animation animation) { }
  @Override
  public void onAnimationEnd(Animation animation) {         //动画结束监听
   RotateYAnimation myBackAnimation = new RotateYAnimation();
   myBackAnimation.setMyOriginDegrees(90);
   myBackAnimation.setMyRotateDegrees(180);
   myImageView.setImageResource(R.mipmap.myicon_back);      //显示反面硬币
   myImageView.startAnimation(myBackAnimation);
   myBackAnimation.setAnimationListener(new Animation.AnimationListener() {
    @Override
    public void onAnimationStart(Animation animation) { }
    @Override
    public void onAnimationEnd(Animation animation) {
     myImageView.setImageResource(R.mipmap.myicon_front);   //显示正面硬币
     myImageView.startAnimation(myFrontAnimation);
    }
    @Override
    public void onAnimationRepeat(Animation animation) { }
   }); }
  @Override
  public void onAnimationRepeat(Animation animation) { }
});
myImageView.startAnimation(myFrontAnimation);               //启动动画
} }
```

上面这段代码在 MyCode\MySample842\app\src\main\java\com\bin\luo\mysample\MainActivity.java 文件中。在这段代码中，RotateYAnimation myBackAnimation = new RotateYAnimation()中的 RotateYAnimation 是自定义类，主要用于在围绕 Y 轴进行旋转时，实现正反面的自动切换。RotateYAnimation 类的主要代码如下：

```java
public class RotateYAnimation extends Animation {
 int myCenterX, myCenterY;                                  //旋转中心坐标
 Camera myCamera = new Camera();
 int myRotateDegrees, myOriginDegrees;                      //旋转前角度和需要旋转的角度
 public void setMyOriginDegrees(int myOriginDegrees) {
  this.myOriginDegrees = myOriginDegrees;
 }
 public void setMyRotateDegrees(int myRotateDegrees) {
  this.myRotateDegrees = myRotateDegrees;
 }
 @Override
 public void initialize(int width, int height, int parentWidth, int parentHeight){
  super.initialize(width, height, parentWidth, parentHeight);
  //设置中心位置坐标
```

```
myCenterX = width / 2;
myCenterY = height / 2;
setFillAfter(true);
setDuration(250);
setInterpolator(new LinearInterpolator());
}
@Override
protected void applyTransformation(float interpolatedTime, Transformation t){
    Matrix myMatrix = t.getMatrix();
    myCamera.save();
    myCamera.rotateY(myOriginDegrees);                          //设置旋转前角度
    myCamera.rotateY(myRotateDegrees * interpolatedTime);       //开始旋转指定角度
    myCamera.getMatrix(myMatrix);                               //获取变换矩阵
    myMatrix.preTranslate( - myCenterX, - myCenterY);
    myMatrix.postTranslate(myCenterX, myCenterY);               //绕中心点进行矩阵变换
    myCamera.restore();
} }
```

上面这段代码在 MyCode\MySample842\app\src\main\java\com\bin\luo\mysample\RotateYAnimation.java 文件中。此实例的完整项目在 MyCode\MySample842 文件夹中。

第 6 章

文件和数据

146 采用 DOM 方式解析 XML 文件的内容

此实例主要通过使用 DocumentBuilderFactory 等 DOM 操作类,从而实现以对象的形式读取 XML 文件的内容。当实例运行之后,单击"采用 DOM 方式解析 XML 文件内容"按钮,将在 ListView 控件中显示 persondom.xml 文件的内容,效果分别如图 146-1 的左图和右图所示。

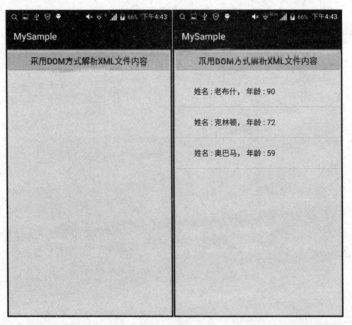

图 146-1

主要代码如下:

```
public void onClickmyBtn1(View v) {//响应单击"采用 DOM 方式解析 XML 文件内容"按钮
    DomHelper myDomHelper = new DomHelper();
    myArrayList = myDomHelper.queryXML(getApplicationContext());
    myArrayAdapter = new ArrayAdapter < Person >(MainActivity.this,
            android.R.layout.simple_expandable_list_item_1, myArrayList);
    myListView.setAdapter(myArrayAdapter);
}
```

上面这段代码在 MyCode\MySample331\app\src\main\java\com\bin\luo\mysample\MainActivity.java 文件中。在这段代码中，DomHelper 是自定义类，它用于以 DOM 方式从 persondom.xml 文件中获取内容，DomHelper 类的主要代码如下：

```java
public class DomHelper {
  public static ArrayList<Person> queryXML(Context context) {
    ArrayList<Person> myPersons = new ArrayList<Person>();
    try {
      DocumentBuilderFactory myDocumentBuilderFactory =
              DocumentBuilderFactory.newInstance();          //获取 DOM 解析器的工厂实例
      DocumentBuilder myDocumentBuilder =
              myDocumentBuilderFactory.newDocumentBuilder();
      Document myDocument =   myDocumentBuilder.parse(
        context.getAssets().open("persondom.xml"));          //将 XML 文件读入 DOM 解析器
      //获取 Person 元素的节点列表
      NodeList myNodes = myDocument.getElementsByTagName("person");
      //遍历该集合,显示集合的元素以及子元素名字
      for (int i = 0; i < myNodes.getLength(); i++) {
        //先从 Person 元素开始解析
        Element myElement = (Element) myNodes.item(i);
        Person myPerson = new Person();
        myPerson.setId(Integer.valueOf(myElement.getAttribute("id")));
        //获取 Person 下的 name 和 age
        NodeList myChildNodes = myElement.getChildNodes();
        for (int j = 0; j < myChildNodes.getLength(); j++) {
          Node myChildNode = myChildNodes.item(j);
          if (myChildNode.getNodeType() == Node.ELEMENT_NODE) {
            Element myChildElement = (Element) myChildNode;
            if ("name".equals(myChildElement.getNodeName()))
              myPerson.setName(myChildElement.getFirstChild().getNodeValue());
            else if ("age".equals(myChildElement.getNodeName()))
              myPerson.setAge(Integer.valueOf(
                      myChildElement.getFirstChild().getNodeValue()));
          }}
        myPersons.add(myPerson);
      }} catch (Exception e) { e.printStackTrace(); }
    return myPersons;
}}
```

上面这段代码在 MyCode\MySample331\app\src\main\java\com\bin\luo\mysample\DomHelper.java 文件中。在这段代码中，首先获得 DOM 解析器的工厂实例，并从工厂实例中获得了 DOM 解析器，最后把要解析的 XML 文件读入 DOM 解析器逐一解析。myDocument = myDocumentBuilder.parse(context.getAssets().open("persondom.xml"))表示解析当前应用的 assets 目录下的 persondom.xml 文件。persondom.xml 文件的主要内容如下：

```xml
<?xml version = "1.0" encoding = "UTF-8"?>
<persons>
 <person id = "1">
  <name>老布什</name>
  <age>90</age></person>
 <person id = "2">
```

```
        <name>克林顿</name>
        <age>72</age></person>
      <person id = "3">
        <name>奥巴马</name>
        <age>59</age></person>
</persons>
```

上面这段代码在 MyCode\MySample331\app\src\main\assets\persondom.xml 文件中。在这段代码中,每对<person></person>标签内的内容表示一个对象,它一定要与 DomHelper 和自定义类 Person 匹配,否则解析会出错。Person 类的主要代码如下:

```
public class Person {
    private int id;
    private String name;
    private int age;
    public Person() { }
    public int getId() { return id; }
    public void setId(int id) { this.id = id; }
    public String getName() { return name; }
    public void setName(String name) { this.name = name; }
    public int getAge() { return age; }
    public void setAge(int age) { this.age = age; }
    @Override
    public String toString() { return "姓名:" + this.name + ",年龄:" + this.age; }
}
```

上面这段代码在 MyCode\MySample331\app\src\main\java\com\bin\luo\mysample\ Person.java 文件中。DOM 是 Document Object Model 文档对象模型的缩写。DOM 是以层次结构组织的节点或信息片断的集合。这个层次结构允许开发人员在树中导航寻找特定信息。分析该结构通常需要加载整个文档和构造层次结构,然后才能做相关工作。DOM 解析方式允许应用对数据和结构做出更改;并且访问是双向的,可以在任何时候在树中上下导航,获取和操作任意部分的数据;缺点是通常需要加载整个 XML 文档来构造层次结构,资源消耗大。此实例的完整项目在 MyCode\ MySample331 文件夹中。

147 采用 Pull 方式解析 XML 文件的内容

此实例主要通过使用 XmlPullParserFactory 等 Pull 操作类,实现以对象的形式读取 XML 文件的内容。当实例运行之后,单击"采用 Pull 方式解析 XML 文件内容"按钮,则将在 ListView 控件中显示 bookpull.xml 文件的内容,效果分别如图 147-1 的左图和右图所示。

主要代码如下:

```
//响应单击"采用 Pull 方式解析 XML 文件内容"按钮
public void onClickmyBtn1(View v) {
    try {
        InputStream myInputStream = getAssets().open("bookpull.xml");
        myArrayList = PullHelper.getBooks(myInputStream);
        myArrayAdapter = new ArrayAdapter<Book>(MainActivity.this,
                android.R.layout.simple_expandable_list_item_1, myArrayList);
```

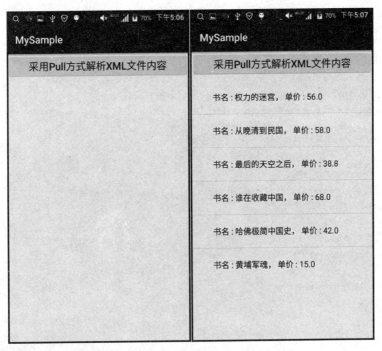

图 147-1

```
    myListView.setAdapter(myArrayAdapter);
  } catch (Exception e) {
    e.printStackTrace();
  }
}
```

上面这段代码在 MyCode\MySample333\app\src\main\java\com\bin\luo\mysample\MainActivity.java 文件中。在这段代码中，PullHelper.getBooks(myInputStream) 是 PullHelper 类的静态方法，用于以流方式读取 bookpull.xml 文件的内容，并将其解析为一个数组列表。PullHelper 类的主要代码如下：

```
public class PullHelper {
  public static ArrayList<Book> getBooks(InputStream xml) throws Exception{
    ArrayList<Book> myBooks = null;
    Book myBook = null;
    XmlPullParserFactory myXmlPullParserFactory =
                                      XmlPullParserFactory.newInstance();
    XmlPullParser myXmlPullParser = myXmlPullParserFactory.newPullParser();
    myXmlPullParser.setInput(xml, "UTF-8");
    int myType = myXmlPullParser.getEventType();
    while (myType != XmlPullParser.END_DOCUMENT) {
      switch (myType) {
        case XmlPullParser.START_DOCUMENT:
          myBooks = new ArrayList<Book>();
          break;
        case XmlPullParser.START_TAG:
          if ("book".equals(myXmlPullParser.getName())) {
            myBook = new Book();
```

```
            int id = Integer.parseInt(myXmlPullParser.getAttributeValue(0));
            myBook.setId(id);
          } else if ("name".equals(myXmlPullParser.getName())) {
            String name = myXmlPullParser.nextText();              //获取该节点的内容
            myBook.setName(name);
          } else if ("price".equals(myXmlPullParser.getName())) {
            float price = Float.parseFloat(myXmlPullParser.nextText());
            myBook.setPrice(price);
          }
          break;
        case XmlPullParser.END_TAG:
          if ("book".equals(myXmlPullParser.getName())) {
            myBooks.add(myBook);
            myBook = null;
          }
          break;
      }
      myType = myXmlPullParser.next();
    }
    return myBooks;
} }
```

上面这段代码在 MyCode\MySample333\app\src\main\java\com\bin\luo\mysample\PullHelper.java 文件中。在这段代码中,Book 是解析对象的类定义,XML 格式的数据文件就是由无数个 Book 对象构成。Book 类的主要代码如下:

```
public class Book {
  private int id;
  private String name;
  private float price;
  public Book() { }
  public void setId(int id) { this.id = id; }
  public void setName(String name) { this.name = name; }
  public void setPrice(float price) { this.price = price; }
  @Override
  public String toString() { return "书名 : " + this.name + ", 单价 : " + this.price; }
}
```

上面这段代码在 MyCode\MySample333\app\src\main\java\com\bin\luo\mysample\ Book.java 文件中。Book 类定义了 XML 文件的数据存储格式,即此实例的数据文件 bookpull.xml 的 <book></book>标签内的内容必须与 Book 类的定义相匹配,否则在解析时可能会出错。bookpull.xml 文件的主要内容如下:

```
<?xml version = "1.0" encoding = "UTF - 8"?>
<books>
  <book id = "1">
    <name>权力的迷宫</name>
    <price>56</price></book>
  <book id = "2">
    <name>从晚清到民国</name>
    <price>58</price></book>
```

```
  < book id = "3" >
    < name >最后的天空之后</name >
    < price > 38.80 </price ></book >
  < book id = "4" >
    < name >谁在收藏中国</name >
    < price > 68 </price ></book >
  < book id = "5" >
    < name >哈佛极简中国史</name >
    < price > 42 </price ></book >
  < book id = "6" >
    < name >黄埔军魂</name >
    < price > 15 </price ></book >
</books >
```

上面这段代码在 MyCode\MySample333\app\src\main\assets\bookpull.xml 文件中。Pull 解析和 SAX 解析一样,也是采用事件驱动进行解析的,当 Pull 解析器开始解析之后,可以调用它的 next()方法获取下一个解析事件(即开始文档,结束文档,开始标签,结束标签),当处于某个元素时可以调用 getAttributte()方法获取属性值。此实例的完整项目在 MyCode\MySample333 文件夹中。

148 使用 JSONArray 解析 JSON 串的多个对象

此实例主要通过使用 JSONArray 和 JSONObject,实现解析 JSON 字符串的多个对象。当实例运行之后,单击"使用 JSONArray 解析 JSON 字符串的多个对象"按钮,将把 JSON 字符串"[{"myName":"Karli Watson","myBook":"C♯入门经典"},{"myName":"Bruce Eckel","myBook":"Java 编程思想"},{"myName":"Stephen Prata","myBook":"C++Primer"}]",解析成三个独立的对象,效果分别如图 148-1 的左图和右图所示。

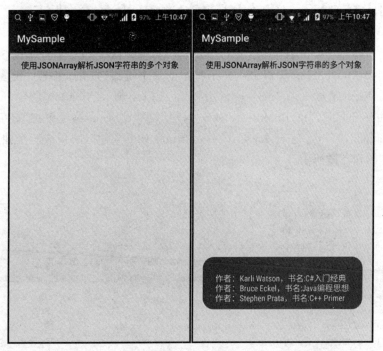

图　148-1

主要代码如下:

```java
//响应单击"使用JSONArray解析JSON字符串的多个对象"按钮
public void onClickmyBtn1(View v) {
    String myJsonData = "[{\"myName\":\"Karli Watson\",\"myBook\":\"C♯入门经典\"},"
        + "{\"myName\":\"Bruce Eckel\",\"myBook\":\"Java编程思想\"},"
        + "{\"myName\":\"Stephen Prata\",\"myBook\":\"C++Primer\"}]";
    try {
        String myInfo = "";
        JSONArray myJSONArray = new JSONArray(myJsonData);
        int myLength = myJSONArray.length();
        for (int i = 0; i < myLength; i++) {
            JSONObject myJSONObject = myJSONArray.getJSONObject(i);
            myInfo += "\n作者:" + myJSONObject.getString("myName");
            myInfo += ",书名:" + myJSONObject.getString("myBook");
        }
        Toast.makeText(this, myInfo, Toast.LENGTH_SHORT).show();
    } catch (Exception e) {
        Toast.makeText(this, e.getMessage().toString(), Toast.LENGTH_SHORT).show();
    }
}
```

上面这段代码在 MyCode\MySample473\app\src\main\java\com\bin\luo\mysample\MainActivity.java 文件中。在这段代码中,myJSONArray = new JSONArray(myJsonData)用于根据JSON格式的字符串myJsonData构造一个JSONArray的实例myJSONArray。myJSONObject = myJSONArray.getJSONObject(i)用于将myJSONArray数组中索引为i的元素转换成JSONObject的实例myJSONObject。myJSONObject.getString("myName")用于从JSON对象myJSONObject中获取属性名称为"myName"的属性值。此实例的完整项目在MyCode\MySample473文件夹中。

149 使用JSONArray解析JSON串的多个键值

此实例主要通过使用JSONArray和JSONObject,实现解析JSON字符串的多个键值。当实例运行之后,单击"使用JSONArray解析JSON字符串的多个键值"按钮,将把一键多值的JSON字符串"[{"myName":"Karli Watson","myBooks":["C♯入门经典","Windows Phone 7入门经典"]},{"myName":"Bruce Eckel","myBooks":["Java编程思想","Scala编程思想","C++编程思想"]},{"myName":"Stephen Prata","myBooks":["C++Primer Plus"]}]",解析成三个独立的对象,效果分别如图149-1的左图和右图所示。

主要代码如下:

```java
//响应单击"使用JSONArray解析JSON字符串的多个键值"按钮
public void onClickmyBtn1(View v) {
    String myJsonData = "[{\"myName\":\"Karli Watson\",\"myBooks\":[\"C♯入门经典\",\"Windows Phone 7入门经典\"]}," + "{\"myName\":\"Bruce Eckel\",\"myBooks\": [\"Java编程思想\",\"Scala编程思想\",\"C++编程思想\"]}," + "{\"myName\":\"Stephen Prata\",\"myBooks\":[\"C++Primer Plus\"]}]";
    try {
        String myInfo = "";
        JSONArray myJSONArray = new JSONArray(myJsonData);
        int myLength = myJSONArray.length();
```

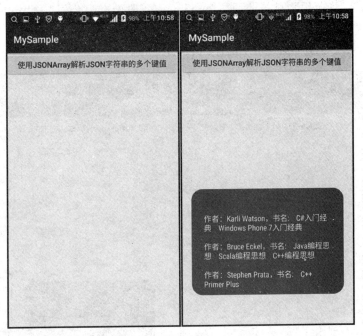

图 149-1

```
for (int i = 0; i < myLength; i++) {
    JSONObject myJSONObject = myJSONArray.getJSONObject(i);
    myInfo += "\n\n作者: " + myJSONObject.getString("myName");
    myInfo += ",书名:";
    JSONArray myBooks = myJSONObject.getJSONArray("myBooks");
    int mySize = myBooks.length();
    for (int j = 0; j < mySize; j++) {
        myInfo += "    " + myBooks.getString(j);
    }
}
Toast.makeText(this, myInfo, Toast.LENGTH_SHORT).show();
} catch (Exception e) {
    Toast.makeText(this, e.getMessage().toString(),Toast.LENGTH_SHORT).show();
}}
```

上面这段代码在 MyCode\MySample475\app\src\main\java\com\bin\luo\mysample\MainActivity.java 文件中。在这段代码中，JSON 字符串的 myBooks 是一键多值，如"myBooks"：["Java 编程思想","Scala 编程思想","C++编程思想"]，myBooks = myJSONObject.getJSONArray("myBooks")用于获取键名 myBooks 对应的多个键值，即 myBooks 数组中包含的"Java 编程思想"、"Scala 编程思想"、"C++编程思想"这三个元素，当使用循环的方式获取 myBooks 数组中的每个元素时，直接使用 myBooks.getString(j)方法即可获得索引 j 对应的元素。此实例的完整项目在 MyCode\MySample475 文件夹中。

150　使用 JSONTokener 获取 JSON 的不同对象

此实例主要通过使用 JSONTokener 的 nextValue()方法，实现获取在 JSON 字符串中的不同类型对象。当实例运行之后，单击"使用 JSONTokener 获取 JSON 的不同对象"按钮，将在弹出的 Toast

中显示"{"myName":"华茂生物科技公司","myType":"股份有限公司"}{"myIncome":"780000万元","myPersons":"3000人","myAssets":"100000万元"}"字符串的"公司概况"和"营收资产"两个对象的键值对信息,效果分别如图 150-1 的左图和右图所示。

图 150-1

主要代码如下：

```java
//响应单击"使用 JSONTokener 获取 JSON 的不同对象"按钮
public void onClickmyBtn1(View v) {
    String myJsonData = "{\"myName\":\"华茂生物科技公司\"," +
            "\"myType\":\"股份有限公司\"}{\"myIncome\":\"780000 万元\"," +
            "\"myPersons\":\"3000 人\",\"myAssets\":\"100000 万元\"}";
    try {
        JSONTokener myJSONTokener = new JSONTokener(myJsonData);
        String myInfo = "\n 公司概况";
        JSONObject myJSONObject1 = (JSONObject) myJSONTokener.nextValue();
        myInfo += "\n 公司名称:" + myJSONObject1.getString("myName");
        myInfo += "\n 公司类别:" + myJSONObject1.getString("myType");
        myInfo += "\n 营收资产";
        JSONObject myJSONObject2 = (JSONObject) myJSONTokener.nextValue();
        myInfo += "\n 年度收入:" + myJSONObject2.getString("myIncome");
        myInfo += "\n 公司员工:" + myJSONObject2.getString("myPersons");
        myInfo += "\n 净资产:" + myJSONObject2.getString("myAssets");
        Toast.makeText(this, myInfo, Toast.LENGTH_SHORT).show();
    } catch (Exception e) {
        Toast.makeText(getApplicationContext(),
                e.getMessage().toString(), Toast.LENGTH_SHORT).show();
    }
}
```

上面这段代码在 MyCode \ MySample479 \ app \ src \ main \ java \ com \ bin \ luo \ mysample \ MainActivity.java 文件中。在这段代码中,{"myName":"华茂生物科技公司","myType":"股份有

限公司"}{"myIncome":"780000万元","myPersons":"3000人","myAssets":"100000万元"}是由{...}{...}构成的平行关系的两个不同对象；myJSONObject1 =（JSONObject）myJSONTokener.nextValue()表示返回第一个{...}所代表的对象，当再次使用nextValue()方法时，则返回下一个{...}所代表的对象，即myJSONObject2 =（JSONObject）myJSONTokener.nextValue()。此实例的完整项目在MyCode\MySample479文件夹中。

151 使用JSONTokener解析JSON非对象文本

此实例主要通过使用JSONTokener的next()方法实现解析JSON字符串的非对象文本。当实例运行之后，单击"使用JSONTokener解析JSON的非对象文本"按钮，将把JSON字符串"名家系列：{"myName":"Karli Watson","myBook":"C#入门经典"}"中的非对象文本"名家系列："解析出来，效果分别如图151-1的左图和右图所示。

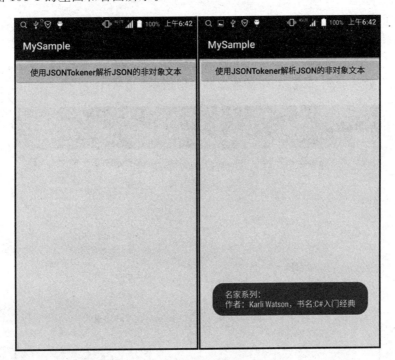

图 151-1

主要代码如下：

```
//响应单击"使用JSONTokener解析JSON的非对象文本"按钮
public void onClickmyBtn1(View v) {
    String myJsonData =
            "名家系列：{\"myName\":\"Karli Watson\",\"myBook\":\"C#入门经典\"}";
    try {
        String myInfo = "";
        JSONTokener myJSONTokener = new JSONTokener(myJsonData);
        myInfo += myJSONTokener.next(5);
        JSONObject myJSONObject = (JSONObject) myJSONTokener.nextValue();
        myInfo += "\n作者：" + myJSONObject.getString("myName");
        myInfo += ",书名：" + myJSONObject.getString("myBook");
        Toast.makeText(this, myInfo, Toast.LENGTH_SHORT).show();
```

```
        } catch (Exception e) {
            Toast.makeText(getApplicationContext(),
                e.getMessage().toString(),Toast.LENGTH_SHORT).show();
        } }
```

上面这段代码在 MyCode\MySample477\app\src\main\java\com\bin\luo\mysample\MainActivity.java 文件中。在这段代码中,myJsonData = "名家系列:{\"myName\":\"Karli Watson\",\"myBook\":\"C♯入门经典\"}"的"名家系列:"在大括号之外,它是一个非对象文本,共有 5 个字符;myJSONTokener.next(5)即是用于获取这 5 个字符。此实例的完整项目在 MyCode\MySample477 文件夹中。

152　使用 Gson 解析 JSON 字符串的单个对象

此实例主要通过在 Gson 的 fromJson()方法中传递类名参数,实现使用 Gson 解析 JSON 字符串的单个对象。当实例运行之后,单击"使用 Gson 解析 JSON 字符串的单个对象"按钮,将把 JSON 字符串"{\"myName\":\"Karli Watson\",\"myBook\":\"C♯入门经典\"}",解析成"作者:Karli Watson 书名:C♯入门经典",效果分别如图 152-1 的左图和右图所示。

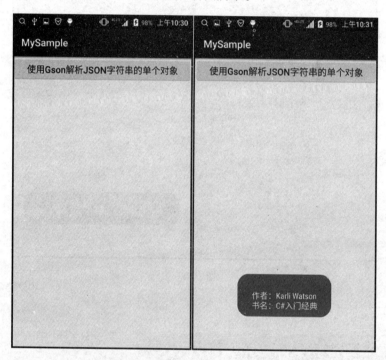

图　152-1

主要代码如下:

```
//响应单击"使用 Gson 解析 JSON 字符串的单个对象"按钮
public void onClickmyBtn1(View v) {
    String myJsonData = "{\"myName\":\"Karli Watson\",\"myBook\":\"C♯入门经典\"}";
    try {
        String myInfo = "";
        Gson myGson = new Gson();
```

```
    User myUser = myGson.fromJson(myJsonData, User.class);
    myInfo += "\n 作者: " + myUser.getMyName();
    myInfo += "\n 书名: " + myUser.getMyBook();
    Toast.makeText(this, myInfo, Toast.LENGTH_SHORT).show();
    } catch (Exception e) {
    Toast.makeText(this, e.getMessage().toString(), Toast.LENGTH_SHORT).show();
    }
}
class User{                         //User 类定义
    public String getMyName() { return myName; }
    public String getMyBook() { return myBook; }
    String myName;
    String myBook;
}
```

上面这段代码在 MyCode\MySample470\app\src\main\java\com\bin\luo\mysample\MainActivity.java 文件中。在这段代码中，myUser = myGson.fromJson(myJsonData, User.class) 用于根据类名从 Json 字符串 myJsonData 中获取单一的对象 myUser。myUser.getMyName() 则是获取类对象的属性值。需要说明的是，使用 Gson 需要在 gradle 中引入 compile 'com.google.code.gson:gson:2.8.1' 依赖项。此实例的完整项目在 MyCode\MySample470 文件夹中。

153 使用 Intent 在 Activity 之间传递基本数据

此实例主要演示了使用 Intent 的 putExtra() 方法和 getStringExtra() 方法，实现在两个 Activity 之间发送和接收 String 等数据类型。当实例运行之后，在 MainActivity 的两个输入框中分别输入"重庆飞达资讯有限公司""20000"，如图 153-1 的左图所示；单击"发送数据"按钮，则将跳转到 SecondActivity。在 SecondActivity 中单击"接收数据"按钮，将在弹出的 Toast 中显示传递的数据，如图 153-1 的右图所示。

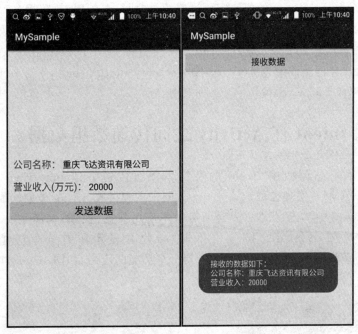

图 153-1

主要代码如下:

```
public void onClickmyBtnSend(View v) {            //响应单击"发送数据"按钮
    Intent myIntent = new Intent(MainActivity.this,SecondActivity.class);
    myIntent.putExtra("myName",myName.getText().toString());
    myIntent.putExtra("myIncome",myIncome.getText().toString());
    startActivity(myIntent);
}
```

上面这段代码在 MyCode\MySample444\app\src\main\java\com\bin\luo\mysample\MainActivity.java 文件中。在这段代码中，myIntent.putExtra("myName",myName.getText().toString())用于将 myName 数据名称及其所代表的数据值以键值对的形式添加到 myIntent 中，putExtra()方法的语法声明如下:

```
public Intent putExtra(String name, String value)
```

其中,参数 String name 表示数据名称,参数 String value 表示数据值。当使用 startActivity(myIntent)方法从 MainActivity 跳转到 SecondActivity 时,myIntent 包含的数据就传递到了 SecondActivity。SecondActivity 的主要代码如下:

```
public void onClickmyBtnReceive(View v) {            //响应单击"接收数据"按钮
    String myData = "接收的数据如下: ";
    Intent myIntent = getIntent();
    myData += "\n 公司名称: " + myIntent.getStringExtra("myName");
    myData += "\n 营业收入: " + myIntent.getStringExtra("myIncome");
    Toast.makeText(getApplicationContext(), myData, Toast.LENGTH_SHORT).show();
}
```

上面这段代码在 MyCode\MySample444\app\src\main\java\com\bin\luo\mysample\SecondActivity.java 文件中。在这段代码中,myIntent.getStringExtra("myName")用于获取在 myIntent 中数据名称为"myName"的值。需要说明的是,如果在传递时,数据名称代表的数据类型为 int 类型,则在接收数据时使用 getIntExtra()方法,其他数据类型以此类推,如 getCharExtra()、getFloatExtra()、getDoubleExtra()等。此外,当在项目中新增了 SecondActivity 之后,则需要在 AndroidManifest.xml 文件中注册该 Activity,如<activity android:name=".SecondActivity"/>。此实例的完整项目在 MyCode\MySample444 文件夹中。

154 使用 Intent 在 Activity 之间传递数组数据

此实例主要演示了如何使用 Intent 的字符串数组传送方法 putStringArrayListExtra()和 getStringArrayListExtra(),实现在两个 Activity 之间发送和接收 String 类型的数组数据。当实例运行之后,在 MainActivity 中单击"获取并发送所有分享应用数据"按钮,如图 154-1 的左图所示,将跳转到 SecondActivity。在 SecondActivity 中单击"接收并显示所有分享应用数据"按钮,将在 TextView 控件中显示当前手机中已经安装的支持分享功能的应用,如图 154-1 的右图所示。

主要代码如下:

```
//响应单击"获取并发送所有分享应用数据"按钮
public void onClickmyBtnSend(View v) {
```

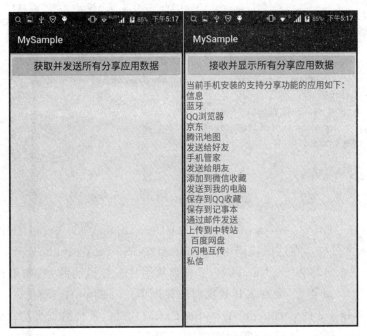

图 154-1

```
//获取当前手机中安装的支持分享功能的应用
List < ResolveInfo > myItems = new ArrayList < ResolveInfo >();
Intent myAppIntent = new Intent(Intent.ACTION_SEND, null);
myAppIntent.addCategory(Intent.CATEGORY_DEFAULT);
myAppIntent.setType("text/plain");
PackageManager myPackageManager = this.getPackageManager();
myItems = myPackageManager.queryIntentActivities(myAppIntent,
                    PackageManager.COMPONENT_ENABLED_STATE_DEFAULT);
ArrayList < String > myNames = new ArrayList < String >();
for (int i = 0; i < myItems.size(); i++) {
 ResolveInfo myResolveInfo = myItems.get(i);
 myNames.add(myResolveInfo.loadLabel(getPackageManager()).toString());
}
//发送所有分享应用数据
Intent mySendIntent = new Intent(MainActivity.this, SecondActivity.class);
mySendIntent.putStringArrayListExtra("myNames", myNames);
startActivity(mySendIntent);
}
```

上面这段代码在 MyCode \ MySample447 \ app \ src \ main \ java \ com \ bin \ luo \ mysample \ MainActivity.java 文件中。在这段代码中，putStringArrayListExtra("myNames"，myNames)用于将 myNames 数据名称及其所代表的字符串数组值以键值对的形式添加到 mySendIntent 中，putStringArrayListExtra()方法的语法声明如下：

putStringArrayListExtra(String name, ArrayList < String > value)

其中，参数 String name 表示数据名称，参数 ArrayList < String > value 表示字符串数组值。当使用 startActivity(mySendIntent)方法从 MainActivity 跳转到 SecondActivity 时，mySendIntent 包含的字符串数组就传递到了 SecondActivity。SecondActivity 的主要代码如下：

```
//响应单击"接收并显示所有分享应用数据"按钮
public void onClickmyBtnReceive(View v) {
    String myText = "当前手机安装的支持分享功能的应用如下：";
    Intent myIntent = getIntent();
    ArrayList < String > myNames = myIntent.getStringArrayListExtra("myNames");
    if (myNames != null) {
        for (int i = 0; i < myNames.size(); i++) {
            String myName = myNames.get(i).toString();
            myText += "\n" + myName;
        }
    }
    myTextView.setText(myText);
}
```

上面这段代码在 MyCode\MySample447\app\src\main\java\com\bin\luo\mysample\SecondActivity.java 文件中。在这段代码中，myIntent.getStringArrayListExtra("myNames")用于获取在 myIntent 中数据名称为"myNames"的字符串数组。需要说明的是，如果在传递数据时，数组名称代表的数据类型为 int 类型，则在发送数据时应使用 Intent 的 putIntegerArrayListExtra()方法，在接收数据时应使用 Intent 的 getIntegerArrayListExtra()方法，其他数据类型以此类推。此外，当在项目中新增了 SecondActivity 之后，则需要在 AndroidManifest.xml 中注册该 Activity，如< activity android:name=".SecondActivity"/>。此实例的完整项目在 MyCode\MySample447 文件夹中。

155 使用 Intent 在 Activity 之间传递图像数据

此实例主要使用 Bitmap 的 compress()方法将图像数据进行压缩，然后通过 Intent 以字节数组的形式实现在两个 Activity 之间传递图像数据。当实例运行之后，将在 MainActivity 的 ImageView 控件中显示一幅图像，如图 155-1 的左图所示；单击"将下面的图像发送到第二个 Activity"按钮，将跳转到 SecondActivity，此时 SecondActivity 的 ImageView 控件是一片空白；单击"接收从第一个 Activity 发送的图像"按钮，将在 ImageView 控件中显示从 MainActivity 发送的图像，如图 155-1 的右图所示。

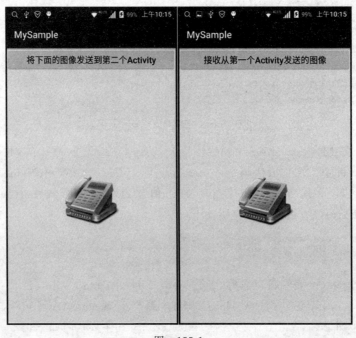

图 155-1

主要代码如下：

```java
//响应单击"将下面的图像发送到第二个Activity"按钮
public void onClickmyBtnSend(View v) {
    Intent myIntent = new Intent(MainActivity.this,SecondActivity.class);
    Bitmap myBitmap = ((BitmapDrawable) myImageView1.getDrawable()).getBitmap();
    ByteArrayOutputStream myByteStream = new ByteArrayOutputStream();
    //如果原始图像是 JPEG 格式,则使用下面的代码
    //myBitmap.compress(Bitmap.CompressFormat.JPEG,100,myByteStream);
    //如果原始图像是 PNG 格式,则使用下面的代码
    myBitmap.compress(Bitmap.CompressFormat.PNG,100,myByteStream);
    byte[] myImageBytes = myByteStream.toByteArray();
    myIntent.putExtra("myBitmap",myImageBytes);
    startActivity(myIntent);
}
```

上面这段代码在 MyCode\MySample443\app\src\main\java\com\bin\luo\mysample\MainActivity.java 文件中。在这段代码中,myBitmap = ((BitmapDrawable) myImageView1.getDrawable()).getBitmap() 表示从 myImageView1 控件中获取 Bitmap。myByteStream = new ByteArrayOutputStream() 用于创建输出流 myByteStream。myBitmap.compress(Bitmap.CompressFormat.PNG,100,myByteStream) 用于按照指定的格式和比例压缩图像,并将数据传送到输出流 myByteStream。myImageBytes=myByteStream.toByteArray() 表示将输出流转换成字节数组。myIntent = new Intent(MainActivity.this,SecondActivity.class) 表示 myIntent 的作用是从 MainActivity 跳转到 SecondActivity。myIntent.putExtra("myBitmap",myImageBytes) 表示在 myIntent 中附加字节数组。SecondActivity 的主要代码如下：

```java
//响应单击"接收从第一个Activity发送的图像"按钮
public void onClickmyBtnReceive(View v) {
    byte[] myImageBytes = getIntent().getByteArrayExtra("myBitmap");
    myImageView2.setImageBitmap(BitmapFactory.decodeByteArray(myImageBytes,
                                0,myImageBytes.length));
}
```

上面这段代码在 MyCode\MySample443\app\src\main\java\com\bin\luo\mysample\SecondActivity.java 文件中。在这段代码中,myImageBytes = getIntent().getByteArrayExtra("myBitmap") 用于从 Intent 中获取 myBitmap 代表的字节数组。BitmapFactory.decodeByteArray(myImageBytes,0,myImageBytes.length) 用于将该字节数组全部解析为一个 Bitmap 对象。此外,当在项目中新增了 SecondActivity 之后,则需要在 AndroidManifest.xml 文件中注册该 Activity ,如 <activity android:name=".SecondActivity"/>。此实例的完整项目在 MyCode\MySample443 文件夹中。

156 使用 Intent 在 Activity 之间传递多幅图像

此实例主要通过使用 Intent 的 putParcelableArrayListExtra()方法和 getParcelableArrayListExtra()方法,实现在两个 Activity 之间传递多幅小图像。当实例运行之后,将在 MainActivity 的 4 个 ImageView 控件中显示 4 幅小图像,如图 156-1 的左图所示；单击"将其中的 3 幅图像发送到第二个 Activity"按钮,将跳转到 SecondActivity,此时 SecondActivity 的 4 个 ImageView 控件是一片空白；

单击"接收从第一个 Activity 发送的 3 幅图像"按钮,将在 ImageView 控件中显示从 MainActivity 传来的 3 幅小图像,如图 156-1 的右图所示。

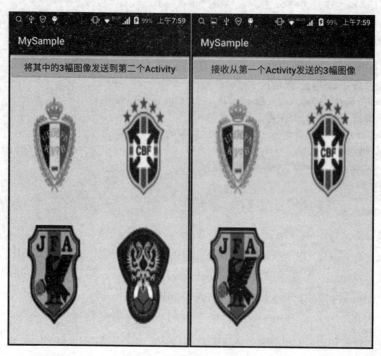

图 156-1

主要代码如下:

```java
//响应单击"将其中的 3 幅图像发送到第二个 Activity"按钮
public void onClickmyBtnSend(View v) {
  try{
    ArrayList<Bitmap> myBitmaps = new ArrayList<>();
    Bitmap myBitmap1 = ((BitmapDrawable) myImageViewA1.getDrawable()).getBitmap();
    Bitmap myBitmap2 = ((BitmapDrawable) myImageViewA2.getDrawable()).getBitmap();
    Bitmap myBitmap3 = ((BitmapDrawable) myImageViewA3.getDrawable()).getBitmap();
    Bitmap myBitmap4 = ((BitmapDrawable) myImageViewA4.getDrawable()).getBitmap();
    myBitmaps.add(myBitmap1);
    myBitmaps.add(myBitmap2);
    myBitmaps.add(myBitmap3);
    //myBitmaps.add(myBitmap4);
    Intent myIntent = new Intent(MainActivity.this, SecondActivity.class);
    myIntent.putParcelableArrayListExtra("myBitmaps", myBitmaps);
    startActivity(myIntent);
  }catch (Exception e){
    Toast.makeText(getApplicationContext(),
              e.getMessage().toString(), Toast.LENGTH_SHORT).show();
  }
}
```

上面这段代码在 MyCode\MySample450\app\src\main\java\com\bin\luo\mysample\MainActivity.java 文件中。在这段代码中,ArrayList<Bitmap> myBitmaps = new ArrayList<>() 用于创建一个数组列表存放多个 Bitmap 对象。myBitmap1=((BitmapDrawable) myImageViewA1.getDrawable()).getBitmap()用于从 myImageViewA1 控件中获取 Bitmap 对象。myBitmaps.

add(myBitmap1)用于将 myBitmap1 添加到 myBitmaps 数组列表中。myIntent=new Intent(MainActivity.this，SecondActivity.class)表示 myIntent 的作用是从 MainActivity 跳转到 SecondActivity。myIntent.putParcelableArrayListExtra("myBitmaps"，myBitmaps)表示在 myIntent 中附加将要传递的图像数组列表 myBitmaps。SecondActivity 的主要代码如下：

```
//响应单击"接收从第一个 Activity 发送的 3 幅图像"按钮
public void onClickmyBtnReceive(View v) {
  try{
    Intent myIntent = getIntent();
    if(myIntent!= null){
      List < Bitmap > myBitmaps =
                 getIntent().getParcelableArrayListExtra("myBitmaps");
      myImageViewB1.setImageBitmap(myBitmaps.get(0));
      myImageViewB2.setImageBitmap(myBitmaps.get(1));
      myImageViewB3.setImageBitmap(myBitmaps.get(2));
      //myImageViewB4.setImageBitmap(myBitmaps.get(3));
    }
  }catch (Exception e){
    Toast.makeText(getApplicationContext(),
            e.getMessage().toString(), Toast.LENGTH_SHORT).show();
  }
}
```

上面这段代码在 MyCode\MySample450\app\src\main\java\com\bin\luo\mysample\SecondActivity.java 文件中。在这段代码中，myIntent = getIntent()用于获取传递的 Intent，myBitmaps = getIntent().getParcelableArrayListExtra("myBitmaps")用于从 Intent 中获取 myBitmaps 代表的图像数组。myImageViewB1.setImageBitmap(myBitmaps.get(0))用于在 myImageViewB1 控件中设置接收的图像。需要说明的是，图像过大或过多可能会导致应用崩溃或者没有响应。此外，当新增了 SecondActivity 之后，则需要在 AndroidManifest.xml 文件中注册该 Activity，如< activity android：name = ".SecondActivity"/>。此实例的完整项目在 MyCode\MySample450 文件夹中。

157 在 Intent 传递数据时使用 Bundle 携带数据

此实例主要实现了当使用 Intent 在两个 Activity 之间传递数据时，使用 Bundle 携带基本类型数据。当实例运行之后，在 MainActivity 的两个输入框中分别输入"无锡宝特软件有限公司""1588"，如图 157-1 的左图所示；单击"发送数据"按钮，将跳转到 SecondActivity。在 SecondActivity 中单击"接收数据"按钮，将在弹出的 Toast 中显示接收的数据，如图 157-1 的右图所示。

主要代码如下：

```
public void onClickmyBtnSend(View v) {     //响应单击"发送数据"按钮
  Intent myIntent =  new Intent(MainActivity.this,SecondActivity.class);
  Bundle myBundle = new Bundle();
  myBundle.putString("myName",myName.getText().toString());
  myBundle.putString("myIncome",myIncome.getText().toString());
  myIntent.putExtras(myBundle);
  startActivity(myIntent);
}
```

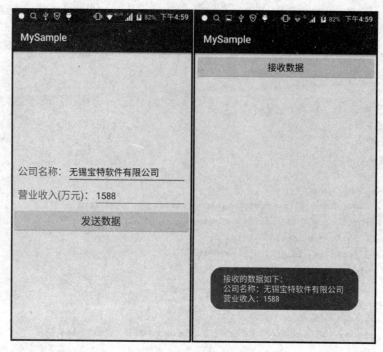

图 157-1

上面这段代码在 MyCode\MySample445\app\src\main\java\com\bin\luo\mysample\MainActivity.java 文件中。在这段代码中，Bundle 是一个简单的数据携带包，Bundle 包含多个基本数据类型方法来存入数据，在传递数据时，首先将数据通过 Bundle 的 putString() 方法附加到 Bundle 对象，再将 Bundle 对象通过 Intent 的 putExtras() 方法附加到 Intent。当使用 startActivity(myIntent)方法从 MainActivity 跳转到 SecondActivity 时，myIntent 包含的 Bundle 数据包就传递到了 SecondActivity。SecondActivity 的主要代码如下：

```java
public void onClickmyBtnReceive(View v) {          //响应单击"接收数据"按钮
    String myData = "接收的数据如下：";
    Intent myIntent = getIntent();
    Bundle myBundle = myIntent.getExtras();
    myData += "\n公司名称：" + myBundle.getString("myName");
    myData += "\n营业收入：" + myBundle.getString("myIncome");
    Toast.makeText(getApplicationContext(), myData, Toast.LENGTH_SHORT).show();
}
```

上面这段代码在 MyCode\MySample445\app\src\main\java\com\bin\luo\mysample\SecondActivity.java 文件中。在这段代码中，myBundle = myIntent.getExtras()用于从获取的 Intent 中取得 Bundle 数据包。myBundle.getString("myName")用于从 Bundle 数据包中获取数据名称为 myName 的数据值。需要说明的是，在使用 Bundle 传递数据包时，如果数据名称代表的数据类型为 int 类型，则在发送数据时使用 putInt()方法，在接收数据时使用 getInt()方法，其他基本数据类型以此类推，如发送数据的 putChar()、putFloat()、putDouble()等方法，接收数据的 getChar()、getFloat()、getDouble()等方法。此外，当在项目中新增了 SecondActivity 之后，则需要在 AndroidManifest.xml 文件中注册该 Activity，如<activity android:name=".SecondActivity"/>。此实例的完整项目在 MyCode\MySample445 文件夹中。

158 使用 Bundle 从 Activity 向 Fragment 传递数据

此实例主要通过使用 Bundle 的 putString()和 getString()方法,实现从 Activity 向 Fragment 传递数据。当实例运行之后,单击底部操作栏上的"搜索"标签,则将显示百度搜索 Fragment,即从 MainActivity 向 WebFragment 通过 Bundle 传递参数 Search,如图 158-1 的左图所示。单击"图片"标签,将显示百度图片 Fragment,如图 158-1 的右图所示。单击其他选项将实现类似的功能。

图 158-1

主要代码如下:

```
public class MainActivity extends Activity implements View.OnClickListener{
  @Override
  protected void onCreate(Bundle savedInstanceState){
    super.onCreate(savedInstanceState);
    setContentView(R.layout.activity_main);
    TextView myTextImage = (TextView)findViewById(R.id.myImage);
    TextView myTextSearch = (TextView)findViewById(R.id.mySearch);
    TextView myTextNews = (TextView)findViewById(R.id.myNews);
    myTextImage.setOnClickListener(this);
    myTextSearch.setOnClickListener(this);
    myTextNews.setOnClickListener(this);
    myTextSearch.performClick();                              //默认显示百度搜索页面
  }
  @Override
  public void onClick(View v){
    FragmentTransaction myFragmentTransaction =
                     getFragmentManager().beginTransaction();  //开启事务
    WebFragment myWebFragment = new WebFragment();             //动态创建 WebFragment
```

```java
Bundle myBundle = new Bundle();
//根据单击标签动态设置传递数据
if(v.getId() == R.id.myImage){
    myBundle.putString("myFlag","Image");
}else if(v.getId() == R.id.myNews){
    myBundle.putString("myFlag","News");
}else if(v.getId() == R.id.mySearch){
    myBundle.putString("myFlag","Search");
}
myWebFragment.setArguments(myBundle);         //向 WebFragment 传递 Bundle
//使用动态创建的 WebFragment 对象替换掉当前正在显示的 WebFragment 对象
myFragmentTransaction.replace(R.id.myFrameLayout,myWebFragment).commit();
}}
```

上面这段代码在 MyCode\MySample811\app\src\main\java\com\bin\luo\mysample\MainActivity.java 文件中。在这段代码中，myBundle.putString("myFlag","Search")用于在 Bundle 中携带 Search 字符串，当新建 WebFragment 时，将在 setArguments(Bundle args)中使用 args.getString("myFlag")读取此值，即 Search 字符串。myWebFragment＝new WebFragment()中的 WebFragment 是以 Fragment 为基类创建的自定义类，用于加载 WebView 控件。WebFragment 类的主要代码如下：

```java
public class WebFragment extends Fragment{
    String myWebFlag;
    @Override
    public void setArguments(Bundle args){             //获取 Bundle 携带的数据内容
        myWebFlag = args.getString("myFlag");
    }
    @Override
    public View onCreateView(LayoutInflater inflater,
                        ViewGroup container,Bundle savedInstanceState){
        WebView myWebView = new WebView(inflater.getContext());   //动态创建 WebView
        myWebView.setWebViewClient(new WebViewClient());
        myWebView.getSettings().setJavaScriptEnabled(true);
        //通过判断传入的 myWebFlag 字符串标志来指定要加载的页面
        myWebView.loadUrl(myWebFlag == "News"?"http://news.sina.com.cn":
                myWebFlag == "Search"?"http://www.baidu.com":
                myWebFlag == "Image"?"http://image.baidu.com":"");
        return myWebView;
}}
```

上面这段代码在 MyCode\MySample811\app\src\main\java\com\bin\luo\mysample\WebFragment.java 文件中。此外，使用 WebView 控件需要在 AndroidManifest.xml 文件中添加 <uses-permission android:name="android.permission.INTERNET"/>权限。此实例的完整项目在 MyCode\MySample811 文件夹中。

159 根据指定网址下载应用安装包到手机 SD 卡

此实例主要通过使用 HttpURLConnection 创建网络连接，实现根据地址下载应用安装包到手机 SD 卡。当实例运行之后，在"下载地址："输入框中输入下载地址，如微信多开安装包下载地址

"http://ftp-apk.pconline.com.cn/3fd55447ae0ddefae8534b1a20ebdf88/pub/download/201010/weixinduokai_v2.4.1.apk"，如图 159-1 的左图所示，单击"开始下载"按钮，则将执行下载操作，下载完成之后该安装包将会保存为 sdcard/update/updata.apk，如图 159-1 的右图所示。

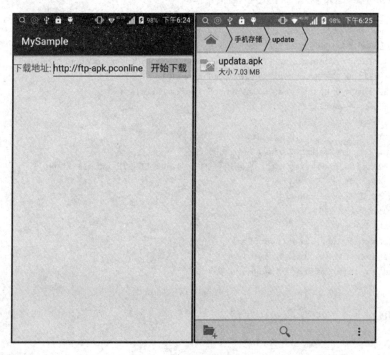

图 159-1

主要代码如下：

```java
public class MainActivity extends Activity {
 EditText myPath;
Handler myHandler = new Handler() {
   @Override
   public void handleMessage(Message msg) {
    if (msg.what == 1) {
     Toast.makeText(MainActivity.this, "开始下载", Toast.LENGTH_SHORT).show();
    } else if (msg.what == 2) {
     Toast.makeText(MainActivity.this, "下载完成", Toast.LENGTH_SHORT).show();
 } } };
@Override
 protected void onCreate(Bundle savedInstanceState) {
   super.onCreate(savedInstanceState);
   setContentView(R.layout.activity_main);
   myPath = (EditText) findViewById(R.id.myPath);
 }
 public void onClickmyBtn1(View v) {                //响应单击"开始下载"按钮
   final String myHttpUrl = myPath.getText().toString();
   new Thread() {
    @Override
    public void run() {   downLoadFile(myHttpUrl); } }.start();
 }
 protected File downLoadFile(String myHttpUrl) {
   final String fileName = "updata.apk";
   File tmpFile = new File("/sdcard/update");
```

```
    if (!tmpFile.exists()) {                    //如果目录不存在
        tmpFile.mkdir();                        //创建目录
    }
    final File myFile = new File("/sdcard/update/"
                              + fileName);      //创建存储文件
    try {
        URL myUrl = new URL(myHttpUrl);
        HttpURLConnection myConnection = (HttpURLConnection) myUrl.openConnection();
        InputStream myInput = myConnection.getInputStream();
        FileOutputStream myOutput = new FileOutputStream(myFile);
        byte[] myBuffer = new byte[256];
        myConnection.connect();
        double myCount = 0;
        if (myConnection.getResponseCode() >= 400) {
            Toast.makeText(MainActivity.this, "连接超时", Toast.LENGTH_SHORT).show();
        } else {
            myHandler.sendEmptyMessage(1);
            while (myCount <= 100) {
                if (myInput != null) {
                    int myData = myInput.read(myBuffer);
                    if (myData <= 0) {    break; }
    else { myOutput.write(myBuffer, 0, myData); }
                } else {   break; }
            }
        }
        myConnection.disconnect();
        myOutput.close();
        myInput.close();
        myHandler.sendEmptyMessage(2);
    } catch (Exception e) { e.printStackTrace(); }
    return myFile;
} }
```

上面这段代码在 MyCode\MySample198\app\src\main\java\com\bin\luo\mysample\MainActivity.java 文件中。在这段代码中，myConnection =（HttpURLConnection）myUrl.openConnection()用于根据安装包的下载地址创建网络连接。myInput = myConnection.getInputStream()用于根据网络连接创建输入流。myOutput = new FileOutputStream（myFile）用于根据存储文件创建输出流。myData = myInput.read(myBuffer)用于将输入流数据读取到缓冲区。myOutput.write(myBuffer, 0，myData)用于将缓冲区数据写到输出流(即存储文件)。此外，访问网络和存储卡需要在AndroidManifest.xml文件中添加< uses-permission android: name = "android.permission.WRITE_EXTERNAL_STORAGE"/>和< uses-permission android: name = "android.permission.INTERNET"/>权限。此实例的完整项目在MyCode\MySample198文件夹中。

160 仅在WiFi时执行DownloadManager下载

此实例主要通过设置DownloadManager的内部类Request的setAllowedNetworkTypes()方法的参数为DownloadManager.Request.NETWORK_WIFI,从而实现仅在WiFi网络连接的情况下才执行下载网络文件的操作。当实例运行之后，在"下载网址："输入框中输入网络地址，如"http://gdown.baidu.com/data/wisegame/0904344dee4a2d92/QQ_718.apk"，如图160-1的左图所示；然后单击"执行下载"按钮，如果当前是WiFi连接，将执行下载操作，下载进度将显示在通知栏上，如图160-1的右图所示；如果当前是移动网络的数据连接，则没有反应。

图 160-1

主要代码如下：

```
public void OnClickBtn1(View v){                                    //响应单击"执行下载"按钮
    DownloadManager myDownloadManager =
        (DownloadManager)getSystemService(DOWNLOAD_SERVICE);        //获取下载管理器
    String myUrl = myEditText.getText().toString();
    DownloadManager.Request myRequest = new DownloadManager.Request(
                    Uri.parse(myUrl));                              //向指定 Url 地址发送下载请求
    myRequest.setNotificationVisibility(DownloadManager.Request.
                    VISIBILITY_VISIBLE);                            //在通知栏上显示当前下载进度
    //仅在 WiFi 连接的情况下执行下载操作
    myRequest.setAllowedNetworkTypes(DownloadManager.Request.NETWORK_WIFI);
    //仅在 NETWORK_MOBILE 连接的情况下执行下载操作
    //myRequest.setAllowedNetworkTypes(DownloadManager.Request.NETWORK_MOBILE);
    myDownloadManager.enqueue(myRequest);                           //将请求加入下载队列
}
```

上面这段代码在 MyCode\MySample894\app\src\main\java\com\bin\luo\mysample\MainActivity.java 文件中。此外，访问网络和 SD 卡需要在 AndroidManifest.xml 文件中添加< uses-permission android:name = "android.permission.INTERNET"/>权限和< uses-permission android:name = "android.permission.WRITE_EXTERNAL_STORAGE"/>权限。此实例的完整项目在 MyCode\MySample894 文件夹中。

161　使用 AsyncTask 实现异步访问网络图像

此实例主要通过使用 AsyncTask 为基类创建异步任务类，从而实现以异步方式访问网络图像。当实例运行之后，在"网址："输入框中输入网络图像地址，如"https://p1.ssl.qhmsg.com/dr/220__/

t01cd98a9eadc0a3a0a.jpg",然后单击"显示指定网址的图像"按钮,将以异步方式加载并显示此网络图像,效果分别如图 161-1 的左图和右图所示。

图 161-1

主要代码如下:

```java
public void onClickmyBtn1(View v) {              //响应单击"显示指定网址的图像"按钮
    String myPath = myEditText.getText().toString();
    //使用 AsyncTask 异步消息处理任务和 Handler 请求网络图像
    new MyTask(myPath, myHandler).execute();
}
Handler myHandler = new Handler(){
    @Override
    public void handleMessage(Message msg) {
        super.handleMessage(msg);
        switch (msg.what){
            case 1:
                Bitmap myBitmap = (Bitmap) msg.obj;
                myImageView.setImageBitmap(myBitmap);
                break;
        }}};
public class MyTask extends AsyncTask<Void,Void,Bitmap> {
    String myUrl;
    Handler myHandler;
    public MyTask(String url, Handler handler) {
        myUrl = url;
        this.myHandler = handler;
    }
    @Override
    protected void onPreExecute() { super.onPreExecute(); }
    @Override
```

```
    protected Bitmap doInBackground(Void... params) {
     Bitmap myBitmap = null;
     try {
      URL newUrl = new URL(myUrl);
      HttpURLConnection myHttpURLConnection =
                         (HttpURLConnection) newUrl.openConnection();
      InputStream myInputStream = myHttpURLConnection.getInputStream();
      myBitmap = BitmapFactory.decodeStream(myInputStream);
     } catch (IOException e) { e.printStackTrace(); }
     return myBitmap;
    }
    @Override
    protected void onPostExecute(Bitmap bitmap) {
     super.onPostExecute(bitmap);
     Message myMessage = myHandler.obtainMessage();
     myMessage.what = 1;
     myMessage.obj = bitmap;
     myHandler.sendMessage(myMessage);
    }}
```

上面这段代码在 MyCode\MySample427\app\src\main\java\com\bin\luo\mysample\MainActivity.java 文件中。在这段代码中，AsyncTask 是 Android 提供的轻量级异步任务类，可以直接继承 AsyncTask 在自定义类中实现异步操作，并提供接口反馈当前异步执行的程度。一个异步任务的执行一般包括以下几个步骤。

（1）execute(Params... params)，执行一个异步任务，需要在代码中调用此方法，触发异步任务的执行。

（2）onPreExecute()，在 execute(Params... params)被调用后立即执行，一般用来在执行后台任务前对 UI 做一些标记。

（3）doInBackground(Params... params)，在 onPreExecute()完成后立即执行，用于执行较为费时的操作，此方法将接收输入参数和返回计算结果。在执行过程中可以调用 publishProgress(Progress...values)来更新进度信息。

（4）onProgressUpdate(Progress... values)，在调用 publishProgress(Progress...values)时，此方法被执行，直接将进度信息更新到 UI 组件上。

（5）onPostExecute(Result result)，当后台操作结束时，此方法将会被调用，结果将作为参数传递到此方法中，直接将结果显示到 UI 组件上。

此外，访问网络需要在 AndroidManifest.xml 文件中添加< uses-permission android：name = "android.permission.INTERNET"/>权限。此实例的完整项目在 MyCode\MySample427 文件夹中。

162 在进度条上显示 AsyncTask 的下载进度

此实例主要通过重写异步任务 AsyncTask 类的 doInBackground()方法执行网络下载操作，从而实现使用异步方式下载网络文件至手机存储卡，并在进度条上显示下载进度。当实例运行之后，在"下载网址："输入框中输入网络地址，如"http://gdown.baidu.com/data/wisegame/0904344dee4a2d92/QQ_718.apk"，然后单击"执行 AsyncTask 下载"按钮，将执行异步下载操作，并在进度条上显示下载进度，效果分别如图 162-1 的左图和右图所示；由于实例下载的是应用安装包，因此在下载完成之后将自动执行安装操作。

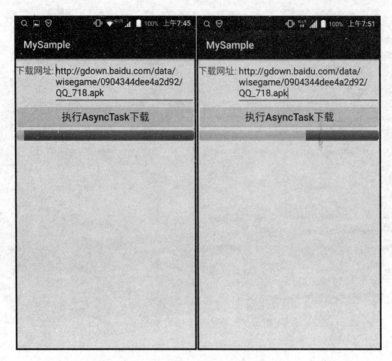

图 162-1

主要代码如下：

```
public void onClickBtn1(View v) {                        //响应单击"执行AsyncTask下载"按钮
    myProgressBar.setVisibility(View.VISIBLE);
    String myHttpUrl = myEditText.getText().toString();
    MyAsyncTask myAsyncTask = new MyAsyncTask(this,
                              myProgressBar,myHttpUrl);   //初始化异步任务对象
    myAsyncTask.execute();                                //执行异步任务
}
```

上面这段代码在 MyCode\MySample877\app\src\main\java\com\bin\luo\mysample\MainActivity.java 文件中。在这段代码中，myAsyncTask = new MyAsyncTask（this, myProgressBar,myHttpUrl）表示使用自定义类 MyAsyncTask 创建异步下载任务。关于 MyAsyncTask 类的详细内容请参考 MyCode\MySample877\app\src\main\java\com\bin\luo\mysample\MyAsyncTask.java 文件。此外，访问网络和 SD 卡需要在 AndroidManifest.xml 中添加 <uses-permission android:name="android.permission.INTERNET"/>和<uses-permission android:name="android.permission.WRITE_EXTERNAL_STORAGE"/>权限。此实例的完整项目在 MyCode\MySample877 文件夹中。

163 以数据流形式加载并显示指定网址的图像

此实例主要通过使用 HttpURLConnection 根据指定网址读取网络图像数据，并将图像数据以二进制形式输出至输出流，然后调用 BitmapFactory 对象的 decodeByteArray()方法将二进制数据解析为 Bitmap，实现以数据流形式动态加载并显示网络图像。当实例运行之后，在"图像网址："输入框中输入代表图像的网络地址，如"https://p1.ssl.qhmsg.com/dr/220__/t013aab5e6d0218b6e2.jpg"，然后单击"加载指定网址的图像"按钮,将在下面显示该网址代表的图像,效果分别如图163-1的左图和右图所示。

图 163-1

主要代码如下:

```
public class MainActivity extends Activity {
 EditText myEditText;
 Handler myHandler = new Handler() {
  @Override
  public void handleMessage(Message msg) {
   if (msg.what == 1) {
    myImageView.setImageBitmap((Bitmap) msg.obj);
    Toast.makeText(MainActivity.this, "加载图像成功!",
                                  Toast.LENGTH_SHORT).show();
   } else if (msg.what == 2) {
    Toast.makeText(MainActivity.this, "加载图像失败!",
                                  Toast.LENGTH_SHORT).show();
} } };
ImageView myImageView;
 @Override
 protected void onCreate(Bundle savedInstanceState) {
  super.onCreate(savedInstanceState);
  setContentView(R.layout.activity_main);
  myEditText = (EditText) findViewById(R.id.myEditText);
  myImageView = (ImageView) findViewById(R.id.myImageView);
 }
 //响应单击"加载指定网址的图像"按钮
 public void onClickBtn1(View v) throws IOException {
  new Thread() {
   @Override
```

```java
public void run() {
    try {
        String myImageUrl = myEditText.getText().toString();
        URL myUrl = new URL(myImageUrl);          //将 URL 字符串转换为 URL 链接对象
        HttpURLConnection myConnection =
                            (HttpURLConnection) myUrl.openConnection();
        myConnection.setRequestMethod("GET");   //以 GET 方式请求图像
        myConnection.setReadTimeout(5 * 1000);//设置请求有效时长
        //通过输入流获得图像数据
        InputStream myInputStream = myConnection.getInputStream();
        byte[] myBuffer = new byte[1024];
        int myLength;
        ByteArrayOutputStream myByteArrayOutputStream =
                                        new ByteArrayOutputStream();
        //输出图像数据
        while ((myLength = myInputStream.read(myBuffer)) != -1) {
            myByteArrayOutputStream.write(myBuffer, 0, myLength);
        }
        myByteArrayOutputStream.close();          //关闭输出流
        //从输出流取出二进制数据
        byte[] myImageData = myByteArrayOutputStream.toByteArray();
        //通过输出流获取图像二进制数据并进行解析
        Bitmap myBitmap = BitmapFactory.decodeByteArray(myImageData,
                                            0, myImageData.length);
        Message myMessage = new Message();
        myMessage.what = 1;
        myMessage.obj = myBitmap;
        myHandler.sendMessage(myMessage);         //通过 Handler 在 UI 线程显示图像
    } catch (Exception e) {                       //向 Handler 发送消息,提示用户图像加载失败
        myHandler.sendEmptyMessage(2);
    }
} }.start();
} }
```

上面这段代码在 MyCode\MySample876\app\src\main\java\com\bin\luo\mysample\MainActivity.java 文件中。在这段代码中,myBitmap = BitmapFactory.decodeByteArray(myImageData,0,myImageData.length)用于将接收的网络图像数据流解析为 Bitmap。decodeByteArray()方法的语法声明如下:

```
public static Bitmap decodeByteArray(byte[] data, int offset, int length)
```

其中,参数 byte[] data 代表图像数据的字节数组。参数 int offset 表示图像数据偏移量,即解码器从哪儿开始解析。参数 int length 代表图像数据的字节长度。

该方法的返回值是解码成功之后的 Bitmap,或者如果图像数据不能被解码则返回值为空。需要注意的是,访问网络需要在 AndroidManifest.xml 文件中添加< uses-permission android:name="android.permission.INTERNET"/>权限。此实例的完整项目在 MyCode\MySample876 文件夹中。

164 使用正则表达式校验在输入框的输入内容

此实例主要通过在重写 TextWatcher 的 afterTextChanged()方法中使用正则表达式校验输入的内容,从而实现及时阻止用户错误输入的效果。当实例运行之后,如果在"账户名称:"输入框中输入

中英文字符,则一切正常,如图 164-1 的左图所示;如果试图在输入框中输入数字、标点符号等非中英文字符,将弹出 Toast 提示"账户名称仅支持中英文!",并且自动删除非中英文字符,如图 164-1 的右图所示。

图 164-1

主要代码如下:

```
public class MainActivity extends Activity {
  EditText myEditText;
  @Override
  protected void onCreate(Bundle savedInstanceState) {
    super.onCreate(savedInstanceState);
    setContentView(R.layout.activity_main);
    myEditText = (EditText) findViewById(R.id.myEditText);
    myEditText.addTextChangedListener(new TextWatcher(){
      //只支持中英文的正则表达式
      String myRegular = "^[\\u4e00-\\u9fa5a-zA-Z]+$";
      String myTemp = "";
      @Override
      //保存符合要求的文本
      public void beforeTextChanged(CharSequence s,int start,int count,int after){
        myTemp = s.toString();
      }
      @Override
    public void onTextChanged(CharSequence s,int start,int before,int count){ }
      @Override
      //检查文本内容是否符合正则表达式规范
      public void afterTextChanged(Editable s){
        //删除不符合正则表达式的部分
        if(!s.toString().matches(myRegular)&&!"".equals(s.toString())){
          //还原之前保存的正确文本
```

```
            myEditText.setText(myTemp);
            //设置光标位置
            myEditText.setSelection(myEditText.getText().toString().length());
            Toast.makeText(MainActivity.this,
                    "账户名称仅支持中英文!",Toast.LENGTH_SHORT).show();
}}});}}
```

上面这段代码在 MyCode\MySample650\app\src\main\java\com\bin\luo\mysample\MainActivity.java 文件中。此实例的完整项目在 MyCode\MySample650 文件夹中。

165 使用随机数生成验证码图像并提交验证

此实例主要通过使用 Random 类的 nextInt() 方法生成随机数,并将随机数使用 StringBuilder 处理之后绘制成验证码图像,实现根据验证码图像显示的字符内容进行验证的功能。当实例运行之后,自动生成一幅包括验证字符的图像,在"请输入验证码(区分大小写):"输入框中输入在图像中的字符,如果输入正确,单击"开始验证"按钮之后,在弹出的 Toast 中将提示"验证成功!",如图 165-1 的右图所示;如果输入错误,单击"开始验证"按钮之后,在弹出的 Toast 中将提示"验证码输入错误,请重试!";如果看不清图像中的验证码字符,则在单击验证码图像之后,将生成新的不同的验证码。

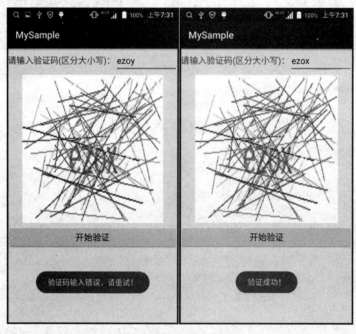

图 165-1

主要代码如下:

```
void onClickButton(View view) {                              //响应单击"开始验证"按钮
    if (myEditText.getText().toString().equals(MyUtils.MyCurrentRandomString)){
        Toast.makeText(MainActivity.this, "验证成功!", Toast.LENGTH_SHORT).show();
        //finish();
    } else {
        Toast.makeText(MainActivity.this,
                "验证码输入错误,请重试!", Toast.LENGTH_SHORT).show();
```

```
    }
}
    void onClickImageView(View view) {         //响应单击验证码图像,即重新生成验证码
        myImageView.setImageBitmap(MyUtils.GetRandomValidatorImage(300, 150));
    }
```

上面这段代码在 MyCode \ MySample732 \ app \ src \ main \ java \ com \ bin \ luo \ mysample \ MainActivity.java 文件中。在这段代码中,MyUtils.GetRandomValidatorImage(300,150)用于生成验证码图像,MyUtils 是自定义工具类,该类的主要内容如下:

```
public class MyUtils {
    static String MyCurrentRandomString = "";              //以字符串形式保存验证码
    private static final String RANDOM_CHARS =
            "123456789QWERTYUIOPASDFGHJKLZXCVBNMqwertyuiopasdfghjklzxcvbnm";
    public static String GetRandomValidatorString() {  //产生随机字符串
        StringBuilder myBuilder = new StringBuilder();
        for (int i = 0; i < 4; i++) {
            myBuilder.append(RANDOM_CHARS.charAt(
                            new Random().nextInt(RANDOM_CHARS.length())));
        }
        return myBuilder.toString();
    }
    //生成随机验证码图像
    public static Bitmap GetRandomValidatorImage(int width, int height) {
        Bitmap myBitmap = Bitmap.createBitmap(width, height, Bitmap.Config.ARGB_8888);
        Canvas myCanvas = new Canvas(myBitmap);
        String myRandomString = GetRandomValidatorString();
        MyCurrentRandomString = myRandomString;
        myCanvas.drawColor(Color.WHITE);
        Paint myPaint = new Paint();
        myPaint.setTextSize(50);
        //绘制验证码
        for (int i = 0; i < myRandomString.length(); i++) {
            int color = Color.rgb(new Random().nextInt(256),
                            new Random().nextInt(256), new Random().nextInt(256));
            myPaint.setColor(color);
            //设置随机字符样式
            myPaint.setFakeBoldText(new Random().nextBoolean());
            myCanvas.drawText(myRandomString, width/2 - 25 * myRandomString.length()/2,
                            height / 2 + 25 * myRandomString.length() / 4, myPaint);
        }
        //绘制噪声(干扰)线条
        for (int i = 0; i < 75; i++) {
            int myColor = Color.rgb(new Random().nextInt(256),
                            new Random().nextInt(256), new Random().nextInt(256));
            int startX = new Random().nextInt(width);
            int startY = new Random().nextInt(height);
            int stopX = new Random().nextInt(width);
            int stopY = new Random().nextInt(height);
            myPaint.setStrokeWidth(1);
            myPaint.setColor(myColor);
            myCanvas.drawLine(startX, startY, stopX, stopY, myPaint);
```

```
        }
        myCanvas.save(Canvas.ALL_SAVE_FLAG);
        myCanvas.restore();
        return myBitmap;
    }}
```

上面这段代码在 MyCode \ MySample732 \ app \ src \ main \ java \ com \ bin \ luo \ mysample \ MyUtils.java 文件中。此实例的完整项目在 MyCode\MySample732 文件夹中。

166　将涂鸦内容在存储卡上保存为图像文件

此实例主要实现了在 ImageView 控件上调用 setOnTouchListener(this)方法监听涂鸦行为,然后使用 Bitmap 的 compress()方法将涂鸦结果作为新图像保存在存储卡上。当实例运行之后,单击"重新绘制"按钮,即可以在 ImageView 控件上进行涂鸦,如图 166-1 的左图所示;单击"保存图像"按钮,则经过涂鸦的图像将被保存在存储卡上(即 1516431917155.jpg、1516432005994.jpg 等图像文件),如图 166-1 的右图所示。

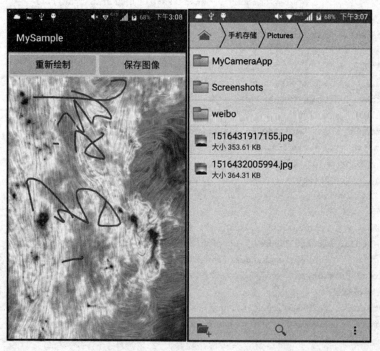

图　166-1

主要代码如下:

```
public class MainActivity extends Activity implements View.OnTouchListener{
    float myLastX = 0, myLastY = 0, myX = 0, myY = 0;
    Canvas myCanvas;
    Bitmap myBitmap, myDrawBitmap;
    Paint myPaint;
    ImageView myImageView;
    @Override
    protected void onCreate(Bundle savedInstanceState) {
```

```java
    super.onCreate(savedInstanceState);
    setContentView(R.layout.activity_main);
    myImageView = (ImageView) findViewById(R.id.myImageView);
}
public void onClickmyBtn1(View v) {                                    //响应单击"重新绘制"按钮
    myBitmap = BitmapFactory.decodeResource(getResources(), R.mipmap.myimage1);
    myDrawBitmap = Bitmap.createBitmap(myBitmap.getWidth(),
                   myBitmap.getHeight(), myBitmap.getConfig());//创建新图像
    myCanvas = new Canvas(myDrawBitmap);                               //初始化画板
    myCanvas.drawBitmap(myBitmap, new Matrix(), myPaint);              //在画板中绘制图像
    myImageView.setImageBitmap(myDrawBitmap);
    //设置触摸监听,即在ImageView控件上实施涂鸦
    myImageView.setOnTouchListener(this);
    myPaint = new Paint();
    myPaint.setColor(Color.RED);                                       //设置画笔颜色
    myPaint.setStrokeWidth(10);                                        //设置画笔粗细
    myPaint.setXfermode(new PorterDuffXfermode(PorterDuff.Mode.CLEAR));
    myCanvas.drawPaint(myPaint);
    myPaint.setXfermode(new PorterDuffXfermode(PorterDuff.Mode.SRC));
    myCanvas.drawBitmap(myBitmap, new Matrix(), myPaint);
}
public void onClickmyBtn2(View v) {                                    //响应单击"保存图像"按钮
    if (myDrawBitmap != null) {
        try {
            //MediaStore.Images.Media.EXTERNAL_CONTENT_URI 表示相册目录
            //ContentValues()用于产生图像文件名
            Uri myImageUri = getContentResolver().insert(
                MediaStore.Images.Media.EXTERNAL_CONTENT_URI, new ContentValues());
            OutputStream myOutStream =
                            getContentResolver().openOutputStream(myImageUri);
            //以JPG格式压缩图像并保存到存储卡
            myDrawBitmap.compress(Bitmap.CompressFormat.JPEG, 90, myOutStream);
            Toast.makeText(this, "图像已保存至存储卡!", Toast.LENGTH_SHORT).show();
        } catch (Exception e) {
            e.printStackTrace();
}}}
@Override
public boolean onTouch(View view, MotionEvent event) {
    int action = event.getAction();
    switch (action) {
    case MotionEvent.ACTION_DOWN:
        myLastX = event.getX();                                        //获取手指按下位置处的X坐标
        myLastY = event.getY();                                        //获取手指按下位置处的Y坐标
        break;
    case MotionEvent.ACTION_UP:
        myX = event.getX();
        myY = event.getY();
        myCanvas.drawLine(myLastX, myLastY, myX, myY, myPaint);        //绘制涂鸦线条
        myImageView.invalidate();
        break;
    case MotionEvent.ACTION_MOVE:
        myX = event.getX();
        myY = event.getY();
```

```
            myCanvas.drawLine(myLastX, myLastY, myX, myY, myPaint);
            myImageView.invalidate();            //将涂鸦后的图像同步至 ImageView 中
            myLastX = myX;
            myLastY = myY;
            break;
        default:
            break;
    }
    return true;
} }
```

上面这段代码在 MyCode \ MySample664 \ app \ src \ main \ java \ com \ bin \ luo \ mysample \ MainActivity.java 文件中。此外，在 SD 卡保存文件需要在 AndroidManifest.xml 文件中添加< uses-permission android:name="android.permission.WRITE_EXTERNAL_STORAGE"/>权限。此实例的完整项目在 MyCode\MySample664 文件夹中。

167　使用 BitmapFactory 读取 SD 卡图像文件

此实例主要通过使用 BitmapFactory 的 decodeStream()方法，实现以文件流方式读取并显示 SD 卡的图像文件的功能。当实例运行之后，在"图像文件："输入框中输入图像文件的路径，然后单击"显示图像"按钮，将在下面显示该文件包含的图像，效果分别如图 167-1 的左图和右图所示。

图　167-1

主要代码如下：

```
public void onClickmyBtn1(View v) {            //响应单击"显示图像"按钮
    try {
        FileInputStream myFileInputStream =
```

```
                new FileInputStream(myEditText.getText().toString());
        BitmapFactory.Options myOptions = new BitmapFactory.Options();
        Bitmap myBitmap =
                BitmapFactory.decodeStream(myFileInputStream,null,myOptions);
        myImageView.setImageBitmap(myBitmap);
    } catch (FileNotFoundException e) { e.printStackTrace(); }
}
```

上面这段代码在 MyCode\MySample221\app\src\main\java\com\bin\luo\mysample\MainActivity.java 文件中。在这段代码中，myFileInputStream = new FileInputStream(myEditText.getText().toString())用于根据指定的文件路径名称创建文件输入流。myOptions = new BitmapFactory.Options()用于创建默认的图像选项。myBitmap = BitmapFactory.decodeStream(myFileInputStream,null,myOptions)用于根据文件输入流和图像选项创建图像。此外，读取在 SD 卡上的文件需要在 AndroidManifest.xml 文件中添加< uses-permission android:name="android.permission.READ_EXTERNAL_STORAGE"/>权限。此实例的完整项目在 MyCode\MySample221 文件夹中。

168 在选择照片窗口中选择图像文件并显示

此实例主要通过使用 MediaStore.Images.Media.EXTERNAL_CONTENT_URI 构造显示所有图像文件的选择照片窗口，从而实现允许用户选择图像文件并显示。当实例运行之后，单击"选择并显示图像文件"按钮，将弹出"选择照片"窗口，如图 168-1 的左图所示。在该窗口中任意选择一个图像文件，将显示图像的预览效果，再单击该预览效果图，将在当前实例中显示此图像文件，如图 168-1 的右图所示。

图 168-1

主要代码如下：

```java
public void onClickmyBtn1(View v) {                    //响应单击"选择并显示图像文件"按钮
    myTextView.setText("显示图像文件全路径");
    Intent myIntent = new Intent(Intent.ACTION_PICK, null);
    //设置在窗口中显示的文件为图像文件
    myIntent.setDataAndType(MediaStore.Images.Media.EXTERNAL_CONTENT_URI, null);
    startActivityForResult(myIntent, 1);
}
@Override
protected void onActivityResult(int myRequestCode,
                int resultCode, Intent myIntent) {    //显示选择的图像文件
    super.onActivityResult(myRequestCode, resultCode, myIntent);
    try {
        switch (myRequestCode) {
        case 1:
            if (resultCode == Activity.RESULT_OK) {
                Uri myUri = myIntent.getData();
                String[] myPathColumns = {MediaStore.Images.Media.DATA};
                Cursor myCursor = getContentResolver().query(myUri,
                                        myPathColumns, null, null, null);
                myCursor.moveToFirst();
                int myIndex = myCursor.getColumnIndex(myPathColumns[0]);
                //获取选择的图像文件
                String myPath = myCursor.getString(myIndex);
                Bitmap myBitmap = BitmapFactory.decodeFile(myPath);
                myImageView.setImageBitmap(myBitmap);
                myTextView.setText(myPath);
            }
            break;
        } } catch (Exception e) { e.printStackTrace(); }
}
```

上面这段代码在 MyCode\MySample381\app\src\main\java\com\bin\luo\mysample\MainActivity.java 文件中。在这段代码中，myIntent = new Intent(Intent.ACTION_PICK，null)用于创建可进行选择的 Intent。myIntent.setDataAndType(MediaStore.Images.Media.EXTERNAL_CONTENT_URI，null)表示此 Intent 的数据类型为图像文件。onActivityResult()用于处理在"选择照片"窗口中选择图像文件之后返回的信息，如图像文件全路径信息等。此外，读取在 SD 卡上的图像文件需要在 AndroidManifest.xml 文件中添加 < uses-permission android：name = " android.permission.READ_EXTERNAL_STORAGE"/> 权限。此实例的完整项目在 MyCode\MySample381 文件夹中。

169 使用 CookieManager 读取和保存数据

此实例主要通过使用 CookieManager 的 setCookie()和 getCookie()方法，实现在存储卡中保存和读取数据。当实例运行之后，在"Cookie 数据内容："输入框中输入自定义数据，如"这是我的 Cookie 数据"，单击"保存数据"按钮，则输入的自定义数据将被保存在存储卡上，如图 169-1 的左图所示。关闭应用，然后重新运行此应用，单击"读取数据"按钮，则将在弹出的 Toast 中显示此前保存的自定义数据，如图 169-1 的右图所示。

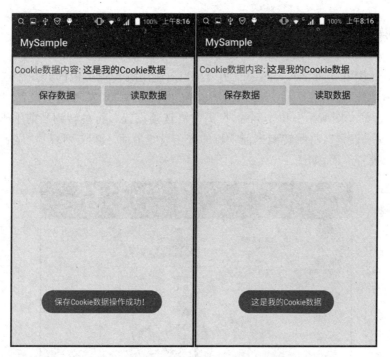

图 169-1

主要代码如下:

```
public void onClickmyBtn1(View v) {          //响应单击"保存数据"按钮
  try {
  CookieManager myCookieManager = CookieManager.getInstance();
  myCookieManager.setCookie("myCookie", myEditText.getText().toString());
  myCookieManager.flush();
  Toast.makeText(getApplicationContext(),
           "保存Cookie数据操作成功!",Toast.LENGTH_SHORT).show();
  } catch (Exception e) {
  Toast.makeText(getApplicationContext(),
          e.getMessage().toString(),Toast.LENGTH_SHORT).show();
} }
public void onClickmyBtn2(View v) {          //响应单击"读取数据"按钮
  try {
  CookieManager myCookieManager = CookieManager.getInstance();
  String myInfo = myCookieManager.getCookie("myCookie");
  Toast.makeText(getApplicationContext(),myInfo,Toast.LENGTH_SHORT).show();
  } catch (Exception e) {
  Toast.makeText(getApplicationContext(),
          e.getMessage().toString(),Toast.LENGTH_SHORT).show();
  }
}
```

上面这段代码在 MyCode\MySample460\app\src\main\java\com\bin\luo\mysample\MainActivity.java 文件中。在这段代码中，myCookieManager＝CookieManager.getInstance()用于获取 CookieManager 的实例。myCookieManager.setCookie("myCookie"，myEditText.getText().toString())用于将自定义数据保存在 CookieManager 的 myCookie 中。myInfo＝myCookieManager.getCookie("myCookie")用于从 CookieManager 的 myCookie 键中读取自定义数据。此实例的完整项目在 MyCode\MySample460 文件夹中。

170　使用 PreferenceScreen 跳转到显示设置

此实例主要通过使用 PreferenceScreen，实现从当前应用跳转到手机的开发者选项、安全、显示等设置界面。当实例运行之后，将显示 PreferenceScreen 的列表项"开发者选项设置""安全设置""显示设置"等，如图 170-1 的左图所示。单击列表项"开发者选项设置"，将跳转到当前手机的开发者选项界面。单击列表项"安全设置"，将跳转到当前手机的安全设置界面。单击列表项"显示设置"，则将跳转到当前手机的显示设置界面，如图 170-1 的右图所示。

图　170-1

主要代码如下：

```java
public class MainActivity extends PreferenceActivity {
 @Override
 protected void onCreate(Bundle savedInstanceState) {
  super.onCreate(savedInstanceState);
  //setContentView(R.layout.activity_main);
  addPreferencesFromResource(R.xml.preferences);
 }}
```

上面这段代码在 MyCode\MySample322\app\src\main\java\com\bin\luo\mysample\MainActivity.java 文件中。在这段代码中，addPreferencesFromResource(R.xml.preferences) 表示 MainActivity 加载的视图是 xml 目录下的 preferences 文件，而不是默认的 setContentView(R.layout.activity_main)。另外，MainActivity 的基类是 PreferenceActivity，而不是默认的 Activity。preferences 文件的主要内容如下：

```xml
<?xml version = "1.0" encoding = "UTF-8"?>
<PreferenceScreen xmlns:android = "http://schemas.android.com/apk/res/android">
<PreferenceScreen android:key = "DevelopmentSettings"
            android:title = "开发者选项设置"
            android:summary = "单击即可跳转到手机的开发者选项界面" >
```

```xml
<intent android:action="android.intent.action.MAIN"
    android:targetPackage="com.android.settings"
    android:targetClass="com.android.settings.DevelopmentSettings" />
</PreferenceScreen>
<PreferenceScreen android:key="SecuritySettings"
        android:title="安全设置"
        android:summary="单击即可跳转到手机的安全设置界面" >
    <intent android:action="android.intent.action.MAIN"
        android:targetPackage="com.android.settings"
        android:targetClass="com.android.settings.SecuritySettings" />
</PreferenceScreen>
<PreferenceScreen android:key="DisplaySettings"
        android:title="显示设置"
        android:summary="单击即可跳转到手机的显示设置界面" >
    <intent android:action="android.intent.action.MAIN"
        android:targetPackage="com.android.settings"
        android:targetClass="com.android.settings.DisplaySettings" />
</PreferenceScreen></PreferenceScreen>
```

上面这段代码在 MyCode\MySample322\app\src\main\res\xml\preferences.xml 文件中。此实例的完整项目在 MyCode\MySample322 文件夹中。

171 使用 PreferenceFragment 实现页面切换

此实例主要通过以 PreferenceFragment 类为基类创建新类，从而实现以 Fragment 页面风格切换包含 CheckBoxPreference 的界面。当实例运行之后，单击"页面风格切换的 Preference"列表项，将在下一个 Fragment 页面中显示多个 CheckBoxPreference，任意选择其中的两个选项，将在弹出的 Toast 中显示两部电影片名，如图 171-1 的左图所示。在关闭应用，再重新启动此应用之后，也将在弹出的 Toast 中显示此前选择的两部电影片名，如图 171-1 的右图所示。

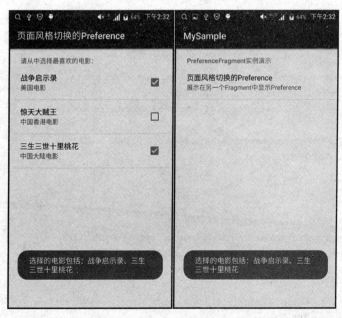

图 171-1

主要代码如下：

```java
public class MainActivity extends Activity {
    @Override
    protected void onCreate(Bundle savedInstanceState) {
        super.onCreate(savedInstanceState);
        getFragmentManager().beginTransaction().replace(
                android.R.id.content, new MyPreferenceFragment()).commit();
    }
    public static class MyPreferenceFragment extends PreferenceFragment {
        CheckBoxPreference myCheckBoxPreference1;
        CheckBoxPreference myCheckBoxPreference2, myCheckBoxPreference3;
        boolean bCheck1, bCheck2, bCheck3;
        String myInfo = "选择的电影包括：";
        @Override
        public void onCreate(Bundle savedInstanceState) {
            super.onCreate(savedInstanceState);
            addPreferencesFromResource(R.xml.preferences);
            myCheckBoxPreference1 =
                    (CheckBoxPreference) findPreference("myCheckBoxPreference1");
            myCheckBoxPreference1.setOnPreferenceClickListener(
                            new Preference.OnPreferenceClickListener(){
                @Override
                public boolean onPreferenceClick(Preference preference) {
                    //这里可以监听到这个CheckBox的单击事件
                    return true;
                } });
            myCheckBoxPreference1.setOnPreferenceChangeListener(
                            new Preference.OnPreferenceChangeListener(){
                @Override
                public boolean onPreferenceChange(Preference arg0, Object newValue){
                    //这里可以监听到CheckBox值是否改变了，并且可以获取新改变的值
                    bCheck1 = (Boolean)newValue;
                    CheckSet();
                    return true;
                } });
            myCheckBoxPreference2 =
                    (CheckBoxPreference) findPreference("myCheckBoxPreference2");
            myCheckBoxPreference2.setOnPreferenceChangeListener(
                            new Preference.OnPreferenceChangeListener(){
                @Override
                public boolean onPreferenceChange(Preference arg0, Object newValue){
                    bCheck2 = (Boolean)newValue;
                    CheckSet();
                    return true;
                } });
            myCheckBoxPreference3 =
                    (CheckBoxPreference) findPreference("myCheckBoxPreference3");
            myCheckBoxPreference3.setOnPreferenceChangeListener(
                            new Preference.OnPreferenceChangeListener(){
                @Override
                public boolean onPreferenceChange(Preference arg0, Object newValue){
                    bCheck3 = (Boolean)newValue;
```

```
        CheckSet();
        return true;
      } });
    bCheck1 = this.myCheckBoxPreference1.isChecked();
    bCheck2 = this.myCheckBoxPreference2.isChecked();
    bCheck3 = this.myCheckBoxPreference3.isChecked();
    CheckSet();
  }
  void CheckSet() {
    String myInfo = "选择的电影包括: ";
    if(bCheck1){ myInfo += "战争启示录、"; }
    if(bCheck2){ myInfo += "惊天大贼王、"; }
    if(bCheck3){ myInfo += "三生三世十里桃花"; }
    Toast.makeText(getActivity(),myInfo, Toast.LENGTH_LONG).show();
} } }
```

上面这段代码在 MyCode\MySample327\app\src\main\java\com\bin\luo\mysample\MainActivity.java 文件中。在这段代码中，public static class MyPreferenceFragment extends PreferenceFragment 用于以 PreferenceFragment 类为基类创建新类 MyPreferenceFragment，此新类的实例将在 getFragmentManager().beginTransaction().replace(android.R.id.content, new MyPreferenceFragment()).commit() 中实现页面切换功能。addPreferencesFromResource(R.xml.preferences) 表示 MyPreferenceFragment 加载的视图是 xml 目录下的 preferences 布局文件。preferences 布局文件的主要内容如下：

```xml
<?xml version = "1.0" encoding = "UTF-8"?>
<PreferenceScreen xmlns:android = "http://schemas.android.com/apk/res/android">
  <PreferenceCategory android:title = "PreferenceFragment 实例演示">
    <PreferenceScreen
        android:key = "myPreferenceFragment"
        android:summary = "展示在另一个 Fragment 中显示 Preference"
        android:title = "页面风格切换的 Preference">
    <!-- 可以在这里放置更多的 Preference,将在下一个 Fragment 呈现出来 -->
    <PreferenceCategory android:title = "请从中选择最喜欢的电影:">
      <CheckBoxPreference android:key = "myCheckBoxPreference1"
              android:title = "战争启示录"
              android:summary = "美国电影"></CheckBoxPreference>
      <CheckBoxPreference android:key = "myCheckBoxPreference2"
              android:title = "惊天大贼王"
              android:summary = "中国香港电影"></CheckBoxPreference>
      <CheckBoxPreference android:key = "myCheckBoxPreference3"
              android:title = "三生三世十里桃花"
              android:summary = "中国大陆电影"></CheckBoxPreference>
    </PreferenceCategory></PreferenceScreen></PreferenceCategory>
</PreferenceScreen>
```

上面这段代码在 MyCode\MySample327\app\src\main\res\xml\preferences.xml 文件中。需要说明的是，在此实例中，没有任何读写存储卡的代码，所有读写存储卡的功能均由 MyPreferenceFragment 和 CheckBoxPreference 自动完成。此实例的完整项目在 MyCode\MySample327 文件夹中。

172 使用 EditTextPreference 实现文本读写

此实例主要通过使用 EditTextPreference，实现自动在存储卡上读写用户编辑的简短文本。当实例运行之后，单击"工作简历"列表项，将显示"工作简历"对话框，在该对话框的输入框中输入简要的文本，如"1991-2000,重庆安定造纸总厂"，再单击"确定"按钮，如图 172-1 的左图所示，输入的文本将被自动保存在存储卡上。然后退出此应用，再重新启动此应用，将在弹出的 Toast 中显示刚才输入的文本内容，如图 172-1 的右图所示。注意：在对话框中单击"取消"按钮将不会保存输入的文本内容。

图 172-1

主要代码如下：

```
public class MainActivity extends PreferenceActivity {
    EditTextPreference myResume;
    @Override
    protected void onCreate(Bundle savedInstanceState) {
        super.onCreate(savedInstanceState);
        //setContentView(R.layout.activity_main);
        addPreferencesFromResource(R.xml.preferences);
        myResume = (EditTextPreference) findPreference("myResume");
        Toast.makeText(MainActivity.this,
                myResume.getText(), Toast.LENGTH_LONG).show();
    }
}
```

上面这段代码在 MyCode\MySample326\app\src\main\java\com\bin\luo\mysample\MainActivity.java 文件中。在这段代码中，myResume = (EditTextPreference) findPreference("myResume")用于在 XML 布局文件 preferences 中查找 key 为 myResume 的 EditTextPreference。myResume.getText()用于获取 EditTextPreference 的文本内容。addPreferencesFromResource

（R. xml. preferences）表示 MainActivity 加载的视图是 xml 目录下的 preferences 布局文件。preferences 布局文件的主要内容如下：

```xml
<?xml version = "1.0" encoding = "UTF-8"?>
<PreferenceScreen xmlns:android = "http://schemas.android.com/apk/res/android">
  <EditTextPreference
      android:dialogTitle = "工作简历"
      android:key = "myResume"
      android:summary = "请在弹出的对话框中输入高中毕业之后的工作经历"
      android:title = "工作简历"></EditTextPreference>
</PreferenceScreen>
```

上面这段代码在 MyCode\MySample326\app\src\main\res\xml\preferences.xml 文件中。在这段代码中，android:key 表示唯一标识符，与 android:id 相类似。android:title 表示标题。android:summary 表示列表选项的简单说明。android:dialogTitle 表示在弹出的对话框中的标题信息。此实例的完整项目在 MyCode\MySample326 文件夹中。

173 使用 SwitchPreference 读写开关状态值

此实例主要通过使用 SwitchPreference，从而实现将开关按钮的选择结果保存在存储卡上或者从存储卡上获取该选择结果。当实例运行之后，如果开关按钮处于开启状态，将在弹出的 Toast 中显示"确认监听来电"，如图 173-1 的左图所示，在关闭应用（实例），再重新启动应用之后，也将在弹出的 Toast 中显示"确认监听来电"。如果开关按钮处于关闭状态，将在弹出的 Toast 中显示"取消监听来电"，如图 173-1 的右图所示，在关闭应用，再重新启动应用之后，也将在弹出的 Toast 中显示"取消监听来电"。

图 173-1

主要代码如下：

```java
public class MainActivity extends PreferenceActivity {
 SwitchPreference mySwitchPreference;
 boolean bCheck;
 @Override
 protected void onCreate(Bundle savedInstanceState) {
  super.onCreate(savedInstanceState);
  addPreferencesFromResource(R.xml.preferences);
  mySwitchPreference = (SwitchPreference) findPreference("mySwitchPreference");
  mySwitchPreference.setOnPreferenceChangeListener(
                    new Preference.OnPreferenceChangeListener() {
   @Override
   public boolean onPreferenceChange(Preference arg0, Object newValue) {
    //这里可以监听到SwitchPreference值是否改变了,并且可以获取新改变的值
    bCheck = (Boolean)newValue;
    CheckSet();
    return true;
   } });
  bCheck = mySwitchPreference.isChecked();
  CheckSet();
 }
 void CheckSet() {
  String myInfo = "";
  if(bCheck){
   myInfo += "确认监听来电!";
  }else{
   myInfo += "取消监听来电!";
  }
  Toast.makeText(MainActivity.this,myInfo, Toast.LENGTH_LONG).show();
 } }
```

上面这段代码在 MyCode\MySample329\app\src\main\java\com\bin\luo\mysample\MainActivity.java 文件中。在这段代码中，mySwitchPreference＝（SwitchPreference）findPreference（"mySwitchPreference"）用于获取在 XML 文件中 key 为 mySwitchPreference 的 SwitchPreference。bCheck＝mySwitchPreference.isChecked()用于获取 SwitchPreference 在改变之前的值。bCheck＝(Boolean)newValue 用于获取 SwitchPreference 在改变之后的值。addPreferencesFromResource(R.xml.preferences)表示 MainActivity 加载的视图是 xml 目录下的 preferences 文件。preferences 文件的主要内容如下：

```xml
<?xml version = "1.0" encoding = "UTF-8"?>
<PreferenceScreen xmlns:android = "http://schemas.android.com/apk/res/android"
                  android:title = "MySample">
  <SwitchPreference android:key = "mySwitchPreference"
                 android:summaryOn = "是"
                 android:summaryOff = "否"
                 android:defaultValue = "true"
                 android:title = "是否开始监听来电"/>
</PreferenceScreen>
```

上面这段代码在 MyCode\MySample329\app\src\main\res\xml\preferences.xml 文件中。需

要说明的是,在此实例中,没有任何读写存储卡的代码,所有读写存储卡的功能均由 PreferenceActivity 和 SwitchPreference 自动完成。此实例的完整项目在 MyCode\MySample329 文件夹中。

174 使用 CheckBoxPreference 实现多选功能

此实例主要通过使用 CheckBoxPreference 实现将多选结果保存在存储卡上。当实例运行之后,如果选择了两部电影,将在弹出的 Toast 中显示两部电影片名,在关闭应用,再重新启动应用之后,也将在弹出的 Toast 中显示两部电影片名。如果选择了 3 部电影,则将在弹出的 Toast 中显示 3 部电影片名,在关闭应用,再重新启动应用之后,也将在弹出的 Toast 中显示 3 部电影片名,效果分别如图 174-1 的左图和右图所示。

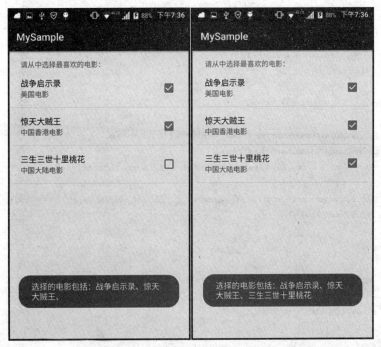

图 174-1

主要代码如下:

```
public class MainActivity extends PreferenceActivity {
  CheckBoxPreference myCheckbox1,myCheckbox2,myCheckbox3;
  boolean isCheck1, isCheck2, isCheck3;
  String myInfo = "选择的电影包括: ";
  @Override
  protected void onCreate(Bundle savedInstanceState) {
    super.onCreate(savedInstanceState);
    //setContentView(R.layout.activity_main);
    addPreferencesFromResource(R.xml.preferences);
    myCheckbox1 = (CheckBoxPreference) findPreference("myCheckbox1");
    myCheckbox1.setOnPreferenceClickListener(new
                Preference.OnPreferenceClickListener() {
      @Override
```

```java
    public boolean onPreferenceClick(Preference preference) {
    //这里可以监听到这个CheckBox的单击事件
     return true;
    } });
 myCheckbox1.setOnPreferenceChangeListener(
               new Preference.OnPreferenceChangeListener() {
  @Override
  public boolean onPreferenceChange(Preference arg0, Object newValue) {
   //这里可以监听到CheckBox值是否改变了,并且可以获取新值
   isCheck1 = (Boolean)newValue;
   CheckSet();
   return true;
  } });
 myCheckbox2 = (CheckBoxPreference) findPreference("myCheckbox2");
 myCheckbox2.setOnPreferenceChangeListener(
                new Preference.OnPreferenceChangeListener() {
  @Override
  public boolean onPreferenceChange(Preference arg0, Object newValue) {
   isCheck2 = (Boolean)newValue;
   CheckSet();
   return true;
  } });
 myCheckbox3 = (CheckBoxPreference) findPreference("myCheckbox3");
 myCheckbox3.setOnPreferenceChangeListener(
                new Preference.OnPreferenceChangeListener() {
  @Override
  public boolean onPreferenceChange(Preference arg0, Object newValue) {
   isCheck3 = (Boolean)newValue;
   CheckSet();
   return true;
  } });
 isCheck1 = this.myCheckbox1.isChecked();
 isCheck2 = this.myCheckbox2.isChecked();
 isCheck3 = this.myCheckbox3.isChecked();
 CheckSet();
}
void CheckSet() {
 String myInfo = "选择的电影包括: ";
 if(isCheck1){ myInfo += "战争启示录、"; }
 if(isCheck2){ myInfo += "惊天大贼王、"; }
 if(isCheck3){ myInfo += "三生三世十里桃花"; }
 Toast.makeText(MainActivity.this,myInfo, Toast.LENGTH_LONG).show();
} }
```

　　上面这段代码在 MyCode\MySample325\app\src\main\java\com\bin\luo\mysample\MainActivity.java 文件中。在这段代码中,myCheckbox1 = (CheckBoxPreference) findPreference("myCheckbox1")用于获取在 XML 文件中 key 为 myCheckbox1 的 CheckBoxPreference。myCheckbox1.isChecked()获取的是改变之前的值。isCheck1 = (Boolean)newValue 获取的是改变之后的值。addPreferencesFromResource(R.xml.preferences)表示 MainActivity 加载的视图是 xml 目录下的 preferences 布局文件。preferences 布局文件的主要内容如下:

```xml
<?xml version = "1.0" encoding = "UTF-8"?>
<PreferenceScreen
        xmlns:android = "http://schemas.android.com/apk/res/android">
<PreferenceCategory android:title = "请从中选择最喜欢的电影:">
<CheckBoxPreference android:key = "myCheckbox1"
                    android:title = "战争启示录"
                    android:summary = "美国电影"></CheckBoxPreference>
<CheckBoxPreference android:key = "myCheckbox2"
                    android:title = "惊天大贼王"
                    android:summary = "中国香港电影"></CheckBoxPreference>
<CheckBoxPreference android:key = "myCheckbox3"
                    android:title = "三生三世十里桃花"
                    android:summary = "中国大陆电影"></CheckBoxPreference>
</PreferenceCategory></PreferenceScreen>
```

上面这段代码在 MyCode\MySample325\app\src\main\res\xml\preferences.xml 文件中。需要说明的是，在此实例中，没有任何读写存储卡的代码，所有读写存储卡的功能均由 PreferenceActivity 和 CheckBoxPreference 自动完成。此实例的完整项目在 MyCode\MySample325 文件夹中。

175　使用 MultiSelectListPreference 实现多选

此实例主要通过使用 MultiSelectListPreference，实现在存储卡上保存用户在多选列表中的选择结果。当实例运行之后，单击"年度大片展播"列表项，将显示一个对话框，在该对话框中任意选择多部影片，如图 175-1 的左图所示，然后单击"确定"按钮返回；将在"年度大片展播"列表项下面显示多选结果，如图 175-1 的右图所示。退出此应用，再重新启动此应用，也将在"年度大片展播"列表项下面显示上次的多选结果，如图 175-1 的右图所示。

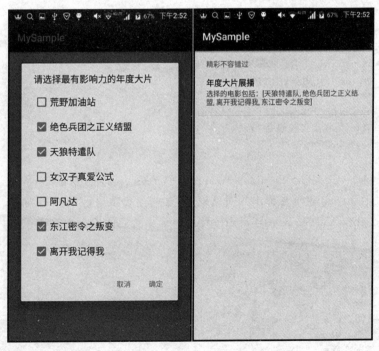

图　175-1

主要代码如下:

```java
public class MainActivity extends PreferenceActivity {
    @Override
    protected void onCreate(Bundle savedInstanceState) {
        super.onCreate(savedInstanceState);
        addPreferencesFromResource(R.xml.preferences);
        MultiSelectListPreference myMultiSelectListPreference =
            (MultiSelectListPreference)findPreference("myMultiSelectListPreference");
        myMultiSelectListPreference.setOnPreferenceChangeListener(
                        new Preference.OnPreferenceChangeListener(){
            @Override
            public boolean onPreferenceChange(Preference pref, Object arg1) {
                pref.setSummary(arg1.toString());
                return true;
            } });
        if(myMultiSelectListPreference.getValues().toString().length()>0){
            myMultiSelectListPreference.setSummary("选择的电影包括: "
                    + myMultiSelectListPreference.getValues().toString());
        }else{ myMultiSelectListPreference.setSummary("这里显示选择结果");}
    } }
```

上面这段代码在 MyCode\MySample328\app\src\main\java\com\bin\luo\mysample\MainActivity.java 文件中。在这段代码中, myMultiSelectListPreference =(MultiSelectListPreference) findPreference("myMultiSelectListPreference")用于在 XML 文件中查找 key 为 myMultiSelectListPreference 的 MultiSelectListPreference。myMultiSelectListPreference.getValues().toString()用于获取用户在多选列表 MultiSelectListPreference 中的选择。addPreferencesFromResource(R.xml.preferences)表示 MainActivity 加载的视图是 xml 目录下的 preferences 文件。preferences 文件的主要内容如下:

```xml
<?xml version="1.0" encoding="UTF-8"?>
<PreferenceScreen  xmlns:android="http://schemas.android.com/apk/res/android">
  <PreferenceCategory android:title="精彩不容错过">
    <MultiSelectListPreference
        android:dialogTitle="请选择最有影响力的年度大片"
        android:entries="@array/mymovies"
        android:entryValues="@array/mymovies"
        android:key="myMultiSelectListPreference"
        android:title="年度大片展播" />
  </PreferenceCategory></PreferenceScreen>
```

上面这段代码在 MyCode\MySample328\app\src\main\res\xml\preferences.xml 文件中。在这段代码中,android:entries 表示在弹出的对话框中,多选列表项显示的文本内容,注意,这里指的是一个数组。android:entryValues 表示与 android:entries 相对应的值。上述数组的内容如下:

```xml
<?xml version="1.0" encoding="utf-8"?>
<resources>
  <string-array name="mymovies">
    <item>荒野加油站</item>
    <item>绝色兵团之正义结盟</item>
    <item>天狼特遣队</item>
    <item>女汉子真爱公式</item>
```

```xml
<item>阿凡达</item>
<item>东江密令之叛变</item>
<item>离开我记得我</item>
</string-array>
</resources>
```

上面这段代码在 MyCode\MySample328\app\src\main\res\values\myarray.xml 文件中。此实例的完整项目在 MyCode\MySample328 文件夹中。

系统和设备

176 使用 ContentResolver 获取手机短信信息

此实例主要通过在 ContentResolver 的 query( )方法中使用"content://sms/inbox"参数,实现查询当前手机接收的所有短信。当实例运行之后,单击"获取当前手机接收的所有短信"按钮,将在下面的列表中显示当前手机接收的所有短信,效果分别如图 176-1 的左图和右图所示。

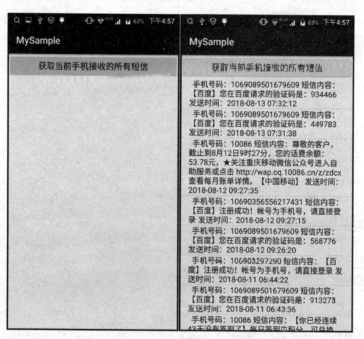

图 176-1

主要代码如下:

```
//响应单击"获取当前手机接收的所有短信"按钮
public void onClickmyBtn1(View v) {
    final String SMS_URI_ALL = "content://sms/";            //所有短信
    final String SMS_URI_INBOX = "content://sms/inbox";     //收信箱
    final String SMS_URI_SEND = "content://sms/sent";       //发信箱
    final String SMS_URI_DRAFT = "content://sms/draft";     //草稿箱
```

```
    String[] myItems = new String[]{"_id", "address",
                        "person", "body", "date", "type"};
    Uri myUri = Uri.parse(SMS_URI_INBOX);
    try {
     ArrayList myArray = new ArrayList<>();
     ContentResolver myContentResolver = getContentResolver();
     Cursor myCursor = myContentResolver.query(myUri, myItems,
                        null, null, "date desc");
     if (myCursor != null) {
      while (myCursor.moveToNext()) {
       String myPhone = myCursor.getString(myCursor.getColumnIndex("address"));
       String myBody = myCursor.getString(myCursor.getColumnIndex("body"));
       SimpleDateFormat myDateFormat = new SimpleDateFormat("yyyy-MM-dd hh:mm:ss");
       Date myTempDate = new Date(Long.parseLong(
                        myCursor.getString(myCursor.getColumnIndex("date"))));
       String myDate = myDateFormat.format(myTempDate);
       String myInfo = " 手机号码:" + myPhone + " 短信内容:"
                        + myBody + " 发送时间:" + myDate;
       myArray.add(myInfo);
      }
      myCursor.close();
     }
     myListView.setAdapter(new ArrayAdapter(this,
                        android.R.layout.simple_list_item_1, myArray));
    } catch (Exception e) {
     Toast.makeText(this, e.getMessage().toString(), Toast.LENGTH_SHORT).show();
    }}
```

上面这段代码在 MyCode\MySample670\app\src\main\java\com\bin\luo\mysample\MainActivity.java 文件中。在这段代码中,myCursor = myContentResolver.query(myUri, myItems, null, null, "date desc")表示根据指定的URI查询对应的内容。如果URI是"content://sms/",则查询结果是所有短信;如果URI是"content://sms/inbox",则查询结果是接收的所有短信;如果URI是"content://sms/sent",则查询结果是本机发送的所有短信。此外,查询短信需要在AndroidManifest.xml文件中添加<uses-permission android:name="android.permission.READ_SMS"/>权限。此实例的完整项目在 MyCode\MySample670 文件夹中。

177 使用 ContentResolver 获取所有联系人信息

此实例主要通过使用 ContentResolver 的 query()方法,实现查询手机通讯录的联系人信息。当实例运行之后,单击"获取手机通讯录的联系人信息"按钮,将在下面的列表中显示当前手机通讯录中的联系人信息,效果分别如图 177-1 的左图和右图所示。

主要代码如下:

```
public void onClickmyBtn1(View v) {     //响应单击"获取手机通讯录的联系人信息"按钮
  List<Contact> myContacts = GetContacts();
  myListView.setAdapter(new ArrayAdapter<Contact>(this,
                    android.R.layout.simple_list_item_1, myContacts));
}
private List GetContacts() {
```

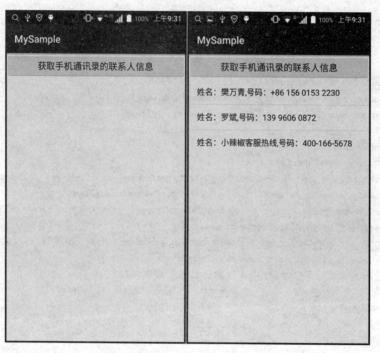

图　177-1

```
    ArrayList myArrayList = new ArrayList<>();
    ContentResolver myContentResolver = getContentResolver();
    Cursor myCursor = myContentResolver.query(
            ContactsContract.CommonDataKinds.Phone.CONTENT_URI,
            new String[]{"contact_id", "display_name",
            ContactsContract.CommonDataKinds.Phone.NUMBER, "sort_key"},
            null, null, "sort_key");
    if (myCursor != null) {
      while (myCursor.moveToNext()) {
        Contact myContact = new Contact();
        String myID = myCursor.getString(0);
        myContact.setName(myCursor.getString(1));
        myContact.setPhone(myCursor.getString(2));
        myArrayList.add(myContact);
      }
      myCursor.close();
    }
    return myArrayList;
}
public class Contact {                    //联系人类定义
    private String name;
    private String phone;
    public String getName() { return name; }
    public void setName(String name) { this.name = name; }
    public String getPhone() { return phone; }
    public void setPhone(String phone) { this.phone = phone; }
    @Override
    public String toString() { return "姓名:" + getName() + ",号码:" + getPhone();}
}
```

上面这段代码在 MyCode\MySample227\app\src\main\java\com\bin\luo\mysample\MainActivity.java 文件中。在这段代码中，Cursor myCursor = myContentResolver.query(ContactsContract.CommonDataKinds.Phone.CONTENT_URI, new String[]{"contact_id", "display_name", ContactsContract.CommonDataKinds.Phone.NUMBER,"sort_key"}, null, null, "sort_key") 用于查询手机联系人信息，参数 ContactsContract.CommonDataKinds.Phone.CONTENT_URI 表示联系人的 URI；参数 new String[]{"contact_id", "display_name", ContactsContract.CommonDataKinds.Phone.NUMBER,"sort_key"} 表示查询结果返回的数据列，此处就是联系人的 ID、姓名和电话号码等，也可以理解为该方法要查询什么；第三个参数名为 selection，表示设置查询条件，相当于 SQL 语句中的 where，null 表示不进行筛选。第四个参数名为 selectionArgs，这个参数需要配合第三个参数使用，如果在第三个参数里面有问号"？"，那么在 selectionArgs 中写的数据就会替换掉"？"，如果第三个参数为 null，此参数也应该为 null；第五个参数名为 sortOrder，表示按照什么要求进行排序，相当于 SQL 语句中的 order by。此外，读取手机联系人需要在 AndroidManifest.xml 文件中添加< uses-permission android:name = "android.permission.READ_CONTACTS"/>权限。此实例的完整项目在 MyCode\MySample227 文件夹中。

178 使用 ContentResolver 查询联系人电话号码

此实例主要通过使用 ContentResolver 的 query() 方法，实现在当前手机的通讯录中根据姓名查询电话号码和电子邮箱的功能。当实例运行之后，在"联系人姓名"输入框中输入将要查询的姓名，如"罗斌"，然后单击"开始查询"按钮，将在弹出的 Toast 中显示查询结果，效果分别如图 178-1 的左图和右图所示。

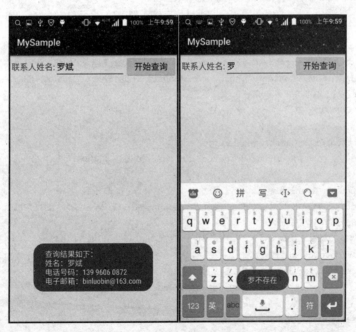

图 178-1

主要代码如下：

```
public void onClickmyBtn1(View v) {                    //响应单击"开始查询"按钮
    Contact myContact = searchContact(myEditText.getText().toString());
```

```java
        String myInfo = "查询结果如下：\n 姓名：" + myEditText.getText().toString()
                +"\n 电话号码：" + myContact.getNumber()
                + "\n 电子邮箱：" + myContact.getEmail();
    Toast.makeText(MainActivity.this, myInfo, Toast.LENGTH_SHORT).show();
}
    public String getContactID(String myName) {              //根据联系人姓名查询 ID
        String myID = "0";
        ContentResolver myContentResolver = getContentResolver();
        Cursor myCursor = myContentResolver.query(
                android.provider.ContactsContract.Contacts.CONTENT_URI,
                new String[]{android.provider.ContactsContract.Contacts._ID},
                android.provider.ContactsContract.Contacts.DISPLAY_NAME +
                " = '" + myName + "'", null, null);
        if (myCursor.moveToNext()) {
         myID = myCursor.getString(myCursor.getColumnIndex(
                    android.provider.ContactsContract.Contacts._ID));
        }
        return myID;
    }
//根据联系人姓名查询电话号码和电子邮箱
public Contact searchContact(String myName) {
    Contact myContact = new Contact();
    myContact.setName(myName);
    String myID = getContactID(myContact.getName());
    myContact.setId(myID);
    if(myID.equals("0")) {
     Toast.makeText(MainActivity.this, myContact.getName()
                          +"不存在", Toast.LENGTH_SHORT).show();
    } else {
     ContentResolver myContentResolver = getContentResolver();
     Cursor myCursor = myContentResolver.query(              //查询电话号码
            android.provider.ContactsContract.CommonDataKinds.Phone.CONTENT_URI,
            new String[]{COLUMN_NUMBER},
            COLUMN_CONTACT_ID + " = '" + myID + "'", null, null);
     while(myCursor.moveToNext()) {
      myContact.setNumber(myCursor.getString(
                            myCursor.getColumnIndex(COLUMN_NUMBER)));
     }
     myCursor = getContentResolver().query(                  //查询电子邮箱
            android.provider.ContactsContract.CommonDataKinds.Email.CONTENT_URI,
            new String[]{COLUMN_EMAIL},
            COLUMN_CONTACT_ID + " = '" + myID + "'", null, null);
     while(myCursor.moveToNext()) {
      myContact.setEmail(myCursor.getString
                            (myCursor.getColumnIndex(COLUMN_EMAIL)));
     }
     myCursor.close();
    }
    return myContact;

    public class Contact {                                   //联系人类定义
     private String email;
     private String id;
```

```
    private String name;
    private String number;
    public Contact() { }
    public Contact(Contact contact) {
     this.name = contact.getName();
     this.number = contact.getNumber();
     this.email = contact.getEmail();
    }
    public String getEmail() {  return email;  }
    public String getId() {  return id;  }
    public String getName() {  return name;  }
    public String getNumber() {  return number;  }
    public void setEmail(String email) { this.email = email; }
    public void setId(String id) { this.id = id; }
    public void setName(String name) { this.name = name; }
    public void setNumber(String number) { this.number = number; }
}
```

上面这段代码在 MyCode\MySample230\app\src\main\java\com\bin\luo\mysample\MainActivity.java 文件中。在这段代码中，Cursor myCursor = myContentResolver.query(android.provider.ContactsContract.CommonDataKinds.Phone.CONTENT_URI, new String[]{COLUMN_NUMBER},COLUMN_CONTACT_ID + "='" + myID + "'", null, null)表示根据联系人 ID 查询联系人电话号码，参数 android.provider.ContactsContract.CommonDataKinds.Phone.CONTENT_URI 表示查询 URI，参数 new String[]{COLUMN_NUMBER}表示返回的数据列是电话号码，参数 COLUMN_CONTACT_ID + "='" + myID + "'"表示查询条件是联系人 ID。此外，查询联系人需要在 AndroidManifest.xml 文件中添加< uses-permission android:name = "android.permission.READ_CONTACTS"/>权限。此实例的完整项目在 MyCode\MySample230 文件夹中。

179　使用 ContentResolver 动态新增联系人信息

此实例主要通过使用 ContentValues 临时存储联系人指定字段键值对集合，然后调用 ContentResolver 的 insert()方法实现插入指定信息至联系人数据库，从而实现动态新增联系人信息。当实例运行之后，在三个输入框中分别输入"联系人姓名""联系人电话""联系人邮箱"等信息，然后单击"在当前手机的通讯录中增加联系人"按钮，如果增加成功，则将在弹出的 Toast 中显示增加成功的信息，如图 179-1 的左图所示。然后即可在手机通讯录中查看刚才增加的联系人信息，如图 179-1 的右图所示。

主要代码如下：

```
public void onClickmyBtn1(View v){          //响应单击"在当前手机的通讯录中增加联系人"按钮
    try {
    ContentValues myContentValues = new ContentValues();
    ContentResolver myContentResolver = getContentResolver();
    Uri myUri = myContentResolver.insert(
            ContactsContract.RawContacts.CONTENT_URI, myContentValues);
    //获取生成的联系人 ID 值
    long myID = ContentUris.parseId(myUri);
```

图 179-1

```
    myContentValues.clear();
//插入联系人姓名
    myContentValues.put(ContactsContract.Data.RAW_CONTACT_ID,myID);
    myContentValues.put(ContactsContract.Data.MIMETYPE,
        ContactsContract.CommonDataKinds.StructuredName.CONTENT_ITEM_TYPE);
    myContentValues.put(ContactsContract.CommonDataKinds.
        StructuredName.GIVEN_NAME,myContactName.getText().toString());
    myContentResolver.insert(ContactsContract.Data.CONTENT_URI,
                                                myContentValues);
    myContentValues.clear();
//插入联系人电话号码
    myContentValues.put(ContactsContract.Data.RAW_CONTACT_ID,myID);
    myContentValues.put(ContactsContract.Contacts.Data.MIMETYPE,
            ContactsContract.CommonDataKinds.Phone.CONTENT_ITEM_TYPE);
    myContentValues.put(ContactsContract.CommonDataKinds.
                Phone.NUMBER,myContactPhone.getText().toString());
    myContentResolver.insert(ContactsContract.Data.CONTENT_URI,
                                                myContentValues);
    myContentValues.clear();
//插入联系人邮箱
    myContentValues.put(ContactsContract.Data.RAW_CONTACT_ID,myID);
    myContentValues.put(ContactsContract.Contacts.Data.MIMETYPE,
            ContactsContract.CommonDataKinds.Email.CONTENT_ITEM_TYPE);
    myContentValues.put(ContactsContract.CommonDataKinds.
            Email.DATA,myContactEmail.getText().toString());
    myContentResolver.insert(ContactsContract.Data.CONTENT_URI,
                                                myContentValues);
    myContentValues.clear();
    Toast.makeText(this,"成功添加联系人!",Toast.LENGTH_SHORT).show();
    } catch (Exception e) {
    Toast.makeText(this, e.getMessage().toString(), Toast.LENGTH_SHORT).show();
    } }
```

上面这段代码在 MyCode\MySample878\app\src\main\java\com\bin\luo\mysample\MainActivity.java 文件中。在这段代码中，getContentResolver（）方法用于返回一个 ContentResolver，ContentResolver 可以访问 ContentProvider 提供的数据；在 Activity 中，直接使用 getContentResolver()方法可以得到当前应用的 ContentResolver 实例。需要注意的是，操作手机联系人需要在 AndroidManifest.xml 文件中添加＜uses-permission android：name＝" android.permission.WRITE_CONTACTS"/＞权限。此实例的完整项目在 MyCode\MySample878 文件夹中。

180　使用 ContentResolver 动态修改联系人信息

此实例主要通过使用 ContentValues 临时存储联系人指定字段键值对集合，然后调用 ContentResolver 的 update()方法，实现修改联系人信息（电话号码）。当实例运行之后，在"联系人姓名："输入框中输入将要查询的联系人姓名，然后单击"查询联系人信息"按钮，如果查询成功，将在"联系人电话："输入框中显示该联系人的电话号码，如图 180-1 的左图所示。如果在"联系人电话："输入框中输入新的电话号码，然后单击"更新联系人信息"按钮，如果更新成功，将在弹出的 Toast 中显示更新成功的信息，如图 180-1 的右图所示。

图　180-1

主要代码如下：

```
public void onClickmyBtn1(View v) {         //响应单击"查询联系人信息"按钮
    String myName = myContactName.getText().toString();
    //根据联系人姓名查询联系人相关信息,并返回游标对象
    ContentResolver myContentResolver = getContentResolver();
    Cursor myCursor = myContentResolver.query(
            ContactsContract.Contacts.CONTENT_URI, null,
            ContactsContract.PhoneLookup.DISPLAY_NAME +
            " = ?", new String[]{myName}, null);
    while (myCursor.moveToNext()) {
        //获取联系人姓名对应列索引值
```

```java
        int myContactNameIndex = myCursor.getColumnIndex(
                        ContactsContract.PhoneLookup.DISPLAY_NAME);
        //获取联系人姓名
        myContactName.setText(myCursor.getString(myContactNameIndex));
        //获取联系人 ID
        myID = myCursor.getString(myCursor.getColumnIndex(
                        ContactsContract.Contacts.NAME_RAW_CONTACT_ID));
        //根据 ID 值查询联系人号码相关信息,并返回游标对象
        Cursor myContactPhoneCursor = myContentResolver.query(
                    ContactsContract.CommonDataKinds.Phone.CONTENT_URI,
                    null,ContactsContract.CommonDataKinds.Phone.
                    NAME_RAW_CONTACT_ID + " = " + myID, null, null);
        while (myContactPhoneCursor.moveToNext()) {
          //获取联系人号码对应列索引值
          int myPhoneIndex = myContactPhoneCursor.getColumnIndex(
                        ContactsContract.CommonDataKinds.Phone.NUMBER);
          //获取指定联系人号码
          myContactPhone.setText(myContactPhoneCursor.getString(myPhoneIndex));
        } }
        myCursor.close();
    }
    public void onClickmyBtn2(View v) {//响应单击"更新联系人信息"按钮
        //获取修改后的手机号码和姓名
        String myPhone = myContactPhone.getText().toString();
        String myName = myContactName.getText().toString();
        //修改 data 表的数据
        Uri myUri = Uri.parse("content://com.android.contacts/data");
        ContentResolver myContentResolver = getContentResolver();
        ContentValues myContentValues = new ContentValues();
        //将手机号码封装至 ContentValues 对象中
        myContentValues.put("data1", myPhone);
        //更新该联系人号码信息
        myContentResolver.update(myUri, myContentValues,
                "mimetype = ? and raw_contact_id = ?",
                new String[]{"vnd.android.cursor.item/phone_v2", myID});
        myContentValues.clear();
        myContentValues.put("data1", myName);
        myContentResolver.update(myUri, myContentValues,
                "mimetype = ? and raw_contact_id = ?",
                new String[]{"vnd.android.cursor.item/name", myID});
        Toast.makeText(this, "更新成功!", Toast.LENGTH_SHORT).show();
    }
```

上面这段代码在 MyCode\MySample883\app\src\main\java\com\bin\luo\mysample\MainActivity.java 文件中。在这段代码中,myContentResolver = getContentResolver()用于返回一个 ContentResolver,ContentResolver 可以访问 ContentProvider 提供的数据。myContentResolver.update(myUri, myContentValues,"mimetype = ? and raw_contact_id = ?", new String[]{"vnd.android.cursor.item/phone_v2", myID})用于根据联系人的 myID 修改联系人的电话号码。此外,操作手机联系人需要在 AndroidManifest.xml 文件中添加< uses-permission android:name = "android.permission.READ_CONTACTS"/>权限和< uses-permission android:name = "android.permission.WRITE_CONTACTS"/>权限。此实例的完整项目在 MyCode\MySample883 文件夹中。

181 使用 ContentResolver 动态删除联系人信息

此实例主要通过使用 ContentResolver 的 delete()方法,实现删除通讯录中的联系人信息。当实例运行之后,在"联系人姓名:"输入框中输入将要查询的联系人姓名,然后单击"查询联系人信息"按钮,如果查询成功,将在"联系人电话:"输入框中显示该联系人的电话号码;单击"删除联系人信息"按钮,将删除刚才查询的联系人信息,如图 181-1 的左图所示。再次在"联系人姓名:"输入框中输入刚才删除的联系人姓名,然后单击"查询联系人信息"按钮,则不会在"联系人电话:"输入框中显示电话号码(因为此人刚才已经被删除),如图 181-1 的右图所示。

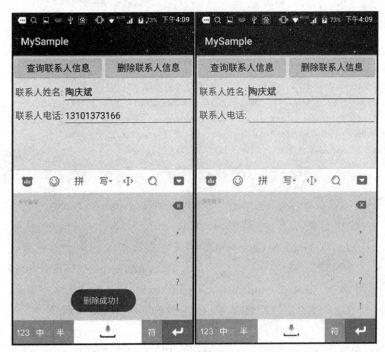

图 181-1

主要代码如下:

```
public void onClickmyBtn1(View v) {           //响应单击"查询联系人信息"按钮
    myContactPhone.setText("");
    String myName = myContactName.getText().toString();
    //根据联系人姓名查询联系人相关信息,并返回游标对象
    ContentResolver myContentResolver = getContentResolver();
    Cursor myCursor = myContentResolver.query(
            ContactsContract.Contacts.CONTENT_URI, null,
            ContactsContract.PhoneLookup.DISPLAY_NAME + " = ?",
            new String[]{myName}, null);
    while (myCursor.moveToNext()) {
        //获取联系人姓名对应列索引值
        int myContactNameIndex = myCursor.getColumnIndex(
                    ContactsContract.PhoneLookup.DISPLAY_NAME);
        //获取联系人姓名
        myContactName.setText(myCursor.getString(myContactNameIndex));
        //获取联系人 ID
```

```java
        myID = myCursor.getString(myCursor.getColumnIndex(
                ContactsContract.Contacts.NAME_RAW_CONTACT_ID));
        //根据 ID 值查询联系人号码相关信息,并返回游标对象
        Cursor myContactPhoneCursor = myContentResolver.query(
                ContactsContract.CommonDataKinds.Phone.CONTENT_URI, null,
                ContactsContract.CommonDataKinds.Phone.NAME_RAW_CONTACT_ID +
                " = " + myID, null, null);
        while (myContactPhoneCursor.moveToNext()) {
            //获取联系人号码对应列索引值
            int myPhoneIndex = myContactPhoneCursor.getColumnIndex(
                    ContactsContract.CommonDataKinds.Phone.NUMBER);
            //获取指定联系人号码
            myContactPhone.setText(myContactPhoneCursor.getString(myPhoneIndex));
        }
        myCursor.close();
    }
    public void onClickmyBtn2(View v) {        //响应单击"删除联系人信息"按钮
        Uri myUri = Uri.parse("content://com.android.contacts/data");
        ContentResolver myContentResolver = getContentResolver();
        myContentResolver.delete(myUri, "raw_contact_id = ?", new String[]{myID});
        Toast.makeText(this, "删除成功!", Toast.LENGTH_SHORT).show();
    }
```

上面这段代码在 MyCode\MySample884\app\src\main\java\com\bin\luo\mysample\MainActivity.java 文件中。在这段代码中,myContentResolver.delete(myUri,"raw_contact_id=?",new String[]{myID})用于根据 myID 删除该联系人对应的信息。此外,操作联系人需要在 AndroidManifest.xml 文件中添加<uses-permission android:name="android.permission.READ_CONTACTS"/>权限和<uses-permission android:name="android.permission.WRITE_CONTACTS"/>权限。此实例的完整项目在 MyCode\MySample884 文件夹中。

182 使用 PhoneStateListener 监听来电号码

此实例主要通过在 PhoneStateListener 的 onCallStateChanged()方法中解析来电信息,从而实现监听手机来电电话号码的功能。当实例运行之后,从其他手机向当前测试手机拨打电话,将在弹出的 Toast 中显示对方的来电电话号码,如图 182-1 所示。

主要代码如下:

```java
public class MainActivity extends Activity {
    @Override
    protected void onCreate(Bundle savedInstanceState) {
        super.onCreate(savedInstanceState);
        setContentView(R.layout.activity_main);
        MyPhoneStateListener myListener = new MyPhoneStateListener();
        TelephonyManager myManager =
                (TelephonyManager)getSystemService(TELEPHONY_SERVICE);
        myManager.listen(myListener, PhoneStateListener.LISTEN_CALL_STATE);
    }
    public class MyPhoneStateListener extends PhoneStateListener {
        @Override
```

图 182-1

```
public void onCallStateChanged(int state,String incomingNumber){
  if(incomingNumber.length()>0&&TelephonyManager.CALL_STATE_RINGING == state){
    Toast.makeText(MainActivity.this,"新来电号码: " +
                    incomingNumber,Toast.LENGTH_SHORT).show();
}}}}
```

上面这段代码在 MyCode\MySample317\app\src\main\java\com\bin\luo\mysample\MainActivity.java 文件中。在这段代码中，myManager.listen（myListener，PhoneStateListener.LISTEN_CALL_STATE）用于监听电话状态，myListener 参数是一个由 PhoneStateListener 类派生的实例，PhoneStateListener.LISTEN_CALL_STATE 表示监听电话状态。此外，监听来电需要在 AndroidManifest.xml 文件中添加< uses-permission android：name = "android.permission.READ_PHONE_STATE"/>权限。此实例的完整项目在 MyCode\MySample317 文件夹中。

183 使用 BroadcastReceiver 监听拨出号码

此实例主要通过使用广播接收者 BroadcastReceiver，实现监听手机向外拨出的电话号码。当实例运行之后，如果使用手机上的拨号器向外拨打电话，将在弹出的 Toast 中显示拨出的电话号码，效果分别如图 183-1 的左图和右图所示。

主要代码如下：

```
public class CallReceiver extends BroadcastReceiver {
  @Override
  public void onReceive(Context context, Intent myIntent) {
    if (myIntent.getAction().equals("android.intent.action.NEW_OUTGOING_CALL")){
      String myPhone =
```

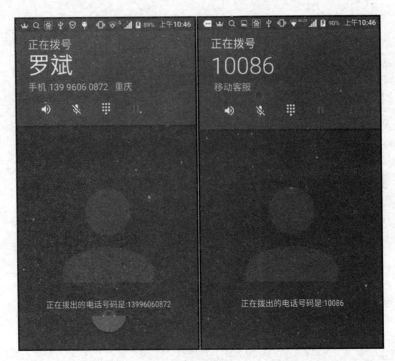

图 183-1

```
        myIntent.getExtras().getString("android.intent.extra.PHONE_NUMBER");
  Toast.makeText(context,"正在拨出的电话号码是:" +
                                   myPhone,Toast.LENGTH_SHORT).show();
}}}
```

上面这段代码在 MyCode\MySample286\app\src\main\java\com\bin\luo\mysample\CallReceiver.java 文件中。在这段代码中，myPhone = myIntent.getExtras().getString("android.intent.extra.PHONE_NUMBER")用于获取手机向外拨出的电话号码，它主要是通过广播接收者 BroadcastReceiver 的 onReceive()方法以参数的形式传递的，因此通常需要以 BroadcastReceiver 为基类创建新的广播接收者，并且需要在 AndroidManifest.xml 文件中进行注册和添加权限，注册 CallReceiver 的代码如下：

```xml
<receiver android:name=".CallReceiver">
  <intent-filter>
    <action android:name="android.intent.action.PHONE_STATE" />
    <action android:name="android.intent.action.NEW_OUTGOING_CALL" />
  </intent-filter>
</receiver>
```

新增权限的代码如下：

```xml
<uses-permission android:name="android.permission.READ_PHONE_STATE" />
<uses-permission android:name="android.permission.PROCESS_OUTGOING_CALLS" />
```

上面这两段代码在 MyCode\MySample286\app\src\main\AndroidManifest.xml 文件中。此实例的完整项目在 MyCode\MySample286 文件夹中。

184 动态注册 BroadcastReceiver 监听网络状态

此实例主要通过使用 registerReceiver() 方法动态注册自定义 BroadcastReceiver，实现监听网络的连接状态。当实例运行之后，在"网址："输入框中输入一个网址，如"http://www.baidu.com"，单击"浏览"按钮，如果当前网络连接是 WiFi，则使用有图模式显示网页，如图 184-1 的左图所示；如果当前网络连接是移动数据连接，则使用无图模式显示网页，如图 184-1 的右图所示。

图　184-1

主要代码如下：

```java
public void onClickBtn1(View v) {                           //响应单击"浏览"按钮
    myWebView.loadUrl(myEditText.getText().toString());
    myBroadcastReceiver = new MyBroadcastReceiver();
    IntentFilter myIntentFilter = new IntentFilter();       //创建过滤器
    //为过滤器设置 Action
    myIntentFilter.addAction(ConnectivityManager.CONNECTIVITY_ACTION);
    registerReceiver(myBroadcastReceiver, myIntentFilter);  //注册广播接收者
}
class MyBroadcastReceiver extends BroadcastReceiver {
    @Override
    public void onReceive(Context context, Intent intent) {
        //获取连接管理器对象
        ConnectivityManager myConnectivityManager = (ConnectivityManager)
                    context.getSystemService(Context.CONNECTIVITY_SERVICE);
        //获取移动网络连接对象
        NetworkInfo myMobileNetInfo =
            myConnectivityManager.getNetworkInfo(ConnectivityManager.TYPE_MOBILE);
        //获取 WiFi 网络连接对象
        NetworkInfo myWifiNetInfo =
```

```
                myConnectivityManager.getNetworkInfo(ConnectivityManager.TYPE_WIFI);
        if (!myMobileNetInfo.isConnected() && !myWifiNetInfo.isConnected()) {
         Toast.makeText(MainActivity.this,
                        "当前网络不可用!", Toast.LENGTH_LONG).show();
        } else if (myMobileNetInfo.isConnected() && !myWifiNetInfo.isConnected()) {
         Toast.makeText(MainActivity.this,
                "正使用移动网络,网页将以无图模式显示!", Toast.LENGTH_LONG).show();
         myWebView.getSettings().setBlockNetworkImage(true);
        } else if (!myMobileNetInfo.isConnected() && myWifiNetInfo.isConnected()) {
         Toast.makeText(MainActivity.this,
                    "正在使用WiFi网络浏览网页!", Toast.LENGTH_LONG).show();
         myWebView.getSettings().setBlockNetworkImage(false);        //禁用无图模式
    } } }
```

上面这段代码在 MyCode\MySample815\app\src\main\java\com\bin\luo\mysample\MainActivity.java 文件中。在这段代码中,registerReceiver(myBroadcastReceiver,myIntentFilter) 用于动态注册自定义广播 BroadcastReceiver。registerReceiver() 方法的语法声明如下:

```
public Intent registerReceiver(BroadcastReceiver receiver,
                                    IntentFilter filter)
```

其中,参数 BroadcastReceiver receiver 表示将要注册的自定义广播。参数 IntentFilter filter 代表一个 IntentFilter,在其中可以过滤需要的(网络连接)动作。

当使用 registerReceiver() 方法注册之后,也可以使用 unregisterReceiver() 方法取消。一般情况下,在 Activity 中注册广播接收者通常在 onResume() 方法中添加代码(也可在 onCreate() 方法中添加代码),在 Activity 中注销广播接收者通常在 onPause() 方法中添加代码(也可在 onDestroy() 方法中添加代码)。需要注意的是,访问网络需要在 AndroidManifest.xml 文件中添加 < uses-permission android:name = "android.permission.ACCESS_NETWORK_STATE"/> 权限和 < uses-permission android:name = "android.permission.INTERNET"/> 权限。此实例的完整项目在 MyCode\MySample815 文件夹中。

185 使用 BroadcastReceiver 实现开机自启动

此实例主要通过监听系统启动广播 BOOT_COMPLETED,实现在手机开机后自动启动当前应用。当应用实例安装在手机上之后,重新启动手机,即可发现此实例会在手机启动结束后自动运行,如图 185-1 所示。

主要代码如下:

```
public class MainActivity extends Activity {
 @Override
 protected void onCreate(Bundle savedInstanceState) {
  super.onCreate(savedInstanceState);
  requestWindowFeature(Window.FEATURE_NO_TITLE);                //去掉标题栏
  getWindow().setFlags(WindowManager.LayoutParams.FLAG_FULLSCREEN,
        WindowManager.LayoutParams.FLAG_FULLSCREEN);            //全屏显示
  setContentView(R.layout.activity_main);
```

图 185-1

```
new Thread() {
  public void run() {
    try {
      sleep(10000);              //10s 后关闭当前应用
    } catch (Exception e) {  e.printStackTrace(); }
    finally { finish(); }
  } }.start();
} }
```

上面这段代码在 MyCode \ MySample224 \ app \ src \ main \ java \ com \ bin \ luo \ mysample \ MainActivity.java 文件中。在这段代码中，Thread 的主要作用就是在当前应用被启动停留 10s 后，自动关闭。监听系统广播 ACTION_BOOT_COMPLETED，并实现开机启动的代码则在新建的 Java 类 BootBroadcastReceiver 中，主要代码如下：

```
public class BootBroadcastReceiver extends BroadcastReceiver {
  static final String myBoot = "android.intent.action.BOOT_COMPLETED";
  @Override
  public void onReceive(Context context, Intent myIntent) {
    if (myIntent.getAction().equals(myBoot)){       //检测手机启动完成
      Intent myStartIntent = new Intent(context,MainActivity.class);
      myStartIntent.addFlags(Intent.FLAG_ACTIVITY_NEW_TASK);
      context.startActivity(myStartIntent);         //启动当前应用
} } }
```

上面这段代码在 MyCode \ MySample224 \ app \ src \ main \ java \ com \ bin \ luo \ mysample \ BootBroadcastReceiver.java 文件中。然后，在 AndroidManifest.xml 文件中注册 BootBroadcastReceiver，主要代码如下：

```xml
<receiver android:name=".BootBroadcastReceiver">
    <intent-filter>
        <action android:name="android.intent.action.BOOT_COMPLETED" />
        <category android:name="android.intent.category.HOME" />
    </intent-filter>
</receiver>
```

同时在该文件中新增下列权限：

```xml
<uses-permission android:name="android.permission.RECEIVE_BOOT_COMPLETED" />
```

上面这两段代码在 MyCode\MySample224\app\src\main\AndroidManifest.xml 文件中。此实例的基本思路是：当 Android 启动时，会发出一个系统广播 ACTION_BOOT_COMPLETED，它的字符串常量为 android.intent.action.BOOT_COMPLETED；因此只要在应用中"捕捉"到这个广播，再启动之即可，即此实例的 BootBroadcastReceiver。此实例的完整项目在 MyCode\MySample224 文件夹中。

186 使用 BroadcastReceiver 获取电量百分比

此实例主要通过使用 registerReceiver()方法向系统注册电池服务请求，实现以广播接收的形式获取当前手机的剩余电量百分比。当实例运行之后，单击"获取当前手机的剩余电量百分比"按钮，则将每隔一段时间，在弹出的 Toast 中显示当前手机的电池剩余电量百分比，效果分别如图 186-1 的左图和右图所示。

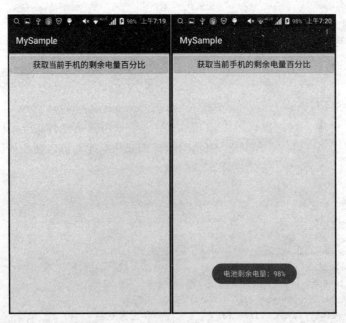

图 186-1

主要代码如下：

```java
//响应单击"获取当前手机的剩余电量百分比"按钮
public void onClickmyBtn1(View v) {
```

```
    IntentFilter myIntentFilter = new IntentFilter(Intent.ACTION_BATTERY_CHANGED);
    registerReceiver(myBroadcastReceiver, myIntentFilter);
}
private BroadcastReceiver myBroadcastReceiver = new BroadcastReceiver() {
    @Override
    public void onReceive(Context context, Intent intent) {
     String myAction = intent.getAction();
     if (myAction.equals(Intent.ACTION_BATTERY_CHANGED)) {
      //电池状态,返回值是一个数字
      //BatteryManager.BATTERY_STATUS_CHARGING 表示当前状态是充电
      //BatteryManager.BATTERY_STATUS_DISCHARGING 表示当前状态是放电中
      //BatteryManager.BATTERY_STATUS_NOT_CHARGING 表示当前状态是未充电
      //BatteryManager.BATTERY_STATUS_FULL 电池满
      int myVoltage = intent.getIntExtra("voltage", 0);        //电池的电压
      int myLevel = intent.getIntExtra("level", 0);            //电池的电量
      if (myVoltage != 0) {
       Toast.makeText(MainActivity.this, "电池剩余电量: "
                           + myLevel + " %", Toast.LENGTH_SHORT).show();
}}}};
```

上面这段代码在 MyCode \ MySample216 \ app \ src \ main \ java \ com \ bin \ luo \ mysample \ MainActivity.java 文件中。在这段代码中,IntentFilter myIntentFilter = new IntentFilter(Intent. ACTION_BATTERY_CHANGED)用于创建过滤电池变化的 IntentFilter。myBroadcastReceiver = new BroadcastReceiver( )用于创建广播接收器。registerReceiver(myBroadcastReceiver,myIntentFilter)则根据过滤器和广播接收器进行接收信息的注册。在 Android 中,手机电池的电量、电压、温度、充电状态等信息,都是由 BatteryService 来提供的;BatteryService 通过广播主动把数据传送给有请求的应用,应用如果想要接收到 BatteryService 发送的电池信息,需要注册 IntentFilter 为 Intent. ACTION_BATTERY_CHANGED 的 BroadcastReceiver。此实例的完整项目在 MyCode \ MySample216 文件夹中。

187 使用 ConnectivityManager 检测数据连接

此实例主要通过以反射的方式调用 getMobileDataEnabled()方法,实现检测移动网络数据连接是否已经打开的功能。当实例运行之后,单击"检测移动网络数据连接是否打开"按钮,将在弹出的 Toast 中显示数据连接状态;如果当前数据连接已经打开,检测结果如图 187-1 的左图所示;如果当前数据连接已经关闭,检测结果如图 187-1 的右图所示。

主要代码如下:

```
//响应单击"检测移动网络数据连接是否打开"按钮
public void onClickmyBtn1(View v) {
  String myInfo = "";
  if (getMobileDataState(this, null)) {
   myInfo = "移动数据连接已经打开!";
  } else {
   myInfo = "移动数据连接已经关闭!";
  }
  Toast.makeText(getApplicationContext(), myInfo, Toast.LENGTH_SHORT).show();
```

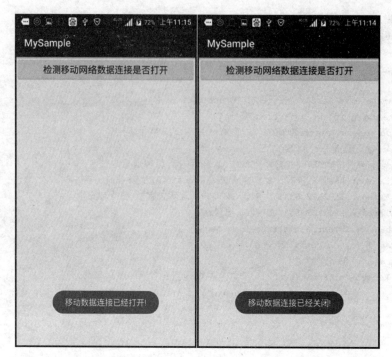

图 187-1

```
}
//返回移动数据连接状态,参数 arg 默认为 null,
//如果该方法返回 true 表示已连接,返回 false 表示未连接
public boolean getMobileDataState(Context myContext, Object[] arg) {
  Boolean bOpen = false;
  try {
    ConnectivityManager myConnectivity = (ConnectivityManager)
myContext.getSystemService(Context.CONNECTIVITY_SERVICE);
    Class myClass = myConnectivity.getClass();
    Class[] myArgs = null;
    if (arg != null) {
     myArgs = new Class[1];
     myArgs[0] = arg.getClass();
    }
    Method myMethod = myClass.getMethod("getMobilcDataEnabled", myArgs);
    bOpen = (Boolean) myMethod.invoke(myConnectivity, arg);
  } catch (Exception e) {
    Toast.makeText(getApplicationContext(),
          e.getMessage().toString(), Toast.LENGTH_SHORT).show();
  }
  return bOpen;
}
```

上面这段代码在 MyCode\MySample264\app\src\main\java\com\bin\luo\mysample\MainActivity.java 文件中。在这段代码中,myMethod = myClass.getMethod("getMobileDataEnabled", myArgs)用于在指定的类中根据方法名称获取方法,因为原始的 getMobileDataEnabled()方法被封装为不可见,这里只能采用反射机制变通一下。然后就可使用 myMethod.invoke(myConnectivity, arg)这种形式调用 getMobileDataEnabled()方法获取移动数据连接的状态。此外,访问移动网络数据

连接需要在 AndroidManifest.xml 文件中添加 <uses-permission android:name=" android.permission.ACCESS_NETWORK_STATE"/> 和 <uses-permission android:name=" android.permission.CHANGE_NETWORK_STATE"/>权限。此实例的完整项目在 MyCode\MySample264 文件夹中。

188 使用 WifiManager 动态打开或关闭 WiFi

此实例主要通过使用系统服务获取 WifiManager，并使用 WifiManager 的 reconnect() 方法和 disconnect() 方法实现动态连接 WiFi 和关闭 WiFi。当实例运行之后，单击"连接 WiFi"按钮，将启动 WiFi，同时在通知栏上显示 WiFi 信号图标，如图 188-1 的左图所示。单击"断开 WiFi"按钮，将关闭 WiFi，同时在通知栏上的 WiFi 信号图标自动消失，如图 188-1 的右图所示。

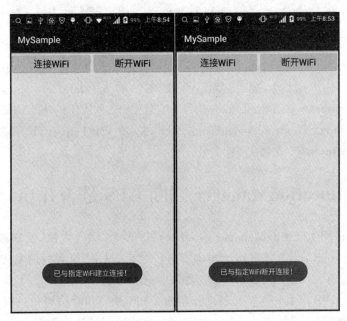

图 188-1

主要代码如下：

```
public class MainActivity extends Activity {
 WifiManager myWifiManager;
 WifiConfiguration myWifiConfiguration;
 @Override
 protected void onCreate(Bundle savedInstanceState) {
  super.onCreate(savedInstanceState);
  setContentView(R.layout.activity_main);
  //获取 WiFi 管理器
  myWifiManager = (WifiManager)getSystemService(Context.WIFI_SERVICE);
  myWifiManager.startScan();                            //开始扫描 WiFi 信号
  myWifiConfiguration = new WifiConfiguration();
  //设置 WiFi 名称(根据测试手机进行设置)
  myWifiConfiguration.SSID = "\"Linksys05506\"";
  myWifiConfiguration.preSharedKey = "\"13996060872\"";   //设置 WiFi 密码
  public void onClickBtn1(View v) {                     //响应单击"连接 WiFi"按钮
```

```
//根据WiFi配置信息获取ID值
    int myWifiId = myWifiManager.addNetwork(myWifiConfiguration);
    myWifiManager.enableNetwork(myWifiId,true);              //启用该WiFi
    myWifiManager.reconnect();                                //重新连接
    Intent myIntent = new Intent(Settings.ACTION_WIFI_SETTINGS);
    startActivity(myIntent);
    Toast.makeText(MainActivity.this,
            "已与指定WiFi建立连接!", Toast.LENGTH_SHORT).show();
}
    public void onClickBtn2(View v) {                        //响应单击"断开WiFi"按钮
    myWifiManager.disableNetwork(myWifiConfiguration.networkId);
    myWifiManager.disconnect();                              //断开连接
    Intent myIntent = new Intent(Settings.ACTION_WIFI_SETTINGS);
    startActivity(myIntent);
    Toast.makeText(MainActivity.this,
        "已与指定WiFi断开连接!", Toast.LENGTH_SHORT).show();
} }
```

上面这段代码在 MyCode\MySample875\app\src\main\java\com\bin\luo\mysample\MainActivity.java 文件中。此外，操作 WiFi 需要在 AndroidManifest.xml 文件中添加 <uses-permission android:name="android.permission.ACCESS_WIFI_STATE"/> 权限和 <uses-permission android:name="android.permission.CHANGE_WIFI_STATE"/> 权限。此实例的完整项目在 MyCode\MySample875 文件夹中。

189 使用 LocationManager 判断 GPS 是否开启

此实例主要通过使用 LocationManager 的 isProviderEnabled() 方法，实现根据该方法的返回值判断当前手机的 GPS 功能是否已经开启。当实例运行之后，单击"检测当前 GPS 状态"按钮，如果当前手机已经开启 GPS，将在弹出的 Toast 中显示"GPS 已经开启!"，如图 189-1 的左图所示；否则将在弹出的 Toast 中显示"GPS 尚未开启!"。手动开启或关闭 GPS 可在"位置信息"中设置，如图 189-1 的右图所示。在 Java 中代码实现开启或关闭 GPS 功能，则需要系统级权限。

图 189-1

主要代码如下：

```java
public class MainActivity extends Activity {
 @Override
 protected void onCreate(Bundle savedInstanceState) {
  super.onCreate(savedInstanceState);
  setContentView(R.layout.activity_main);
 }
 public void onClickmyBtn1(View v) {//响应单击"检测当前GPS状态"按钮
if (isOpen()) {
  Toast.makeText(this, "GPS 已经开启!", Toast.LENGTH_SHORT).show();
  } else {
  Toast.makeText(this, "GPS 尚未开启!", Toast.LENGTH_SHORT).show();
}}
 public boolean isOpen() {
  LocationManager myLocationManager =
                 (LocationManager) getSystemService(Context.LOCATION_SERVICE);
  boolean isGPS =
         myLocationManager.isProviderEnabled(LocationManager.GPS_PROVIDER);
  boolean myNetwork =
         myLocationManager.isProviderEnabled(LocationManager.NETWORK_PROVIDER);
  if (isGPS || myNetwork) {
   return true;
  }
  return false;
}}
```

上面这段代码在 MyCode\MySample464\app\src\main\java\com\bin\luo\mysample\MainActivity.java 文件中。在这段代码中，LocationManager myLocationManager = (LocationManager) getSystemService(Context.LOCATION_SERVICE)用于创建位置服务对象 myLocationManager。isGPS = myLocationManager.isProviderEnabled(LocationManager.GPS_PROVIDER)用于判断当前 GPS 是否已经开启，如果该方法的返回值为 true，表示已经开启 GPS；如果该方法的返回值为 false，表示已经关闭 GPS。此实例的完整项目在 MyCode\MySample464 文件夹中。

190 使用 TelephonyManager 获取运营商等信息

此实例主要通过使用 TelephonyManager 类的成员方法，从而获取手机网络运营商信息和 SIM 卡信息。当实例运行之后，单击"获取运营商信息"按钮，将在弹出的 Toast 中显示手机网络运营商信息，如图 190-1 的左图所示；单击"获取 SIM 卡信息"按钮，将在弹出的 Toast 中显示手机 SIM 卡信息，如图 190-1 的右图所示。出于安全考虑，最近新办理的 SIM 卡上都没有存储手机号码，因此在手机上测试此实例无法显示手机号码，但是在 Android 模拟器上测试可以显示模拟器预置的手机号码。

主要代码如下：

```java
public void onClickmyBtn1(View v) {       //响应单击"获取运营商信息"按钮
 //获得系统提供的 TelephonyManager 实例
 TelephonyManager myManager =
                 (TelephonyManager) getSystemService(Context.TELEPHONY_SERVICE);
```

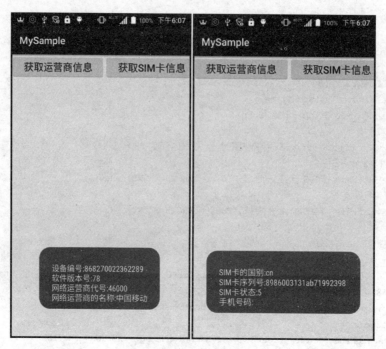

图 190-1

```
    String myInfo = "";
    myInfo += "\n 设备编号:" + myManager.getDeviceId();
    myInfo += "\n 软件版本号:" + myManager.getDeviceSoftwareVersion();
    myInfo += "\n 网络运营商代号:" + myManager.getNetworkOperator();
    myInfo += "\n 网络运营商的名称:" + myManager.getNetworkOperatorName();
    Toast.makeText(getApplicationContext(), myInfo, Toast.LENGTH_SHORT).show();
}
public void onClickmyBtn2(View v) { //响应单击"获取 SIM 卡信息"按钮
    TelephonyManager myManager =
                    (TelephonyManager) getSystemService(Context.TELEPHONY_SERVICE);
    String myInfo = "";
    myInfo += "\nSIM 卡的国别:" + myManager.getSimCountryIso();
    myInfo += "\nSIM 卡序列号:" + myManager.getSimSerialNumber();
    myInfo += "\nSIM 卡状态:" + myManager.getSimState();
    myInfo += "\n 手机号码:" + myManager.getLine1Number();
    Toast.makeText(getApplicationContext(), myInfo, Toast.LENGTH_SHORT).show();
}
```

上面这段代码在 MyCode\MySample189\app\src\main\java\com\bin\luo\mysample\MainActivity.java 文件中。在这段代码中，TelephonyManager myManager =（TelephonyManager）getSystemService（Context.TELEPHONY_SERVICE）用于获取一个 TelephonyManager 对象。在 Android 中，TelephonyManager 主要用于管理手机通话状态、设备信息和 SIM 卡信息以及网络信息，侦听电话呼叫状态、信号强度状态以及可以调用电话拨号器拨打电话等。此外，当使用 TelephonyManager 的成员方法时，通常需要在 AndroidManifest.xml 文件中添加< uses-permission android:name = "android.permission.READ_PHONE_STATE"/>和< uses-permission android:name = "android.permission.ACCESS_COARSE_LOCATION"/>权限。此实例的完整项目在 MyCode\MySample189 文件夹中。

191 使用 TelephonyManager 检测卡槽类型

此实例主要通过使用 TelephonyManager 类的 getPhoneCount()方法获取手机的卡槽数量,实现检测当前手机是双卡槽手机还是单卡槽手机。当实例运行之后,单击"检测当前手机是双卡槽还是单卡槽"按钮,将在弹出的 Toast 中显示手机卡槽类型,效果分别如图 191-1 的左图和右图所示。

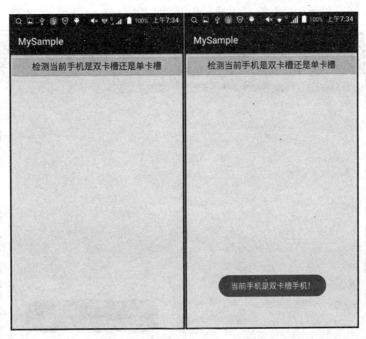

图 191-1

主要代码如下:

```java
//响应单击"检测当前手机是双卡槽还是单卡槽"按钮
public void onClickmyBtn1(View v) {
    TelephonyManager myTelephonyManager =
                (TelephonyManager) getSystemService(TELEPHONY_SERVICE);
    String myInfo = "";
    int myCount = myTelephonyManager.getPhoneCount();
    if (myCount == 1) {
        myInfo = "当前手机是单卡槽手机!";
    } else if (myCount == 2) {
        myInfo = "当前手机是双卡槽手机!";
    } else {
        myInfo = "当前手机卡槽检测出现错误!";
    }
    Toast.makeText(getApplicationContext(), myInfo, Toast.LENGTH_SHORT).show();
}
```

上面这段代码在 MyCode\MySample220\app\src\main\java\com\bin\luo\mysample\MainActivity.java 文件中。在这段代码中,myTelephonyManager =(TelephonyManager)getSystemService(TELEPHONY_SERVICE)用于获取电话管理器 TelephonyManager 的实例。myCount = myTelephonyManager.getPhoneCount()用于获取手机卡槽的数量。此实例的完整项目

在 MyCode\MySample220 文件夹中。

192 使用 PackageManager 获取包名版本等信息

此实例主要通过使用 PackageManager 的成员方法,实现获取当前应用的包名和版本等信息。当实例运行之后,单击"获取当前应用的包名和版本等信息"按钮,将在弹出的 Toast 中显示实例的包名和版本等信息,效果分别如图 192-1 的左图和右图所示。

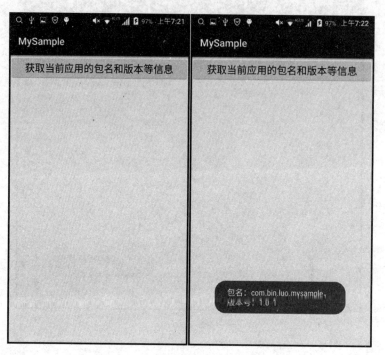

图 192-1

主要代码如下：

```
//响应单击"获取当前应用的包名和版本等信息"按钮
public void onClickmyBtn1(View v) {
  try {
    String myPackageName = this.getPackageName();
    PackageManager myPackageManager = this.getPackageManager();
    String myVersion =
            myPackageManager.getPackageInfo(myPackageName, 0).versionName;
    int myCode = myPackageManager.getPackageInfo(myPackageName, 0).versionCode;
    Toast.makeText(this, "包名：" + myPackageName + ",\n版本号：" +
                   myVersion + "   " + myCode, Toast.LENGTH_SHORT).show();
  } catch (Exception e) { }
}
```

上面这段代码在 MyCode\MySample207\app\src\main\java\com\bin\luo\mysample\MainActivity.java 文件中。在这段代码中,myPackageName = this.getPackageName()用于获取当前应用的包名。myVersion = myPackageManager.getPackageInfo(myPackageName, 0).versionName 用于获取当前应用的版本信息。此实例的完整项目在 MyCode\MySample207 文件夹中。

193 使用 WallpaperManager 随机更换壁纸

此实例主要通过在自定义 Service 服务组件中获取壁纸管理器对象,并在该组件的 onStartCommand() 方法中随机指定图像更换壁纸,然后通过 AlarmManager 对象实现定时执行服务操作,从而实现动态随机更换壁纸的效果。当实例运行之后,单击按钮"启用壁纸更换服务",手机壁纸将动态随机更换,同时当前应用隐藏到通知栏,效果分别如图 193-1 的左图和右图所示。

图 193-1

主要代码如下:

```
public class MainActivity extends Activity {
  private PendingIntent myPendingIntent;
  AlarmManager myAlarmManager;
  Intent myIntent;
  @Override
  protected void onCreate(Bundle savedInstanceState) {
    super.onCreate(savedInstanceState);
    setContentView(R.layout.activity_main);
    myAlarmManager = (AlarmManager) getSystemService(Service.ALARM_SERVICE);
    myIntent = new Intent(MainActivity.this, MyWallpaperService.class);
    myPendingIntent = PendingIntent.getService(MainActivity.this, 0, myIntent, 0);
    //在通知栏显示当前应用
    Intent myBarIntent = new Intent(this, MainActivity.class);
    PendingIntent myPending = PendingIntent.getActivity(MainActivity.this,
                                                        0, myBarIntent, 0);
    Notification myNotification = new Notification.Builder(this)
        .setAutoCancel(true)
        .setContentTitle("壁纸更换服务控制器")
```

```
                .setContentText("")
                .setContentIntent(myPending)
                .setSmallIcon(R.mipmap.ic_launcher)
                .setWhen(System.currentTimeMillis())
                .build();
    NotificationManager myManager =
                (NotificationManager) getSystemService(NOTIFICATION_SERVICE);
    myManager.notify(110, myNotification);
}
public void onClickBtn1(View v) {        //响应单击"启用壁纸更换服务"按钮
    //此时间不准确,耐心等待即可有效果
    myAlarmManager.setRepeating(AlarmManager.RTC_WAKEUP,
                                                0, 2500, myPendingIntent);
    Toast.makeText(MainActivity.this,
                "已启动壁纸更换服务!", Toast.LENGTH_SHORT).show();
    finish();                           //在后台运行服务
}
public void onClickBtn2(View v) {        //响应单击"停止壁纸更换服务"按钮
    myAlarmManager.cancel(myPendingIntent);
    Toast.makeText(MainActivity.this,
                "已停用壁纸更换服务!", Toast.LENGTH_SHORT).show();
}}
```

上面这段代码在 MyCode\MySample755\app\src\main\java\com\bin\luo\mysample\MainActivity.java 文件中。在这段代码中,myAlarmManager.setRepeating(AlarmManager. RTC_WAKEUP,0,2500,myPendingIntent)表示在指定时间内多次重复调用 myPendingIntent 的壁纸更换服务,该方法的语法声明如下:

```
setRepeating(int type, long triggerAtMillis,
                    long intervalMillis, PendingIntent operation)
```

其中,参数 int type 表示闹钟(AlarmManager)的类型,如 ELAPSED_REALTIME、RTC、RTC_WAKEUP、ELAPSED_REALTIME_WAKEUP 等;ELAPSED_REALTIME 表示闹钟在手机睡眠状态下不可用,该状态下闹钟使用相对时间(相对于系统启动时间);ELAPSED_REALTIME_WAKEUP 表示闹钟在手机睡眠状态下会唤醒系统并执行提示功能,该状态下闹钟也使用相对时间;RTC 表示闹钟在手机睡眠状态下不可用,该状态下闹钟使用绝对时间,即当前系统时间;RTC_WAKEUP 表示闹钟在手机睡眠状态下会唤醒系统并执行提示功能,该状态下闹钟使用绝对时间。

参数 long triggerAtMillis 表示闹钟的第一次执行时间,以毫秒为单位,可以自定义时间,不过一般使用当前时间;需要注意的是,此参数与第一个参数 type 密切相关。如果第一个参数对应的闹钟使用的是相对时间(ELAPSED_REALTIME 和 ELAPSED_REALTIME_WAKEUP),那么此参数就得使用相对时间(相对于系统启动时间来说)。

参数 long intervalMillis 表示两次闹钟执行的间隔时间,以毫秒为单位。

myIntent = new Intent(MainActivity.this,MyWallpaperService.class)中的 MyWallpaperService.class 表示自定义更换壁纸服务,该类的主要代码如下:

```
public class MyWallpaperService extends Service {
    int[] myWallpapers = new int[]{R.mipmap.myimage1, R.mipmap.myimage2,
        R.mipmap.myimage3, R.mipmap.myimage4, R.mipmap.myimage5};        //定义壁纸资源
```

```
private WallpaperManager myWallpaperManager;
@Override
public IBinder onBind(Intent intent) {   return null; }
@Override
public int onStartCommand(Intent intent, int flags, int startId) {
 try {                          //更换壁纸
  myWallpaperManager.setResource(
            myWallpapers[new Random().nextInt(myWallpapers.length)]);
 } catch (Exception e) { e.printStackTrace(); }
 return START_STICKY;
}
@Override
public void onCreate() {
 super.onCreate();
 myWallpapaerManager = WallpaperManager.getInstance(this);
} }
```

上面这段代码在 MyCode\MySample755\app\src\main\java\com\bin\luo\mysample\MyWallpaperService.java 文件中。一般情况下，自定义服务应该在 MyCode\MySample755\app\src\main\AndroidManifest.xml 文件中进行注册，如下面的粗体字所示：

```
<?xml version = "1.0" encoding = "utf-8"?>
< manifest xmlns:android = "http://schemas.android.com/apk/res/android"
         package = "com.bin.luo.mysample">
 < application
     android:allowBackup = "true"
     android:icon = "@mipmap/ic_launcher"
     android:label = "@string/app_name"
     android:supportsRtl = "true"
     android:theme = "@style/AppTheme">
  < activity android:name = ".MainActivity">
   < intent - filter >
    < action android:name = "android.intent.action.MAIN"/>
    < category android:name = "android.intent.category.LAUNCHER"/>
   </ intent - filter >
  </ activity >
  < service android:name = ".MyWallpaperService"/>
 </ application >
 < uses - permission android:name = "android.permission.SET_WALLPAPER"/>
</ manifest >
```

此实例的完整项目在 MyCode\MySample755 文件夹中。

194　使用 RingtoneManager 自定义来电铃声

此实例主要通过使用 RingtoneManager 的 setActualDefaultRingtoneUri() 方法，实现将指定的音乐文件设置为手机来电铃声。当实例运行之后，在"音乐文件："输入框中输入在 SD 卡上存储的音乐文件名称，然后单击"播放音乐文件"按钮，将播放指定的音乐；单击"设置为来电铃声"按钮，该音乐将被设置为来电铃声，如图 194-1 的左图所示；当手机有来电时，则自动播放此音乐，如图 194-1 的右图所示。

图 194-1

主要代码如下：

```java
public void onClickmyBtn1(View v) {              //响应单击"播放音乐文件"按钮
  try {
    MediaPlayer myMediaPlayer = new MediaPlayer();
    myMediaPlayer.setDataSource(myEditText.getText().toString());
    myMediaPlayer.prepare();
    myMediaPlayer.start();
  } catch (Exception e) { e.printStackTrace();}
}
public void onClickmyBtn2(View v) {              //响应单击"设置为来电铃声"按钮
  try {
    setUserCustomVoice(myEditText.getText().toString(),
                                    RingtoneManager.TYPE_RINGTONE);
    Toast.makeText(this, "设置来电铃声操作成功!", Toast.LENGTH_SHORT).show();
  } catch (Exception e) { e.printStackTrace(); }
}
private void setUserCustomVoice (String path, int type){
  File myFile = new File(path);
  String myMimeType = "audio/*";
  ContentValues myContentValues = new ContentValues();
  myContentValues.put(MediaStore.MediaColumns.DATA, myFile.getAbsolutePath());
  myContentValues.put(MediaStore.MediaColumns.TITLE, myFile.getName());
  myContentValues.put(MediaStore.MediaColumns.MIME_TYPE, myMimeType);
  myContentValues.put(MediaStore.MediaColumns.SIZE, myFile.length());
  myContentValues.put(MediaStore.Audio.Media.IS_RINGTONE, false);
  myContentValues.put(MediaStore.Audio.Media.IS_NOTIFICATION, false);
  myContentValues.put(MediaStore.Audio.Media.IS_ALARM, false);
  myContentValues.put(MediaStore.Audio.Media.IS_MUSIC, false);
  switch (type){
    case RingtoneManager.TYPE_NOTIFICATION:
      myContentValues.put(MediaStore.Audio.Media.IS_NOTIFICATION, true);
      break;
    case RingtoneManager.TYPE_ALARM:
```

```
        myContentValues.put(MediaStore.Audio.Media.IS_ALARM, true);
        break;
      case RingtoneManager.TYPE_RINGTONE:
        myContentValues.put(MediaStore.Audio.Media.IS_RINGTONE, true);
        break;
      default:
        type = RingtoneManager.TYPE_ALL;
        myContentValues.put(MediaStore.Audio.Media.IS_MUSIC, true);
        break;
    }
    Uri myUri =
        MediaStore.Audio.Media.getContentUriForPath(myFile.getAbsolutePath());
    getContentResolver().delete(myUri, null, null);
    final Uri newUri = getContentResolver().insert(myUri, myContentValues);
    RingtoneManager.setActualDefaultRingtoneUri(MainActivity.this, type, newUri);
}
```

上面这段代码在 MyCode\MySample665\app\src\main\java\com\bin\luo\mysample\MainActivity.java 文件中。在这段代码中，setUserCustomVoice（String path，int type）自定义方法除了可以设置来电铃声 TYPE_RINGTONE 之外，也可以设置闹钟铃声 TYPE_ALARM 和通知铃声 TYPE_NOTIFICATION 等。此外，访问外部存储介质和设置手机铃声需要在 AndroidManifest.xml 文件中添加 < uses-permission android：name = " android. permission. READ _ EXTERNAL _ STORAGE"/>权限、< uses-permission android：name = " android. permission. WRITE_ EXTERNAL_ STORAGE"/>权限、< uses-permission android：name= "android. permission. WRITE_SETTINGS"/>权限。此实例的完整项目在 MyCode\MySample665 文件夹中。

195 通过重力传感器控制飞行器的轨迹和速度

此实例主要通过使用重力传感器感应手机放置位置的变化，实现控制飞行器的轨迹和速度。当实例运行之后，在蓝天白云背景下面有一个飞行器，如果不断晃动手机，此飞行器将不断改变移动轨迹和速度，效果分别如图 195-1 的左图和右图所示。

图　195-1

主要代码如下:

```java
public class MainActivity extends Activity
        implements SensorEventListener, SurfaceHolder.Callback {
  float mPosX, mPosY;
  private Bitmap myAirCraft;
  private Bitmap mySky;
  private int myScreenWidth;
  private int myScreenHeight;
  private SurfaceHolder myHolder;
  private SurfaceView mySurfaceView;
  @Override
  protected void onCreate(Bundle savedInstanceState) {
    super.onCreate(savedInstanceState);
    setContentView(R.layout.activity_main);
    mySurfaceView = (SurfaceView) findViewById(R.id.mySurfaceView);
    myHolder = mySurfaceView.getHolder();
    myHolder.addCallback(this);
    myScreenWidth = getWindowManager().getDefaultDisplay().getWidth();
    myScreenHeight = getWindowManager().getDefaultDisplay().getHeight();
    mySky = Bitmap.createScaledBitmap(BitmapFactory.decodeResource(getResources(), R.mipmap.myimage1),
myScreenWidth, myScreenHeight, false);    //背景图像
    myAirCraft = BitmapFactory.decodeResource(getResources(),
              R.mipmap.myimage2);     //飞行器图像
  }
  @Override
  //根据重力传感器的 X、Y 值变化绘制飞行器
  public void onSensorChanged(final SensorEvent event) {
 new Thread() {
    @Override
    public void run() {
      float currentX = event.values[SensorManager.DATA_X];
      float currentY = event.values[SensorManager.DATA_Y];
      mPosX -= currentX;
      mPosY += currentY;
      if(mPosX < 0)
        mPosX = myScreenWidth;
      if(mPosY > myScreenHeight)
        mPosY = 0;
      Canvas myCanvas = myHolder.lockCanvas();
      myCanvas.drawBitmap(mySky, 0, 0, new Paint());
      myCanvas.drawBitmap(myAirCraft, mPosX, mPosY, new Paint());
      myHolder.unlockCanvasAndPost(myCanvas);
    } }.start();
}
  @Override
  public void onAccuracyChanged(Sensor sensor, int accuracy) { }
  @Override
  public void surfaceCreated(SurfaceHolder holder) {
    Canvas myCanvas = holder.lockCanvas();
    myCanvas.drawBitmap(mySky, 0, 0, new Paint());
    holder.unlockCanvasAndPost(myCanvas);
    SensorManager mySensorManager =
              (SensorManager) getSystemService(SENSOR_SERVICE);
    //表示当前操作的传感器是重力(加速度)传感器
```

```
    Sensor mySensor = mySensorManager.getDefaultSensor(Sensor.TYPE_ACCELEROMETER);
    //注册重力传感器的监听器
    mySensorManager.registerListener(this, mySensor,
                                    SensorManager.SENSOR_DELAY_GAME);
}
@Override
public void surfaceChanged(SurfaceHolder holder,
                          int format, int width, int height){ }
@Override
public void surfaceDestroyed(SurfaceHolder holder) { }
}
```

上面这段代码在 MyCode \ MySample763 \ app \ src \ main \ java \ com \ bin \ luo \ mysample \ MainActivity.java 文件中。在这段代码中，mySensorManager.registerListener(this，mySensor，SensorManager.SENSOR_DELAY_GAME)用于注册重力传感器的监听功能，在此可以设置传感器的数据频率等参数。registerListener()方法的语法声明如下：

```
    public boolean registerListener(SensorEventListener listener,
                                    Sensor sensor, int samplingPeriodUs)
```

其中，SensorEventListener listener 参数表示传感器的监听器，该监听器需要实现 SensorEventListener 接口，实现该接口需重写 onSensorChanged() 和 onAccuracyChanged() 方法。Sensor sensor 参数表示传感器对象。int samplingPeriodUs 参数用于指定传感器的数据频率，它支持以下几个频率值。

（1）SensorManager.SENSOR_DELAY_FASTEST，该值表示最快。即延迟小，只有特别依赖传感器的应用才推荐使用，因为这会造成手机耗电量增大。

（2）SensorManager.SENSOR_DELAY_NORMAL，该值适合正常频率。

（3）SensorManager.SENSOR_DELAY_GAME，该值表示游戏的频率。

（4）SensorManager.SENSOR_DELAY_UI，该值适合普通用户界面的频率。

onSensorChanged(final SensorEvent event)事件响应方法用于获取重力传感器实时测量出来的 X、Y、Z 的变化值，实例即根据这些值绘制飞行器图像，X、Y、Z 的三维坐标如图 195-2 所示。

此实例的完整项目在 MyCode \ MySample763 文件夹中。

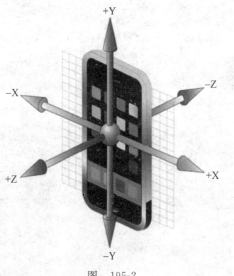

图 195-2

196 使用加速度传感器实现微信的摇一摇功能

此实例主要通过使用自定义加速度（重力）传感器类 SensorHelper，从而实现类似于微信的摇一摇功能，即摇一摇手机就可生成一个新的随机验证码的功能。当实例运行之后，每次摇一摇手机，即可在弹出的 Toast 中显示不同的随机验证码，效果分别如图 196-1 的左图和右图所示。

图 196-1

主要代码如下：

```java
public class MainActivity extends Activity {
 Handler myHandler = new Handler() {
  @Override
  public void handleMessage(Message msg) {
   if (msg.what == 1) {
    Toast.makeText(MainActivity.this,
           "新的随机验证码是：" + msg.obj, Toast.LENGTH_LONG).show();
   } } };
 @Override
 protected void onCreate(Bundle savedInstanceState) {
  super.onCreate(savedInstanceState);
  setContentView(R.layout.activity_main);
  final SensorHelper mySensorHelper = new SensorHelper(this);
  mySensorHelper.setOnShakeListener(new SensorHelper.OnShakeListener() {
   @Override
   public void onShake() {   //响应摇  摇事件
    GetNewRandomChars(myHandler);
   } });
 }
 public static void GetNewRandomChars(final Handler handler) {
  new Thread() {
   @Override
   public void run() {
    try {
     String RANDOM_CHARS =
         "123456789QWERTYUIOPASDFGHJKLZXCVBNMqwertyuiopasdfghjklzxcvbnm";
     StringBuilder myBuilder = new StringBuilder();
```

```
        //产生随机验证码
        for (int i = 0; i < 4; i++) {
          myBuilder.append(RANDOM_CHARS.charAt(
                             new Random().nextInt(RANDOM_CHARS.length())));
        }
        Message myMessage = new Message();
        myMessage.what = 1;
        myMessage.obj = myBuilder.toString();
        handler.sendMessage(myMessage);
      } catch (Exception e) { e.printStackTrace(); }
    } }.start();
} }
```

上面这段代码在 MyCode \ MySample759 \ app \ src \ main \ java \ com \ bin \ luo \ mysample \ MainActivity.java 文件中。在这段代码中，mySensorHelper = new SensorHelper(this)中的 SensorHelper 是一个检测重力传感器变化的自定义类，该类的主要代码如下：

```
public class SensorHelper implements SensorEventListener {
  //重力感应三维坐标
  private float lastX;
  private float lastY;
  private float lastZ;
  //两次摇一摇的时间间隔阈值
  private static final int UPDATE_INTERVAL_TIME = 50;
  //上次检测时间
  private long lastUpdateTime;
  //速度阈值,当摇晃速度达到该值后产生作用
  private static final int SPEED_THRESHOLD = 5000;
  //重力感应监听器
  private OnShakeListener onShakeListener;
  //设置重力感应监听器
  public void setOnShakeListener(OnShakeListener listener) {
    onShakeListener = listener;
  }
  public SensorHelper(Context context) {
    SensorManager myManager =
         (SensorManager) context.getSystemService(Context.SENSOR_SERVICE);
    //获取手机上的重力(加速度)传感器
    Sensor mySensor = myManager.getDefaultSensor(Sensor.TYPE_ACCELEROMETER);
    myManager.registerListener(this, mySensor, SensorManager.SENSOR_DELAY_GAME);
  }
  @Override
  public void onSensorChanged(SensorEvent event) {
    long currentUpdateTime = System.currentTimeMillis();        //当前检测时间
    long timeInterval = currentUpdateTime - lastUpdateTime;     //两次检测时间间隔
    //判断是否达到了检测时间间隔阈值
    if (timeInterval < UPDATE_INTERVAL_TIME) return;
    lastUpdateTime = currentUpdateTime;                         //当前时间变成上次时间
    float x = event.values[0];                                  //获得 X、Y、Z 坐标
    float y = event.values[1];
    float z = event.values[2];
    float deltaX = x - lastX;                                   //获得 X、Y、Z 的变化值
```

```
            float deltaY = y - lastY;
            float deltaZ = z - lastZ;
            lastX = x;                      //将当前坐标变成上次坐标
            lastY = y;
            lastZ = z;
            double speed = Math.sqrt(deltaX * deltaX
                    + deltaY * deltaY + deltaZ * deltaZ)/ timeInterval * 10000;
            if (speed >= SPEED_THRESHOLD)   //达到速度阈值,发出提示
                onShakeListener.onShake();
        }
        @Override
        public void onAccuracyChanged(Sensor sensor, int accuracy) { }
        interface OnShakeListener { void onShake(); }
    }
```

上面这段代码在 MyCode\MySample759\app\src\main\java\com\bin\luo\mysample\SensorHelper.java 文件中。此实例的完整项目在 MyCode\MySample759 文件夹中。

197　使用传感器监测手机周围光线亮度变化

此实例主要通过在 SensorEventListener 的 onSensorChanged() 事件响应方法中监听光线传感器的值,实现使用手机监测周围的光线亮度变化的功能。当实例运行之后,如果把手机放在光线较弱的地方,则显示的提示文字字体自动放大,如图 197-1 的左图所示。如果把手机放在光线较强的地方,则显示的提示文字字体自动缩小,如图 197-1 的右图所示。

图　197-1

主要代码如下：

```
public class MainActivity extends Activity implements SensorEventListener {
    TextView myTextView;
    private SensorManager mySensorManager;
    private Sensor mySensor;
```

```java
@Override
protected void onCreate(Bundle savedInstanceState) {
  super.onCreate(savedInstanceState);
  setContentView(R.layout.activity_main);
  myTextView = (TextView) findViewById(R.id.myTextView);
  mySensorManager = (SensorManager)getSystemService(SENSOR_SERVICE);
  mySensor = mySensorManager.getDefaultSensor(Sensor.TYPE_LIGHT);
}
@Override
protected void onResume() {
  mySensorManager.registerListener(this, mySensor,
                        SensorManager.SENSOR_DELAY_NORMAL);
  super.onResume();
}
@Override
protected void onPause() {
  mySensorManager.unregisterListener(this);
  super.onPause();
}
public void onAccuracyChanged(Sensor sensor, int accuracy) { }
public void onSensorChanged(SensorEvent event) {
  if (event.sensor.getType() == Sensor.TYPE_LIGHT) {
    float myLight = event.values[0];
    myTextView.setTextSize(30 - myLight/10);
    myTextView.setText("当前手机周围的亮度值是:" + myLight);
} } }
```

上面这段代码在 MyCode\MySample272\app\src\main\java\com\bin\luo\mysample\MainActivity.java 文件中。在这段代码中，mySensor = mySensorManager.getDefaultSensor(Sensor.TYPE_LIGHT)表示当前传感器的类型是光线传感器，该传感器通过 onSensorChanged(SensorEvent event)的 event.values[0]获取手机周围的光线亮度。需要说明的是，当使用光线传感器监测手机周围的光线亮度时，反应有点慢，有时可能需要几秒钟才能检测出来，这主要与手机硬件有关。此实例的完整项目在 MyCode\MySample272 文件夹中。

198 使用方向传感器实现自制指南针

此实例主要通过在 SensorEventListener 的 onSensorChanged()事件响应方法中监听方向传感器的值，实现自制指南针。当实例运行后，将手机对着任意方向，指南针立即指示当前的方位，效果分别如图 198-1 的左图和右图所示。

主要代码如下：

```java
public class MainActivity extends Activity implements SensorEventListener{
  ImageView myImageView;
  TextView myTextView;
  float myDegree = 0f;                        //指南针图像的旋转角度
  SensorManager mySensorManager;
  @Override
  public void onCreate(Bundle savedInstanceState) {
    super.onCreate(savedInstanceState);
    setContentView(R.layout.activity_main);
    myImageView = (ImageView)findViewById(R.id.myImageView);
```

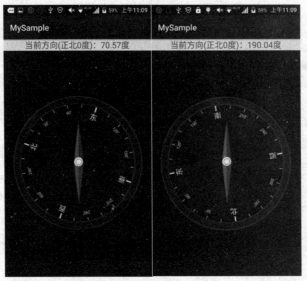

图 198-1

```
    myTextView = (TextView)findViewById(R.id.myTextView);
    mySensorManager = (SensorManager)getSystemService(SENSOR_SERVICE);
}
@Override
protected void onResume(){
    super.onResume();
    mySensorManager.registerListener(this,
        mySensorManager.getDefaultSensor(Sensor.TYPE_ORIENTATION),
        SensorManager.SENSOR_DELAY_GAME);            //注册传感器
}
@Override
protected void onPause(){
    mySensorManager.unregisterListener(this);        //取消注册
    super.onPause();
}
@Override
protected void onStop(){
    mySensorManager.unregisterListener(this);
    super.onStop();
}
@Override                                             //响应精度值改变
public void onAccuracyChanged(Sensor sensor, int accuracy) { }
@Override                                             //响应传感器值改变
public void onSensorChanged(SensorEvent event) {
    //获取触发 event 的传感器类型
    int mySensorType = event.sensor.getType();
    switch(mySensorType){
        case Sensor.TYPE_ORIENTATION:                 //方向传感器
            float myDegree = event.values[0];         //获取转过的角度
            myTextView.setText("当前方向(正北 0 度): " + myDegree + "度");
            RotateAnimation myRotateAnimation = new RotateAnimation(this.myDegree,
                - myDegree,Animation.RELATIVE_TO_SELF,
                0.5f,Animation.RELATIVE_TO_SELF,0.5f);  //旋转指南针时产生动画
            myRotateAnimation.setDuration(100);
            myImageView.startAnimation(myRotateAnimation);
            this.myDegree = - myDegree;
            break;
}}}
```

上面这段代码在 MyCode\MySample270\app\src\main\java\com\bin\luo\mysample\MainActivity.java 文件中。在这段代码中，Sensor.TYPE_ORIENTATION 表示当前传感器是方向传感器，event.values[0] 表示方向传感器转过的角度。方向传感器提供三个数据，分别为 azimuth、pitch 和 roll。azimuth 表示方位，返回磁北极和 Y 轴的夹角，范围为 0°～360°，0°=北，90°=东，180°=南，270°=西。pitch 表示 X 轴和水平面的夹角，范围为−180°～180°；当 Z 轴向 Y 轴转动时，角度为正值。roll 表示 Y 轴和水平面的夹角，由于历史原因，此值范围为−90°～90°；当 X 轴向 Z 轴移动时，角度为正值。此外，监听传感器值变化的监听器必须经过注册才有效，如 mySensorManager.registerListener(this, mySensorManager.getDefaultSensor(Sensor.TYPE_ORIENTATION), SensorManager.SENSOR_DELAY_GAME)。此实例的完整项目在 MyCode\MySample270 文件夹中。

199　使用 DisplayMetrics 获取屏幕分辨率

此实例主要通过使用显示参数 DisplayMetrics，实现获取当前手机的屏幕分辨率。当实例运行之后，单击"获取手机屏幕分辨率"按钮，将在弹出的 Toast 中显示当前手机的屏幕分辨率，如图 199-1 所示。

图　199-1

主要代码如下：

```
//响应单击"获取手机屏幕分辨率"按钮
public void onClickmyBtn1(View v) {
    DisplayMetrics myMetrics = new DisplayMetrics();
    getWindowManager().getDefaultDisplay().getMetrics(myMetrics);
    String myInfo = "当前手机的屏幕分辨率是：";
    myInfo += myMetrics.widthPixels;
    myInfo += " * " + myMetrics.heightPixels;
```

```
         Toast.makeText(getApplicationContext(), myInfo,Toast.LENGTH_SHORT).show();
}
```

上面这段代码在 MyCode\MySample458\app\src\main\java\com\bin\luo\mysample\MainActivity.java 文件中。在这段代码中,myMetrics = new DisplayMetrics()用于创建显示参数的实例。getWindowManager().getDefaultDisplay().getMetrics(myMetrics)用于获取默认的显示参数。myMetrics.widthPixels 表示水平像素。myMetrics.heightPixels 表示垂直像素。此实例的完整项目在 MyCode\MySample458 文件夹中。

200　使用 StatFs 获取存储卡的空间大小信息

此实例主要通过使用 StatFs 的 getBlockCount( )方法、getAvailableBlocks( )方法和 getBlockSize( )方法,实现获取存储卡的总空间和可用空间大小信息。当实例运行之后,单击"获取 SD 卡总空间大小"按钮,将在弹出的 Toast 中显示 SD 卡的总空间大小,如图 200-1 的左图所示;单击"获取 SD 卡可用空间大小"按钮,将在弹出的 Toast 中显示 SD 卡的可用空间大小,如图 200-1 的右图所示。

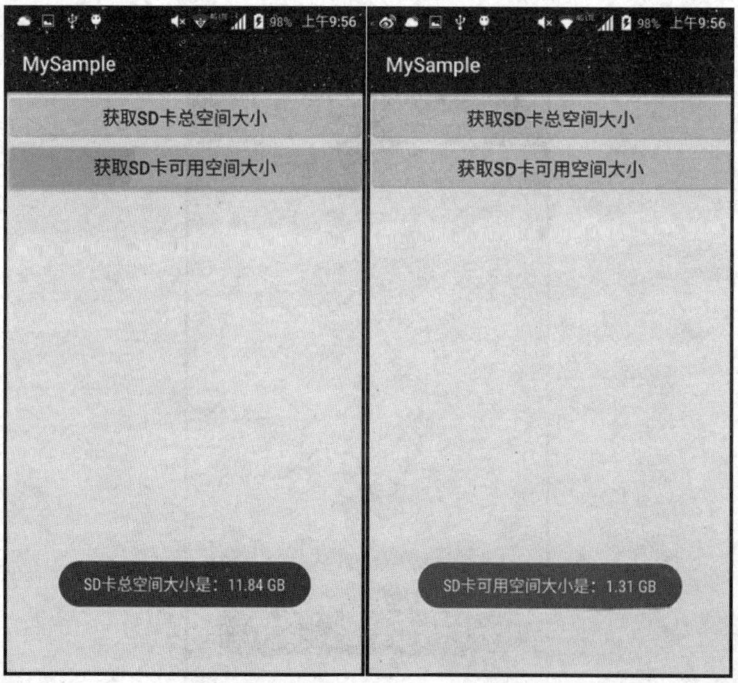

图　200-1

主要代码如下:

```
public void onClickmyBtn1(View v) {               //响应单击"获取 SD 卡总空间大小"按钮
    if(isSdSafe()){
      Toast.makeText(this, "SD 卡总空间大小是: "
                 + getTotalSize(this), Toast.LENGTH_SHORT).show();
}}
public void onClickmyBtn2(View v) {               //响应单击"获取 SD 卡可用空间大小"按钮
    if(isSdSafe()){
      Toast.makeText(this, "SD 卡可用空间大小是: "
```

```java
                                + getAvailableSize(this), Toast.LENGTH_SHORT).show();
    }}
    public boolean isSdSafe() {                    //判断SD卡是否可用
        //当返回值为true时,表示SD卡可正常使用
        return Environment.getExternalStorageState().equals(
                                Environment.MEDIA_MOUNTED);
    }
    public String getTotalSize(Context context) {     //获取SD卡总空间大小
        //获取SD卡根目录
        File myPath = Environment.getExternalStorageDirectory();
        //获取SD卡根目录的存储信息
        StatFs myStatFs = new StatFs(myPath.getPath());
        //获取单个扇区大小
        long myBlockSize = myStatFs.getBlockSize();
        //获取扇区的数量
        long myBlockCount = myStatFs.getBlockCount();
        //总空间 = 扇区数量 * 扇区大小
        long myTotalSize = myBlockSize * myBlockCount;
        return Formatter.formatFileSize(context, myTotalSize);
    }
    public String getAvailableSize(Context context){ //获取SD卡可用空间大小
        //获取SD卡目录
        File path = Environment.getExternalStorageDirectory();
        //获取SD卡目录的存储信息
        StatFs myStatFs = new StatFs(path.getPath());
        //获取单个扇区大小
        long myBlockSize = myStatFs.getBlockSize();
        //获取可用扇区数量
        long myAvailableBlocks = myStatFs.getAvailableBlocks();
        //可用空间 = 扇区大小 * 可用扇区数量
        long myAvailableSize = myBlockSize * myAvailableBlocks;
        return Formatter.formatFileSize(context, myAvailableSize);
    }
```

上面这段代码在 MyCode \ MySample666 \ app \ src \ main \ java \ com \ bin \ luo \ mysample \ MainActivity.java 文件中。在上面这段代码中,myBlockSize = myStatFs.getBlockSize()用于获取单个扇区的大小。myBlockCount = myStatFs.getBlockCount()用于获取全部扇区的数量。myAvailableBlocks = myStatFs.getAvailableBlocks()用于获取可用扇区的数量。扇区大小 * 扇区数量即为空间大小。此实例的完整项目在 MyCode\MySample666 文件夹中。

201 使用 Camera 实现打开或关闭手电筒

此实例主要通过使用 Camera.Parameters 的 setFlashMode()方法,实现打开或关闭手机的闪光灯(手电筒)。当实例运行之后,单击"开启闪光灯"按钮,将打开手机的闪光灯;单击"关闭闪光灯"按钮,将关闭手机的闪光灯,效果分别如图 201-1 的左图和右图所示。

主要代码如下:

```java
public void onClickmyBtn1(View v) {              //响应单击"开启闪光灯"按钮
    Camera.Parameters myParameters = myCamera.getParameters();
    myParameters.setFlashMode(Camera.Parameters.FLASH_MODE_TORCH);
    myCamera.setParameters(myParameters);
```

图 201-1

```
}
public void onClickmyBtn2(View v) {          //响应单击"关闭闪光灯"按钮
    Camera.Parameters myParameters = myCamera.getParameters();
    myParameters.setFlashMode(Camera.Parameters.FLASH_MODE_OFF);
    myCamera.setParameters(myParameters);
}
```

　　上面这段代码在 MyCode\MySample223\app\src\main\java\com\bin\luo\mysample\MainActivity.java 文件中。在这段代码中，Camera.Parameters myParameters = myCamera.getParameters()用于获取手机的照相机参数信息。myParameters.setFlashMode（Camera.Parameters.FLASH_MODE_TORCH）用于设置手机的闪光灯为打开状态。myParameters.setFlashMode(Camera.Parameters.FLASH_MODE_OFF)用于设置手机的闪光灯为关闭状态。此外，操作手机闪光灯需要在 AndroidManifest.xml 文件中添加< uses-permission android:name="android.permission.FLASHLIGHT"/> 权限、< uses-permission android:name="android.permission.WAKE_LOCK"/> 权限、< uses-permission android:name="android.permission.CAMERA"/>权限和< uses-feature android:name="android.hardware.camera"/>权限。此实例的完整项目在 MyCode\MySample223 文件夹中。

202　使用 Camera 捕捉前置和后置摄像头画面

　　此实例主要通过扫描手机上的所有摄像头，并通过 Camera 的 facing 属性，从而确定是否以前置摄像头或后置摄像头捕捉画面。当实例运行之后，单击"后置摄像头"按钮，将显示后置摄像头捕捉的画面，如图 202-1 的左图所示；单击"前置摄像头"按钮，将显示前置摄像头捕捉的画面，如图 202-1 的右图所示。

图 202-1

主要代码如下:

```java
public class MainActivity extends Activity implements SurfaceHolder.Callback{
  private SurfaceView mySurfaceView;
  Camera myCamera;
  private SurfaceHolder mySurfaceHolder;
  @Override
  protected void onCreate(Bundle savedInstanceState) {
    super.onCreate(savedInstanceState);
    setContentView(R.layout.activity_main);
    mySurfaceView = (SurfaceView) findViewById(R.id.mySurfaceView);
    mySurfaceHolder = mySurfaceView.getHolder();
    mySurfaceHolder.addCallback(this);
  }
  @Override
  public void surfaceCreated(SurfaceHolder holder) {
    if (myCamera == null) {
      myCamera = Camera.open();                         //开启相机
      try {
        myCamera.setDisplayOrientation(90);
        myCamera.setPreviewDisplay(holder);             //通过 SurfaceView 显示取景画面
        myCamera.startPreview();                        //开始预览
      } catch (IOException e) { e.printStackTrace(); }
    }
  }
  @Override
  public void surfaceChanged(SurfaceHolder holder,
                             int format, int width, int height){ }
  @Override
  public void surfaceDestroyed(SurfaceHolder holder) {
    myCamera.stopPreview();
```

```
        myCamera.release();
        myCamera = null;
    }
    public void onClickBtn1(View v) {                    //响应单击"后置摄像头"按钮
      Camera.CameraInfo myCameraInfo = new Camera.CameraInfo();
      int myCameraCount = myCamera.getNumberOfCameras();
      for (int i = 0; i < myCameraCount; i++) {
        Camera.getCameraInfo(i, myCameraInfo);           //获取每个摄像头的信息
        //代表摄像头的方位,CAMERA_FACING_BACK 表示后置摄像头
        if (myCameraInfo.facing == Camera.CameraInfo.CAMERA_FACING_BACK) {
          myCamera.stopPreview();                        //停止原来摄像头的预览
          myCamera.release();                            //释放资源
          myCamera = null;                               //取消原来摄像头
          myCamera = Camera.open(i);                     //打开当前选中的摄像头
          myCamera.setDisplayOrientation(90);
          try {
            myCamera.setPreviewDisplay(mySurfaceHolder); //显示取景画面
          } catch (IOException e) { e.printStackTrace(); }
          myCamera.startPreview();                       //开始预览
    } } }
    public void onClickBtn2(View v) {                    //响应单击"前置摄像头"按钮
      Camera.CameraInfo myCameraInfo = new Camera.CameraInfo();
      int myCameraCount = myCamera.getNumberOfCameras();
      for (int i = 0; i < myCameraCount; i++) {          //获取每个摄像头的信息
        Camera.getCameraInfo(i, myCameraInfo);
        //代表摄像头方位,CAMERA_FACING_FRONT 表示前置摄像头
        if (myCameraInfo.facing == Camera.CameraInfo.CAMERA_FACING_FRONT) {
          myCamera.stopPreview();                        //停止原来摄像头的预览
          myCamera.release();                            //释放资源
          myCamera = null;                               //取消原来摄像头
          myCamera = Camera.open(i);                     //打开当前选中的摄像头
          myCamera.setDisplayOrientation(90);
          try {
            myCamera.setPreviewDisplay(mySurfaceHolder); //显示取景画面
          } catch (IOException e) { e.printStackTrace(); }
          myCamera.startPreview();                       //开始预览
    } } }
```

上面这段代码在 MyCode\MySample750\app\src\main\java\com\bin\luo\mysample\MainActivity.java 文件中。在这段代码中,Camera 的 getNumberOfCameras()方法用于获取摄像头数量。myCameraInfo.facing == Camera.CameraInfo.CAMERA_FACING_BACK 表示当前是后置摄像头。myCameraInfo.facing == Camera.CameraInfo.CAMERA_FACING_FRONT 表示当前是前置摄像头。此外,操作相机需要在 AndroidManifest.xml 文件中添加< uses-permission android:name = " android.permission.CAMERA "/> 权限。此实例的完整项目在 MyCode\MySample750 文件夹中。

203 使用 TextureView 实现照相机的预览功能

此实例主要通过以 TextureView 类为基类创建新类 CameraPreview,实现预览功能。当实例运行之后,摄像头获取的预览图像通过 CameraPreview 显示出来,如图 203-1 的左图所示;单击"开始照

相"按钮,当前的图像将以照片的形式保存在"手机存储\Pictures\MyCameraApp"文件夹中,如图 203-1 的右图所示。

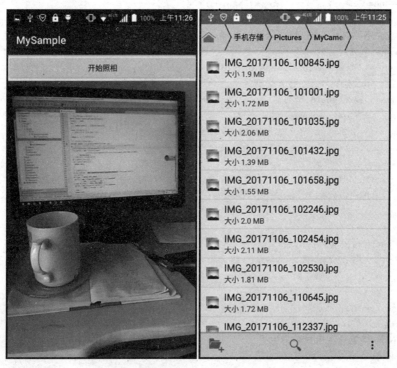

图 203-1

主要代码如下:

```
public class MainActivity   extends Activity {
 public static final int MEDIA_TYPE_IMAGE = 1;
 public static final int MEDIA_TYPE_VIDEO = 2;
 private Camera myCamera;
 private CameraPreview myPreview;
 private Camera.PictureCallback myPicture = new Camera.PictureCallback() {
  @Override
  public void onPictureTaken(byte[] data, Camera camera) {
   File myFile = getOutputMediaFile(MEDIA_TYPE_IMAGE);
   if (myFile == null) { return; }
   try {
    FileOutputStream myStream = new FileOutputStream(myFile);
    myStream.write(data);
    myStream.close();
    Toast.makeText(MainActivity.this,
                   "图像保存成功!",Toast.LENGTH_LONG).show();
    camera.startPreview();                     //重新开始预览
   } catch (Exception e) { } } };
 private FrameLayout myFrameLayout;
 @Override
 protected void onCreate(Bundle savedInstanceState) {
  super.onCreate(savedInstanceState);
  setContentView(R.layout.activity_main);
```

```
    myCamera = getCameraInstance();
    myCamera.setDisplayOrientation(90);
    myPreview = new CameraPreview(this,myCamera);
    myPreview.setSurfaceTextureListener(myPreview);
    //设置透明度;不要这行代码,则完全清晰
    //myPreview.setAlpha(0.5f);
    myFrameLayout = (FrameLayout) findViewById(R.id.myFrameLayout);
    //设置透明度;不要这行代码,则完全清晰
    //myFrameLayout.setAlpha(0.5f);
    myFrameLayout.addView(myPreview);
    Button myButton = (Button) findViewById(R.id.myButton);
    myButton.setOnClickListener(new View.OnClickListener() {
      @Override
      public void onClick(View v) {                      //响应单击"开始照相"按钮
        myCamera.takePicture(null, null, myPicture);     //从摄像头获取照片
      }
    });
}
//安全获取 Camera 对象实例的方法
public static Camera getCameraInstance() {
    Camera c = null;
    try {
      c = Camera.open();
    } catch (Exception e) {
      //摄像头不可用(正被占用或不存在)
    }
    return c; //不可用则返回 null
}
//为保存照片或视频创建 File
private static File getOutputMediaFile(int type) {
    //可以用 Environment.getExternalStorageState()检查 SD 卡是否已装入
    //"手机存储\Pictures\MyCameraApp"是自定义的存放照片的目录
    File myPath = new File(Environment
        .getExternalStoragePublicDirectory(Environment.DIRECTORY_PICTURES),
        "MyCameraApp");
    if (!myPath.exists()) {                              //如果不存在,则创建存储目录
      if (!myPath.mkdirs()) { return null; }
    }
    //创建媒体文件名
    String myStamp = new SimpleDateFormat("yyyyMMdd_HHmmss").format(new Date());
    File myFile;
    if (type == MEDIA_TYPE_IMAGE) {
      myFile = new File(myPath.getPath() + File.separator
                        + "IMG_" + myStamp + ".jpg");
    } else if (type == MEDIA_TYPE_VIDEO) {
      myFile = new File(myPath.getPath() + File.separator
                        + "VID_" + myStamp + ".mp4");
    } else { return null; }
    return myFile;
}
}
```

上面这段代码在 MyCode\MySample421\app\src\main\java\com\bin\luo\mysample\MainActivity.java 文件中。在这段代码中,myFrameLayout.addView(myPreview)用于将预览控件(CameraPreview 的实例 myPreview)放置在布局控件中实现照相前的图像预览功能。

CameraPreview 是一个以 TextureView 为基类的自定义类，CameraPreview 类的主要代码如下：

```java
public class CameraPreview extends TextureView implements
        TextureView.SurfaceTextureListener {
    private Camera myCamera;
    public CameraPreview(Context context , Camera camera) {
        super(context);
        myCamera = camera;
    }
    public void onSurfaceTextureAvailable(SurfaceTexture surface,
                                          int width, int height) {
        try {
            myCamera.setPreviewTexture(surface);
            myCamera.startPreview();
        } catch (IOException ioe) { }
    }
    public void onSurfaceTextureSizeChanged(SurfaceTexture surface,
                                            int width, int height) { }
    public boolean onSurfaceTextureDestroyed(SurfaceTexture surface) {
        myCamera.setPreviewCallback(null);
        myCamera.stopPreview();
        myCamera.lock();
        myCamera.release();
        myCamera = null;
        return true;
    }
    public void onSurfaceTextureUpdated(SurfaceTexture surface) { }
}
```

上面这段代码在 MyCode\MySample421\app\src\main\java\com\bin\luo\mysample\CameraPreview.java 文件中。在这段代码中，基类 TextureView 主要用于展示内容流，如视频或者 OpenGL 场景的内容流，内容流可以来自本应用以及其他进程，可以实现平移、动画等效果，也可使用该类的 setAlpha(0.5f) 方法将 TextureView（预览图像）设置成半透明。此外，进行照相需要在 AndroidManifest.xml 文件中添加 < uses-permission android: name = " android. permission. CAMERA"/>权限 和< uses-permission android: name = " android. permission. WRITE_EXTERNAL_STORAGE"/>权限。此实例的完整项目在 MyCode\MySample421 文件夹中。

204　通过处理按键实现双击后退键退出应用

此实例主要通过在 onKeyDown()事件响应方法中检测后退键 KEYCODE_BACK 前后两次被单击的时间间隔(500ms)，实现双击后退键才能退出应用的效果。当实例运行之后，如果仅单击一次后退键，将在弹出的 Toast 中提示"双击后退键才能退出应用！"，如图 204-1 所示；如果快速两次单击(双击)后退键，则直接退出当前应用。

主要代码如下：

```java
public class MainActivity extends Activity {
    private long myTime = 0;
    @Override
    protected void onCreate(Bundle savedInstanceState) {
```

图 204-1

```
    super.onCreate(savedInstanceState);
    setContentView(R.layout.activity_main);
}
public boolean onKeyDown(int keyCode, KeyEvent event) {
    if (keyCode == KeyEvent.KEYCODE_BACK &&
            event.getAction() == KeyEvent.ACTION_DOWN) {
        if ((System.currentTimeMillis() - myTime) > 500) {
            Toast.makeText(getApplicationContext(),
                    "双击后退键才能退出应用!", Toast.LENGTH_SHORT).show();
            myTime = System.currentTimeMillis();
        } else {
            finish();
            System.exit(0);
        }
        return true;
    }
    return super.onKeyDown(keyCode, event);
}
```

上面这段代码在 MyCode\MySample501\app\src\main\java\com\bin\luo\mysample\MainActivity.java 文件中。在这段代码中，onKeyDown()方法用于响应手机按键操作。keyCode == KeyEvent.KEYCODE_BACK && event.getAction() == KeyEvent.ACTION_DOWN 表示后退键被按下。此实例的完整项目在 MyCode\MySample501 文件夹中。

205 使用 GestureDetector 实现横向滑动切换

此实例主要通过使用 GestureDetector 动态监听控件操作，并重写 onFling()方法监听滑动事件，实现在 ViewFlipper 控件上左右滑动时切换图像。当实例运行之后，向左滑动手指，则 ViewFlipper

控件滑入下一幅图像，向右滑动手指，则 ViewFlipper 控件滑入上一幅图像，效果分别如图 205-1 的左图和右图所示。

图 205-1

主要代码如下：

```java
public class MainActivity extends Activity {
    int[] myImages = {R.mipmap.myimage1, R.mipmap.myimage2,
            R.mipmap.myimage3, R.mipmap.myimage4, R.mipmap.myimage5};
    ViewFlipper myViewFlipper;
    @Override
    protected void onCreate(Bundle savedInstanceState) {
        super.onCreate(savedInstanceState);
        setContentView(R.layout.activity_main);
        myViewFlipper = (ViewFlipper) findViewById(R.id.myViewFlipper);
        for (int i = 0; i < myImages.length; i++) {
            ImageView myImageView = new ImageView(MainActivity.this);
            myImageView.setImageResource(myImages[i]);
            myImageView.setScaleType(ImageView.ScaleType.FIT_XY);
            myViewFlipper.addView(myImageView);
        }
        myViewFlipper.setClickable(true);                        //设置 ViewFlipper 控件可单击
        final GestureDetector myGestureDetector =
                new GestureDetector(new GestureDetector.SimpleOnGestureListener() {
            @Override
            //检测手势滑动事件
            public boolean onFling(MotionEvent e1,
                           MotionEvent e2, float velocityX, float velocityY) {
                if (e2.getX() > e1.getX()) showNextView();       //向右滑动
                else if (e1.getX() > e2.getX()) showPreviousView();  //向左滑动
                return false;
            }
        });
        myViewFlipper.setOnTouchListener(new View.OnTouchListener() {
```

```
    @Override
    public boolean onTouch(View v, MotionEvent event) {
      //传递事件至手势识别器对象
      return myGestureDetector.onTouchEvent(event);
    } });
  }
  public void showPreviousView() {
    //设置右滑时 ViewFlipper 子视图进入和退出动画
    myViewFlipper.setInAnimation(AnimationUtils.loadAnimation(this,
                              R.anim.slide_right_in));
    myViewFlipper.setOutAnimation(AnimationUtils.loadAnimation(this,
                              R.anim.slide_left_out));
    myViewFlipper.showPrevious(); //显示上一幅图像
  }
  public void showNextView() {
    //设置左滑时 ViewFlipper 子视图进入和退出动画
    myViewFlipper.setInAnimation(AnimationUtils.loadAnimation(this,
                              R.anim.slide_left_in));
    myViewFlipper.setOutAnimation(AnimationUtils.loadAnimation(this,
                              R.anim.slide_right_out));
    myViewFlipper.showNext(); //显示下一幅图像
  } }
```

上面这段代码在 MyCode\MySample834\app\src\main\java\com\bin\luo\mysample\MainActivity.java 文件中。在这段代码中，myViewFlipper.setInAnimation（AnimationUtils.loadAnimation(this，R.anim.slide_right_in)）用于设置右滑时图像的进场动画，R.anim.slide_right_in 进场动画的主要代码如下：

```
<?xml version = "1.0" encoding = "utf-8"?>
<set xmlns:android = "http://schemas.android.com/apk/res/android">
  <translate android:duration = "500"
             android:fromXDelta = "100 % p"
             android:toXDelta = "0 % p"/>
</set>
```

上面这段代码在 MyCode\MySample834\app\src\main\res\anim\slide_right_in.xml 文件中。在 MainActivity.java 文件中，myViewFlipper.setOutAnimation（AnimationUtils.loadAnimation(this，R.anim.slide_left_out)）用于设置右滑时图像的退场动画，R.anim.slide_left_out 退场动画的主要代码如下：

```
<?xml version = "1.0" encoding = "utf-8"?>
<set xmlns:android = "http://schemas.android.com/apk/res/android">
  <translate android:duration = "500"
             android:fromXDelta = "0 % p"
             android:toXDelta = " - 100 % p"/>
</set>
```

上面这段代码在 MyCode\MySample834\app\src\main\res\anim\slide_left_out.xml 文件中。在 MainActivity.java 文件中，myViewFlipper.setInAnimation（AnimationUtils.loadAnimation(this，R.anim.slide_left_in)）用于设置左滑时图像的进场动画，R.anim.slide_left_in 进场动画的主

要代码如下:

```xml
<?xml version = "1.0" encoding = "utf-8"?>
<set xmlns:android = "http://schemas.android.com/apk/res/android">
<translate android:duration = "500"
           android:fromXDelta = "-100%p"
           android:toXDelta = "0%p"/>
</set>
```

上面这段代码在 MyCode\MySample834\app\src\main\res\anim\slide_left_in.xml 文件中。在 MainActivity.java 文件中,myViewFlipper.setOutAnimation(AnimationUtils.loadAnimation(this,R.anim.slide_right_out))用于设置左滑时图像的退场动画,R.anim.slide_right_out 退场动画的主要代码如下:

```xml
<?xml version = "1.0" encoding = "utf-8"?>
<set xmlns:android = "http://schemas.android.com/apk/res/android">
<translate android:duration = "500"
           android:fromXDelta = "0%p"
           android:toXDelta = "100%p"/>
</set>
```

上面这段代码在 MyCode\MySample834\app\src\main\res\anim\slide_right_out.xml 文件中。此实例的完整项目在 MyCode\MySample834 文件夹中。

206 使用锁屏标志实现在锁屏时是否显示窗口

此实例主要通过使用 addFlags()和 clearFlags()方法,实现允许或禁止在锁屏时显示指定的应用窗口。当实例运行之后,单击"允许锁屏时显示此应用"按钮,然后按下电源键关闭手机,再按下电源键开启手机,将显示当前应用的窗口;单击"禁止锁屏时显示此应用"按钮,然后按下电源键关闭手机,再按下电源键开启手机,将不显示当前应用的窗口,效果分别如图 206-1 的左图和右图所示。

图 206-1

主要代码如下:

```java
public class MainActivity extends Activity {
 @Override
 protected void onCreate(Bundle savedInstanceState) {
  super.onCreate(savedInstanceState);
  boolean isScreenShotEnabled = getIntent().getBooleanExtra("isLocked", true);
  if(!isScreenShotEnabled){                //允许锁屏时显示此应用
   getWindow().addFlags(WindowManager.LayoutParams.FLAG_SHOW_WHEN_LOCKED);
  }else{                                   //禁止锁屏时显示此应用
   getWindow().clearFlags(WindowManager.LayoutParams.FLAG_SHOW_WHEN_LOCKED);
  }
  setContentView(R.layout.activity_main);
 }
 public void onClickmyBtn1(View v) {       //响应单击"允许锁屏时显示此应用"按钮
  Toast.makeText(this, "已经启用在锁屏时显示此应用,可以锁屏测试一下",
                                          Toast.LENGTH_SHORT).show();
  Intent myIntent = getIntent();
  myIntent.putExtra("isLocked",false);
  finish();
  startActivity(myIntent);
 }
 public void onClickmyBtn2(View v) {       //响应单击"禁止锁屏时显示此应用"按钮
 Toast.makeText(this, "已经禁止在锁屏时显示此应用,可以锁屏测试一下",
                                          Toast.LENGTH_SHORT).show();
  Intent myIntent = getIntent();
  myIntent.putExtra("isLocked",true);
  finish();
  startActivity(myIntent);
} }
```

上面这段代码在 MyCode\MySample249\app\src\main\java\com\bin\luo\mysample\MainActivity.java 文件中。在这段代码中,getWindow().addFlags(WindowManager.LayoutParams.FLAG_SHOW_WHEN_LOCKED)用于实现在锁屏时允许显示此应用的窗口,getWindow().clearFlags(WindowManager.LayoutParams.FLAG_SHOW_WHEN_LOCKED)用于实现在锁屏时禁止显示此应用的窗口,即恢复正常状态。此实例的完整项目在 MyCode\MySample249 文件夹中。

207 在当前应用中实现关机和重启功能

此实例主要通过执行 Android 系统的关机和重启命令行,实现在当前应用中实现关机和重启功能。当实例运行之后,单击"关机"按钮,将直接关机;单击"重启"按钮,将直接重启,不会出现手动关机或重启时的选择对话框,如图 207-1 所示。注意:请直接在手机中测试,大多数模拟器在测试此实例时无法实现其功能。

主要代码如下:

```java
public void onClickmyBtn1(View v) {        //响应单击"关机"按钮
 execCmd("reboot -p");
}
```

图　207-1

```
public void onClickmyBtn2(View v) {          //响应单击"重启"按钮
   execCmd("reboot");
}
public static boolean execCmd(String myCommand) {
   Process myProcess = null;
   DataOutputStream myStream = null;
   try {
     myProcess = Runtime.getRuntime().exec("su");
     myStream = new DataOutputStream(myProcess.getOutputStream());
     myStream.writeBytes(myCommand + "\n");
     myStream.writeBytes("exit\n");
     myStream.flush();
     myProcess.waitFor();
   } catch (Exception e) {
     return false;
   } finally {
     try {
       if (myStream != null) { myStream.close(); }
       if (myProcess != null) { myProcess.destroy(); }
     } catch (Exception e) { e.printStackTrace(); }
   }
   return true;
}
```

上面这段代码在 MyCode\MySample356\app\src\main\java\com\bin\luo\mysample\MainActivity.java 文件中。此实例的完整项目在 MyCode\MySample356 文件夹中。

第 8 章

Intent

208　使用 Intent 启动百度地图进行骑行导航

此实例主要通过使用 Intent 调用百度地图,实现根据出发地和目的地对骑行线路进行导航。当实例运行之后,在"出发地:"和"目的地:"输入框中分别输入"29.6904466666667,106.5939533333332"和"29.716311,106.638184",然后单击"启动百度地图进行骑行导航"按钮,将在百度地图中显示从出发地(长福西路)到达目的地(重庆江北机场)的骑行线路,并通过 GPS 进行实时定位导航,效果分别如图 208-1 的左图和右图所示。如果在"出发地:"和"目的地:"输入框中分别输入"29.6904466666667,106.5939533333332"和"29.722556,106.58752",然后单击"启动百度地图进行骑行导航"按钮,将在百度地图中显示从出发地(长福西路)到达目的地(重庆中央公园)的骑行线路,并通过 GPS 进行实时定位导航,效果分别如图 208-2 的左图和右图所示。注意:在测试此应用之前一定要安装百度地图 App。

图　208-1

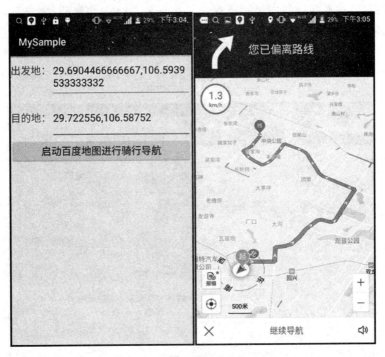

图 208-2

主要代码如下：

```
//响应单击"启动百度地图进行骑行导航"按钮
public void onClickmyBtn1(View v) {
    Intent myIntent = new Intent();
    myIntent.setData(Uri.parse("baidumap://map/bikenavi?origin = "
            + myEditBegin.getText().toString() + "&destination = "
            + myEditEnd.getText().toString()));
    startActivity(myIntent);
}
```

上面这段代码在 MyCode\MySample347\app\src\main\java\com\bin\luo\mysample\MainActivity.java 文件中。在这段代码中，myIntent.setData(Uri.parse("baidumap://map/bikenavi?origin = " + myEditBegin.getText().toString() + "&destination = " + myEditEnd.getText().toString()))表示百度地图App根据出发地和目的地的纬度和经度值查询骑行线路并在骑行过程中进行导航，在导航过程中，百度地图会自动显示骑行线路附近的地形状况。测试表明，origin 参数和 destination 参数仅支持纬度和经度值，输入中文地名无效。此实例的完整项目在MyCode\MySample347 文件夹中。

209 使用Intent启动百度地图查询公交线路

此实例主要通过使用 Intent 调用百度地图，实现在百度地图中查询指定城市的公交线路。当实例运行之后，在"城市名称："和"公交线路："输入框中分别输入相应的内容，如"重庆市"和"877 路"，然后单击"开始通过百度地图查询该公交线路"按钮，将在百度地图中显示重庆主城区的877公交线路，并在下面显示该公交线路的开收班时间，效果分别如图 209-1 的左图和右图所示。如果在"城市名称："和"公交线路："输入框中分别输入"重庆市涪陵区"和"112 路"，然后单击"开始通过百度地图查询该公交线路"按钮,将在百度地图中显示重庆市涪陵区的 112 公交线路,并在下面显示该公交线路的开收班时间,效果分别如图 209-2 的左图和右图所示。需要说明的是:重庆市涪陵区不在重庆主城九区范围内,因此它的公交线路编号独立于重庆主城九区,相当于一个单独的城市,其他城市的公交线路可以参考这两个城市的方式进行查询。注意:在测试此应用之前一定要安装百度地图 App。

图 209-1

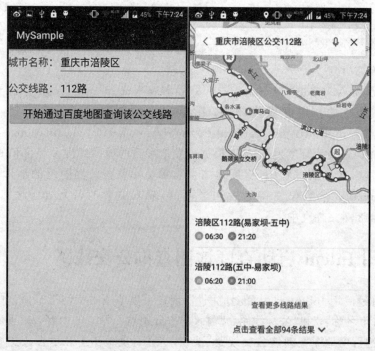

图 209-2

主要代码如下：

```
//响应单击"开始通过百度地图查询该公交线路"按钮
public void onClickmyBtn1(View v) {
    try {
        //baidumap://map/line?region = 不起作用 &name = 重庆市轨道交通 3 号线
        //baidumap://map/line?region = 不起作用 &name = 重庆市公交 877 路
        //baidumap://map/line?region = 不起作用 &name = 北京市地铁一号线
        Intent myIntent = Intent.getIntent(
                "baidumap://map/line?region = 不起作用 &name = "
                + myEditCity.getText().toString() + "公交"
                + myEditLine.getText().toString());
        startActivity(myIntent);
    } catch (Exception e) { Log.e("intent", e.getMessage()); }
}
```

上面这段代码在 MyCode\MySample343\app\src\main\java\com\bin\luo\mysample\MainActivity.java 文件中。在这段代码中，myIntent = Intent.getIntent("baidumap://map/line?region=不起作用 &name = " + myEditCity.getText().toString() + "公交" + myEditLine.getText().toString())表示在百度地图 App 中查询指定城市的公交线路。需要说明的是：region 参数原本是用来指定城市，包括百度地图的 API 也是这样举例的，但是实际测试表明，它不起作用（但不能为空，可以输入任意的字符），城市名称应合并在 name 参数中才能得到正确的结果。此实例的完整项目在 MyCode\MySample343 文件夹中。

210 使用 Intent 启动百度地图查询步行线路

此实例主要通过使用 Intent 调用百度地图，实现根据出发地和目的地查询步行线路。当实例运行之后，在"城市名称："""出发地：""目的地："3 个输入框中分别输入"重庆市渝中区""朝天门"和"解放碑"，然后单击"开始通过百度地图查询步行路线"按钮，则将在百度地图中显示从朝天门到解放碑的步行线路（路线），效果分别如图 210-1 的左图和右图所示。如果在"城市名称：""出发地：""目的地："3 个输入框中分别输入"重庆市渝中区""三峡博物馆"和"解放碑"，然后单击"开始通过百度地图查询步行路线"按钮，将在百度地图中显示从三峡博物馆到解放碑的步行线路，效果分别如图 210-2 的左图和右图所示。注意：在测试此应用之前一定要安装百度地图 App。

主要代码如下：

```
public void onClickmyBtn1(View v) {        //响应单击"开始通过百度地图查询步行路线"按钮
    try {
        Intent myIntent = Intent.getIntent("intent://map/direction?origin = "
                + myEditBegin.getText().toString() + "&destination = "
                + myEditEnd.getText().toString() + "&mode = walking&region = "
                + myEditCity.getText().toString() +
                " # Intent;scheme = bdapp;package = com.baidu.BaiduMap;end");
        startActivity(myIntent);                //启动百度地图 App
    } catch (Exception e) {
        Log.e("intent", e.getMessage());
    }
}
```

上面这段代码在 MyCode\MySample341\app\src\main\java\com\bin\luo\mysample\MainActivity.java 文件中。在这段代码中，myIntent = Intent.getIntent("intent://map/direction?

图 210-1

图 210-2

origin="+myEditBegin.getText().toString()+"&destination="+myEditEnd.getText().toString()+"&mode=walking&region="+myEditCity.getText().toString()+"#Intent;scheme=bdapp;package=com.baidu.BaiduMap;end")表示在百度地图中根据出发地和目的地查询步行线路。此实例的完整项目在 MyCode\MySample341 文件夹中。

211 使用 Intent 启动百度地图查询兴趣点

此实例主要通过使用 Intent 调用百度地图，实现根据指定的要求查询指定位置附近的学校、超市、宾馆等兴趣点信息（POI）。当实例运行之后，在"纬度,经度："""查询类别：""范围半径（米）："输入框中分别输入对应的值,如"29.6904466666667,106.5939533333332""学校"和"2000",然后单击"开始在百度地图中查询"按钮,将在百度地图中显示对应的学校内容,效果分别如图 211-1 的左图和右图所示。如果在"纬度,经度:""查询类别:""范围半径(米):"输入框中分别输入对应的值,如"29.6904466666667,106.5939533333332""超市"和"2000",然后单击"开始在百度地图中查询"按钮,将在百度地图中显示对应的超市内容,效果分别如图 211-2 的左图和右图所示。需要说明的是:在"纬度,经度:"输入框中一定要按照顺序输入纬度和经度值,并且小数点后面的位数越多,地图内容越详细。注意:在测试此应用之前一定要安装百度地图 App。

图 211-1

图 211-2

主要代码如下:

```
public void onClickmyBtn1(View v) {          //响应单击"开始在百度地图中查询"按钮
    Intent myIntent = new Intent();
    myIntent.setData(Uri.parse("baidumap://map/place/search?query = "
            + myEditType.getText().toString()
            + "&location = " + myEditPos.getText().toString()
            + "&radius = " + myEditRadius.getText().toString() + ""));
    startActivity(myIntent);
}
```

上面这段代码在 MyCode\MySample340\app\src\main\java\com\bin\luo\mysample\MainActivity.java 文件中。在这段代码中,myIntent.setData(Uri.parse("baidumap://map/place/search?query="+myEditType.getText().toString()+"&location="+myEditPos.getText().toString()+"&radius="+myEditRadius.getText().toString()+""))用于在百度地图中根据指定的位置在指定半径范围内查询指定要求的超市、学校等信息。需要注意的是:location=后面的数据是纬度和经度,而不是经度和纬度,并且两者之间有一个半角逗号。此实例的完整项目在 MyCode\MySample340 文件夹中。

212 使用 Intent 启动百度地图根据地名定位

此实例主要通过使用 Intent 调用百度地图,实现在百度地图中显示指定地名位置附近的地图的功能。当实例运行之后,在"地名:"输入框中输入地名,如"清华大学",然后单击"在百度地图中显示指定地名附近的地图"按钮,将启用百度地图显示该位置(清华大学)附近的地图,效果分别如图 212-1 的左图和右图所示。注意:在测试此应用之前一定要安装百度地图 App。

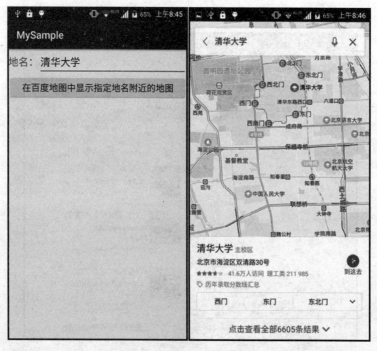

图 212-1

主要代码如下：

```
//响应单击"在百度地图中显示指定地名附近的地图"按钮
public void onClickmyBtn1(View v) {
    if (!isAvailable(this, "com.baidu.BaiduMap")) {
        Toast.makeText(this, "当前手机中没有安装百度地图,请安装之后再测试!", Toast.LENGTH_SHORT).show();
        return;
    }
    Intent myIntent = new Intent();
    myIntent.setData(Uri.parse(
            "baidumap://map/geocoder?src=openApiDemo&address="
            + myEditText.getText().toString()));
    startActivity(myIntent);
}
public static boolean isAvailable(Context context, String packageName) {
    final PackageManager myPackageManager = context.getPackageManager();
    //获取所有已安装应用的包信息
    List<PackageInfo> myPackageInfos = myPackageManager.getInstalledPackages(0);
    //保存所有已安装应用的包名
    List<String> myPackageNames = new ArrayList<String>();
    //将包名字逐一取出
    if (myPackageInfos != null) {
        for (int i = 0; i < myPackageInfos.size(); i++) {
            String myPackageName = myPackageInfos.get(i).packageName;
            myPackageNames.add(myPackageName);
        }
    }
    //判断是否有目标应用的包名,如有返回true,没有返回false
    return myPackageNames.contains(packageName);
}
```

上面这段代码在 MyCode\MySample338\app\src\main\java\com\bin\luo\mysample\MainActivity.java 文件中。在这段代码中，myIntent.setData(Uri.parse("baidumap://map/geocoder?src=openApiDemo&address="+myEditName.getText().toString()))用于在百度地图中显示指定地名位置附近的地图。需要说明的是：地名可以是门牌号码、公司名称等，百度地图首先会返回一个与地名相匹配的可选列表，然后则可以在该可选列表中选择与查询要求最接近的选项对应的地图。此实例的完整项目在 MyCode\MySample338 文件夹中。

213 使用 Intent 启动百度地图助手搜索地点

此实例主要通过在 Intent 的 setData()方法中使用"baidumap://map/trip"参数，实现启动百度地图的行程助手实现搜地点、查公交、找路线等功能。当实例运行之后，单击"启动百度地图的行程助手"按钮，将通过百度地图启动行程助手。在行程助手操作界面中，在顶部的输入框中输入内容，即可搜索想要的信息，效果分别如图 213-1 的左图和右图所示。注意：在测试时，手机一定要安装百度地图 App。

主要代码如下：

```
public void onClickBtn1(View v) {    //响应单击"启动百度地图的行程助手"按钮
    try {
        Intent myIntent = new Intent();
```

图 213-1

```
    myIntent.setData(Uri.parse("baidumap://map/trip"));
    startActivity(myIntent);
} catch (Exception e) {
Toast.makeText(MainActivity.this,
        e.getMessage().toString(), Toast.LENGTH_SHORT).show();
} }
```

上面这段代码在 MyCode\MySampleX07\app\src\main\java\com\bin\luo\mysample\MainActivity.java 文件中。此实例的完整项目在 MyCode\MySampleX07 文件夹中。

214 使用 Intent 在百度地图中展示详情页

此实例主要通过在 Intent 的 setData()方法的参数中使用"baidumap://map/place/detail"字符串组合某地的 POI 的 UID,从而实现在百度地图中展示该 UID 代表的某地详细信息。当实例运行之后,在"POI 的 ID:"输入框中输入重庆市人民大礼堂的 UID 值,如"cdb116dbf944882397f08c4c",然后单击"在百度地图中显示详细信息"按钮,则将在百度地图中显示该 UID 代表的重庆市人民大礼堂的详细信息,效果分别如图 214-1 的左图和右图所示。如果在"POI 的 ID:"输入框中输入 UID 值,如"53a0dad38a3f5ad64d776e92",然后单击"在百度地图中显示详细信息"按钮,则将在百度地图中显示该 UID 代表的重庆江北国际机场的详细信息,效果分别如图 214-2 的左图和右图所示。需要说明的是,某个地方的 POI 的 UID 可以在百度全景选择工具(http://quanjing.baidu.com/apipickup)中获得,即在"设置全景 UID"输入框中的内容。注意:在测试此应用之前一定要安装百度地图 App。

主要代码如下:

```
public void onClickBtn1(View v) {     //响应单击"在百度地图中显示详细信息"按钮
    try {
```

图　214-1

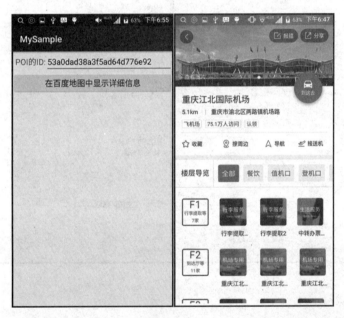

图　214-2

```
Intent myIntent = new Intent();
//鑫城名都 UID = e3082848c0cf1cbfb24c498b
//重庆市人民大礼堂 UID = cdb116dbf944882397f08c4c
//重庆江北国际机场 UID = 53a0dad38a3f5ad64d776e92
myIntent.setData(Uri.parse("baidumap://map/place/detail?uid = "
        + myEditUID.getText().toString() + "&show_type = detail_page"));
startActivity(myIntent);
} catch (Exception e) {
Toast.makeText(MainActivity.this,
        e.getMessage().toString(), Toast.LENGTH_SHORT).show();
} }
```

上面这段代码在 MyCode\MySampleX03\app\src\main\java\com\bin\luo\mysample\MainActivity.java 文件中。在这段代码中，myIntent.setData(Uri.parse("baidumap：//map/place/detail?uid="+ myEditUID.getText().toString() + "&show_type=detail_page"))的 detail 表示在百度地图上展示详细信息，uid 后面的内容表示某地的 POI 的 UID，detail_page 是 POI 详情页。此实例的完整项目在 MyCode\MySampleX03 文件夹中。

215　使用 Intent 启动百度地图查询实时公交

此实例主要通过使用 Intent 调用百度地图，实现在百度地图中查询当前位置附近的实时公交信息。当实例运行之后，单击"在百度地图中查询附近的实时公交"按钮，则将显示当前位置附近的实时公交信息，效果分别如图 215-1 的左图和右图所示。注意：在测试此应用之前一定要安装百度地图 App。

图　215-1

主要代码如下：

```java
//响应单击"在百度地图中查询附近的实时公交"按钮
public void onClickmyBtn1(View v) {
    Intent myIntent = new Intent();
    myIntent.setData(Uri.parse(
            "baidumap://map/page/realtimebus?mode = NORMAL_MAP_MODE"));
    startActivity(myIntent);
}
```

上面这段代码在 MyCode\MySample349\app\src\main\java\com\bin\luo\mysample\MainActivity.java 文件中。在这段代码中，myIntent.setData(Uri.parse("baidumap：//map/page/realtimebus?mode=NORMAL_MAP_MODE"))表示在百度地图中查询当前位置附近的实时公交信息。此实例的完整项目在 MyCode\MySample349 文件夹中。

216 使用Intent启动百度地图查询实时路况

此实例主要通过使用Intent调用百度地图的"出行早晚报",实现根据当前手机位置查询实时路况。当实例运行之后,单击"启动百度地图查询当地的实时路况"按钮,则将显示当地的实时路况,效果分别如图216-1的左图和右图所示。注意:在测试此应用之前一定要安装百度地图App。

图 216-1

主要代码如下:

```
public static boolean isAvailable(Context context, String packageName) {
    final PackageManager myPackagcManager = context.getPackageManager();
    //获取所有已安装应用的包信息
    List<PackageInfo> myPackageInfos = myPackageManager.getInstalledPackages(0);
    //保存所有已安装应用的包名
    List<String> myPackageNames = new ArrayList<String>();
    //将包名字逐一取出
    if (myPackageInfos != null) {
        for (int i = 0; i < myPackageInfos.size(); i++) {
            String myPackageName = myPackageInfos.get(i).packageName;
            myPackageNames.add(myPackageName);
        }
    }
    //判断是否有目标应用的包名,如有返回true,没有返回false
    return myPackageNames.contains(packageName);
}
//响应单击"启动百度地图查询当地的实时路况"按钮
public void onClickmyBtn1(View v) {
    if (!isAvailable(this, "com.baidu.BaiduMap")) {
        Toast.makeText(this, "当前手机中没有安装百度地图,请安装之后再测试!", Toast.LENGTH_SHORT).show();
```

```
    return;
}
Intent myIntent = new Intent();
myIntent.setData(Uri.parse("baidumap://map/newsassistant"));
startActivity(myIntent);
}
```

上面这段代码在 MyCode\MySample336\app\src\main\java\com\bin\luo\mysample\MainActivity.java 文件中。在这段代码中，myIntent.setData(Uri.parse("baidumap://map/newsassistant"))表示通过 Intent 跳转到百度地图的 newsassistant，从而实现在 startActivity(myIntent)之后，跳转到百度地图的实时路况。此实例的完整项目在 MyCode\MySample336 文件夹中。

217 使用 Intent 启动百度地图显示实时汇率

此实例主要通过在 Intent 的 setData()方法中使用"baidumap://map/component?comName=international&target=international_exchangerate_page"参数，实现启动百度地图的实时汇率页面。当实例运行之后，单击"启动百度地图的实时汇率页面"按钮，将通过百度地图启动实时汇率页面，效果分别如图 217-1 的左图和右图所示。注意：在测试时，手机一定要安装百度地图 App。

图 217-1

主要代码如下：

```
public void onClickBtn1(View v) {    //响应单击"启动百度地图的实时汇率页面"按钮
    try {
Intent myIntent = new Intent();
    myIntent.setData(Uri.parse("baidumap://map/component?" +
```

```
            "comName = international&target = international_exchangerate_page"));
    startActivity(myIntent);
  } catch (Exception e) {
    Toast.makeText(MainActivity.this,
          e.getMessage().toString(), Toast.LENGTH_SHORT).show();
} }
```

上面这段代码在 MyCode\MySampleX08\app\src\main\java\com\bin\luo\mysample\MainActivity.java 文件中。此实例的完整项目在 MyCode\MySampleX08 文件夹中。

218　使用 Intent 直接跳转到百度地图 App 界面

此实例主要通过使用百度地图 URI 创建 Intent 对象，实现从当前应用直接跳转到百度地图的主操作界面。当实例运行之后，单击"直接跳转到百度地图的主操作界面"按钮，则将显示百度地图 App 的主操作界面，效果分别如图 218-1 的左图和右图所示；如果百度地图尚未启动，则直接启动百度地图。注意：测试手机一定要安装百度地图 App。

图　218-1

主要代码如下：

```
//响应单击"直接跳转到百度地图的主操作界面"按钮
public void onClickBtn1(View v) {
  try {
    Intent myIntent = new Intent("android.intent.action.VIEW",
                         android.net.Uri.parse("baidumap://map"));
    startActivity(myIntent);
  }catch (Exception e) {
    Toast.makeText(MainActivity.this,
          e.getMessage().toString(), Toast.LENGTH_SHORT).show();
} }
```

上面这段代码在 MyCode\MySample988\app\src\main\java\com\bin\luo\mysample\MainActivity.java 文件中。在这段代码中，myIntent= new Intent("android.intent.action.VIEW",android.net.Uri.parse("baidumap://map"))用于根据百度地图 URI 创建 Intent，百度地图 URI 支持复杂的参数，如使用 android.net.Uri.parse("baidumap://map/geocoder?location="+myPoint.getLatitude()+","+myPoint.getLongitude()),则可以进行地图定位。startActivity(myIntent)用于启动该 Intent。此实例的完整项目在 MyCode\MySample988 文件夹中。

219 使用 Intent 启动腾讯地图查询驾车线路

此实例主要通过在 Intent 的 setData()方法的参数中使用"qqmap://map/routeplan?type=drive"字符串组合起点和终点的名称和纬度及经度值，实现在腾讯地图中查询该起点和终点之间的驾车线路。当实例运行之后，在"起点站名称："输入框中输入起点站名称，如"重庆江北国际机场"，在"起点站纬度、经度："输入框中输入纬度和经度值，如"29.7177702,106.638030"，在"终点站名称："输入框中输入终点站名称，如"重庆抗战遗址博物馆"，在"终点站纬度、经度："输入框中输入终点站的纬度和经度值，如"29.564190,106.618710"，然后单击"在腾讯地图中显示驾车线路"按钮，将在腾讯地图中显示重庆江北国际机场到重庆抗战遗址博物馆的驾车线路，效果分别如图 219-1 的左图和右图所示。如果在"起点站名称："输入框中输入"重庆江北国际机场"，在"起点站纬度、经度："输入框中输入"29.7177702,106.638030"，在"终点站名称："输入框中输入"长寿古镇"，在"终点站纬度、经度："输入框中输入"29.851020,107.067560"，然后单击"在腾讯地图中显示驾车线路"按钮，将在腾讯地图中显示重庆江北国际机场到长寿古镇的驾车线路，效果分别如图 219-2 的左图和右图所示。需要说明的是：可以在腾讯的坐标拾取器(http://lbs.qq.com/tool/getpoint/Index.html)中查询某地的纬度和经度值。注意：在测试此应用之前一定要安装腾讯地图 App。

图 219-1

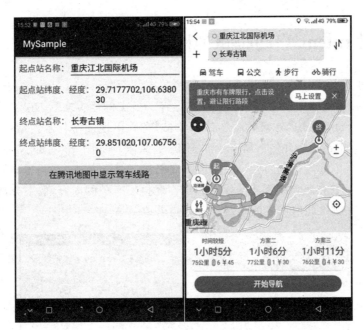

图 219-2

主要代码如下：

```
public void onClickBtn1(View v) {        //响应单击"在腾讯地图中显示驾车线路"按钮
  try {
    Intent myIntent = new Intent();
//    myIntent.setData(Uri.parse("qqmap://map/routeplan?type = drive&" +
//        "from = 重庆江北国际机场 &fromcoord = 29.7177702,106.638030&" +
//        "to = 重庆抗战遗址博物馆 &tocoord = 29.564190,106.618710&policy = 1"));
//    myIntent.setData(Uri.parse("qqmap://map/routeplan?type = drive&" +
//        "from = 重庆江北国际机场 &fromcoord = 29.7177702,106.638030&" +
//        "to = 长寿古镇 &tocoord = 29.851020,107.067560&policy = 1"));
    myIntent.setData(Uri.parse("qqmap://map/routeplan?type = drive&" +
        "from = " + myEditFromName.getText().toString() +
        "&fromcoord = " + myEditFromLatlgt.getText().toString() +
        "&to = " + myEditToName.getText().toString() +
        "&tocoord = " + myEditToLatlgt.getText().toString() + "&policy = 1"));
    startActivity(myIntent);
  } catch (Exception e) {
    Toast.makeText(MainActivity.this, e.getMessage().toString(),
    Toast.LENGTH_SHORT).show();
  } }
```

上面这段代码在 MyCode\MySample995\app\src\main\java\com\bin\luo\mysample\MainActivity.java 文件中。在这段代码中，myIntent.setData(Uri.parse("qqmap://map/routeplan?type＝drive&"＋"from＝"＋myEditFromName.getText().toString()＋"&fromcoord＝"＋myEditFromLatlgt.getText().toString()＋"&to＝"＋myEditToName.getText().toString()＋"&tocoord＝"＋myEditToLatlgt.getText().toString()＋"&policy＝1"))的 type＝drive 表示查询驾车路线，from 后面的内容表示起点站的名称，fromcoord 后面的内容表示起点站的纬度经度值，to 后面的内容表示终点站的名称，tocoord 后面的内容表示终点站的纬度经度值，policy 表示策略方式，可以有以下取值：0 表示较快捷，1 表示无高速，2 表示距离优先。此实例的完整项目在 MyCode\MySample995 文件夹中。

220 使用 Intent 启动腾讯地图搜索感兴趣内容

此实例主要通过在 Intent 的 setData() 方法的参数中使用"qqmap://map/search?keyword"构造搜索字符串,实现在腾讯地图中搜索感兴趣的特定内容。当实例运行之后,在"搜索内容:"输入框中输入特定内容,如"重庆市渝北区永辉超市",然后单击"在腾讯地图中显示搜索结果"按钮,将启动腾讯地图,并在腾讯地图中显示渝北区的所有永辉超市,效果分别如图 220-1 的左图和右图所示。如果在"搜索内容:"输入框中输入特定内容,如"重庆江北机场",然后单击"在腾讯地图中显示搜索结果"按钮,将启动腾讯地图,并在腾讯地图中显示与重庆江北机场有关的内容,效果分别如图 220-2 的左图和右图所示。注意:在测试此应用之前一定要安装腾讯地图 App。

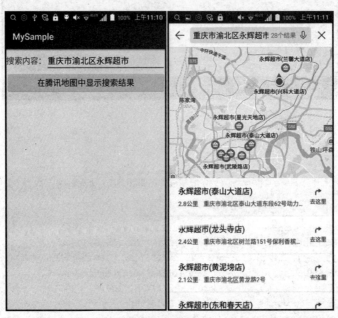

图 220-1

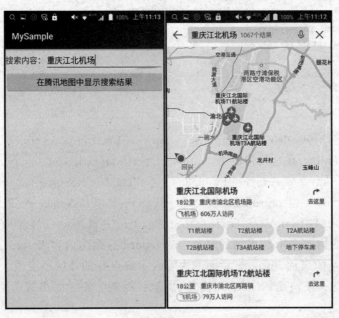

图 220-2

主要代码如下：

```java
public void onClickBtn1(View v) {         //响应单击"在腾讯地图中显示搜索结果"按钮
    try {
        Intent myIntent = new Intent();
        myIntent.setData(Uri.parse("qqmap://map/search?keyword = "
                        + myEditText.getText().toString()));
        startActivity(myIntent);
    } catch (Exception e) {
        Toast.makeText(MainActivity.this,
                e.getMessage().toString(), Toast.LENGTH_SHORT).show();
    } }
```

上面这段代码在 MyCode\MySample990\app\src\main\java\com\bin\luo\mysample\MainActivity.java 文件中。在这段代码中，myIntent.setData(Uri.parse("qqmap://map/search?keyword=" + myEditText.getText().toString()))用于设置将要在腾讯地图中显示的与搜索内容相关的信息。startActivity(myIntent)用于启动腾讯地图显示搜索结果。此实例的完整项目在 MyCode\MySample990 文件夹中。

221　使用 Intent 启动腾讯地图显示指定位置

此实例主要通过在 Intent 的 setData()方法的参数中使用"qqmap://map/geocoder?coord"字符串组合纬度和经度值，实现在腾讯地图中显示指定纬度和经度位置附近的地图。当实例运行之后，在"纬度："输入框中输入纬度值，如"39.904956"，在"经度："输入框中输入经度值，如"116.389449"，然后单击"在腾讯地图中显示此位置"按钮，将启动腾讯地图 App，并在腾讯地图中显示此位置附近的地图，效果分别如图 221-1 的左图和右图所示。注意：在测试此应用之前一定要安装腾讯地图 App。

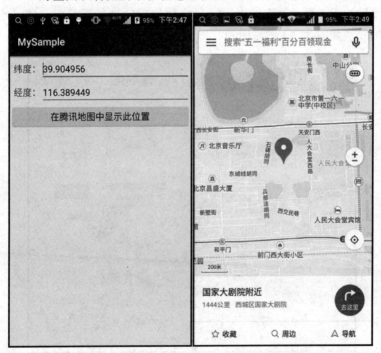

图　221-1

主要代码如下：

```
public void onClickBtn1(View v) {      //响应单击"在腾讯地图中显示此位置"按钮
  try {
    Intent myIntent = new Intent();
    myIntent.setData(Uri.parse("qqmap://map/geocoder?coord = "
        + myEditLatitude.getText().toString() + ","
        + myEditLongitude.getText().toString()));
    startActivity(myIntent);
  } catch (Exception e) {
    Toast.makeText(MainActivity.this,
             e.getMessage().toString(), Toast.LENGTH_SHORT).show();
  } }
```

上面这段代码在 MyCode\MySample991\app\src\main\java\com\bin\luo\mysample\MainActivity.java 文件中。在这段代码中，myIntent.setData(Uri.parse("qqmap://map/ geocoder?coord＝"＋myEditLatitude.getText().toString()＋","＋myEditLongitude.getText().toString()))用于在腾讯地图中根据指定纬度和经度值显示该位置附近的地图。需要注意的是：coord＝后面的数据是纬度和经度，而不是经度和纬度，并且两者之间有一个半角逗号。此实例的完整项目在 MyCode\MySample991 文件夹中。

222 使用 Intent 启动 QQ 浏览器显示腾讯地图

此实例主要通过使用 getLaunchIntentForPackage("com.tencent.mtt")创建 Intent，同时在 Intent 的 setData()方法的参数中使用"http://apis.map.qq.com/uri/v1/search"字符串组合搜索关键词、目标城市，实现在腾讯地图中搜索指定城市的指定类别（关键词）的信息。当实例运行之后，在"城市名称："输入框中输入城市名称，如"重庆市"，在"关键词："输入框中输入分类信息，如"国际机场"，然后单击"在腾讯地图中显示搜索结果"按钮，将使用 QQ 浏览器显示在腾讯地图中的重庆市的国际机场信息，效果分别如图 222-1 的左图和右图所示。如果在"城市名称："输入框中输入"重庆市

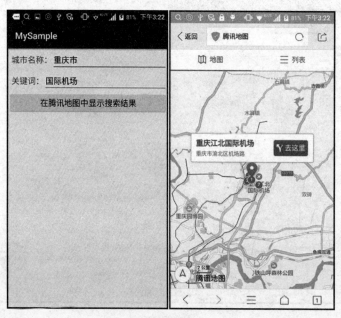

图 222-1

渝北区",在"关键词："输入框中输入"沃尔玛超市",然后单击"在腾讯地图中显示搜索结果"按钮,将使用 QQ 浏览器显示在腾讯地图中的重庆市渝北区的沃尔玛超市分布信息,效果分别如图 222-2 的左图和右图所示。

图　222-2

主要代码如下：

```
public void onClickBtn1(View v) {        //响应单击"在腾讯地图中显示搜索结果"按钮
    try {
        PackageManager myPackageManager = getPackageManager();
        //使用 QQ 浏览器查询腾讯地图
        Intent myIntent =
                myPackageManager.getLaunchIntentForPackage("com.tencent.mtt");
        //myIntent.setData(Uri.parse("http://apis.map.qq.com/uri/v1/
        //search?keyword = 沃尔玛超市 &region = 重庆市渝北区"));
        //myIntent.setData(Uri.parse("http://apis.map.qq.com/uri/v1/
        //search?keyword = 国际机场 &region = 重庆市"));
        myIntent.setData(Uri.parse("http://apis.map.qq.com/uri/v1/search?keyword = "
                + myEditKeyword.getText().toString() + "&region = "
                + myEditCity.getText().toString()));
        startActivity(myIntent);
    } catch (Exception e) {
        Toast.makeText(MainActivity.this,
                e.getMessage().toString(), Toast.LENGTH_SHORT).show();
    }
}
```

上面这段代码在 MyCode \ MySample999 \ app \ src \ main \ java \ com \ bin \ luo \ mysample \ MainActivity. java 文件中。在这段代码中，Intent myIntent = myPackageManager. getLaunchIntentForPackage("com. tencent. mtt")表示使用 QQ 浏览器应用包"com. tencent. mtt"创建 myIntent 对象。myIntent. setData (Uri. parse (" http://apis. map. qq. com/uri/v1/search?

keyword="+myEditKeyword.getText().toString()+"&region="+myEditCity.getText().toString()))这行代码可以简化为myIntent.setData(Uri.parse("http://apis.map.qq.com/uri/v1/search?keyword=国际机场&region=重庆市")),表示在腾讯地图中搜索重庆市范围内与国际机场相关的信息。此实例的完整项目在MyCode\MySample999文件夹中。

223 使用Intent将文本内容仅分享到微信

此实例主要通过在Intent中指定微信包名,实现将文本内容仅分享到微信。当实例运行后,在输入框中输入分享内容,然后单击"分享到我的微信"按钮,将弹出"分享方式"窗口,选择"发送给朋友"选项,如图223-1的左图所示;然后在微信通讯录中选择联系人,将把分享内容发送给联系人,如图223-1的右图所示。注意:在测试此应用之前一定要安装微信App。

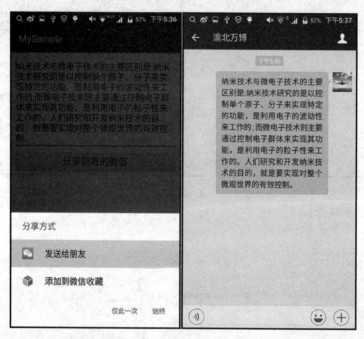

图 223-1

主要代码如下:

```
public void onClickmyBtn1(View v) {              //响应单击"分享到我的微信"按钮
    try{
     Intent myIntent = new Intent(Intent.ACTION_SEND);
     myIntent.setPackage("com.tencent.mm");      //微信应用包名
     myIntent.setType("text/plain");
     myIntent.putExtra(Intent.EXTRA_TEXT, myEditText.getText().toString());
     startActivity(myIntent);
    }catch (Exception e){
     Toast.makeText(getApplicationContext(),
           "请确认手机已经安装了微信!", Toast.LENGTH_SHORT).show();
    }}
```

上面这段代码在 MyCode\MySample448\app\src\main\java\com\bin\luo\mysample\MainActivity.java文件中。在这段代码中,myIntent = new Intent(Intent.ACTION_SEND)用于创

建分享内容的 Intent。myIntent.setPackage("com.tencent.mm")表示将内容分享到微信。myIntent.putExtra(Intent.EXTRA_TEXT，myEditText.getText().toString())用于设置分享内容。此实例的完整项目在 MyCode\MySample448 文件夹中。

224　使用 Intent 将本地图像发送到微信朋友圈

此实例主要通过在 Intent 的 setComponent()方法的参数中使用"com.tencent.mm"和"com.tencent.mm.ui.tools.ShareToTimeLineUI"创建 ComponentName 对象，实现将本地图像仅发送到微信的朋友圈。当实例运行之后，将在 ImageView 控件中显示本地的图像资源，如图 224-1 的左图所示，单击"发送到微信朋友圈"按钮，将弹出发送窗口，在发送窗口中单击"发送"按钮，如图 224-1 的右图所示，该图像将立即出现在微信的朋友圈中。注意：在测试此应用之前一定要安装微信 App。

图　224-1

主要代码如下：

```java
public void onClickmyBtn1(View v) {                  //响应单击"发送到微信朋友圈"按钮
  try {
   Bitmap myBitmap =
            BitmapFactory.decodeResource(getResources(), R.mipmap.myimage1);
   String filePrefix = null;
   try {
    filePrefix = saveImageToGallery(this,myBitmap);
   } catch (IOException e) { e.printStackTrace(); }
   Intent myIntent = new Intent();
   ComponentName myComponentName = new ComponentName("com.tencent.mm",
                    "com.tencent.mm.ui.tools.ShareToTimeLineUI");
   myIntent.setComponent(myComponentName);
   myIntent.setAction(Intent.ACTION_SEND);
```

```java
        myIntent.setFlags(Intent.FLAG_ACTIVITY_NEW_TASK);
        myIntent.setType("image/jpg");
        Uri myUri = Uri.parse(filePrefix);
        myIntent.putExtra(Intent.EXTRA_STREAM,myUri);
        startActivity(myIntent);
    } catch (Exception e) {
        Toast.makeText(getApplicationContext(),
                "请确认手机已经安装了微信!", Toast.LENGTH_SHORT).show();
    } }
    public static String saveImageToGallery(Context context,
                                Bitmap bmp) throws IOException {
    File myFile = new File(Environment.getExternalStorageDirectory(),
                    "luoGallery"); //保存图像文件
    if (!myFile.exists()) { myFile.mkdir(); }
    String myFileName = System.currentTimeMillis() + ".jpg";
    String myNameTemp;
    File myNewFile = new File(myFile, myFileName);
    myNewFile.createNewFile();
    try {
        FileOutputStream myStream = new FileOutputStream(myNewFile);
        bmp.compress(Bitmap.CompressFormat.JPEG, 100, myStream);
        myStream.flush();
        myStream.close();
    } catch (FileNotFoundException e) { e.printStackTrace(); }
    catch (IOException e) {   e.printStackTrace(); }
    //把图像文件插入到系统图库
    try {
        myNameTemp = MediaStore.Images.Media.insertImage(
        context.getContentResolver(),myNewFile.getAbsolutePath(), myFileName, null);
    } catch (FileNotFoundException e) {
        e.printStackTrace();
        myNameTemp = "";
    }
    //通知图库更新
    context.sendBroadcast(new Intent(Intent.ACTION_MEDIA_SCANNER_SCAN_FILE,
            Uri.parse(Environment.getExternalStorageDirectory().toString()
            + "/luoGallery")));
    return myNameTemp;
    }
```

上面这段代码在 MyCode\MySample954\app\src\main\java\com\bin\luo\mysample\MainActivity.java 文件中。在这段代码中，myComponentName = new ComponentName("com.tencent.mm","com.tencent.mm.ui.tools.ShareToTimeLineUI")用于指定将图像仅发送到微信的朋友圈。myIntent.putExtra(Intent.EXTRA_STREAM,myUri)表示以流的形式传递图像资源。自定义方法 saveImageToGallery()用于在发送图像前，首先将图像保存在存储卡上。此外，访问存储卡需要在 AndroidManifest.xml 文件中添加<uses-permission android:name="android.permission.WRITE_EXTERNAL_STORAGE"/>权限。此实例的完整项目在 MyCode\MySample954 文件夹中。

225 使用 Intent 将图像发送到微信我的收藏

此实例主要通过在 Intent 的 setComponent()方法中使用"com.tencent.mm"和"com.tencent.mm.ui.tools.AddFavoriteUI"两个参数创建 ComponentName 对象,实现将本地图像文件通过 Intent 发送到微信的"我的收藏"。当实例运行之后,单击"选择图像文件"按钮,然后在弹出的窗口中选择本地图像文件并返回,该图像文件的路径信息将显示在"文件路径:"输入框中,如图 225-1 的左图所示,单击"发送到微信收藏"按钮,该图像文件将立即出现在微信的"我的收藏"中,如图 225-1 的右图所示。注意:在测试此应用之前一定要安装微信 App。

图 225-1

主要代码如下:

```
public void onClickBtn1(View v) {           //响应单击"选择图像文件"按钮
    Intent myIntent = new Intent(Intent.ACTION_GET_CONTENT);
    myIntent.setType("image/*");
    myIntent.addCategory(Intent.CATEGORY_OPENABLE);
    try {
        startActivityForResult(Intent.createChooser(myIntent, null), 1);
    } catch (Exception ex) { }
}
public void onClickBtn2(View v) {           //响应单击"发送到微信收藏"按钮
    try {
        Intent myIntent = new Intent();
        ComponentName myComponentName = new ComponentName("com.tencent.mm",
                "com.tencent.mm.ui.tools.AddFavoriteUI");
        myIntent.setComponent(myComponentName);
        myIntent.setAction(Intent.ACTION_SEND);
        myIntent.setFlags(Intent.FLAG_ACTIVITY_NEW_TASK);
        myIntent.setType("*/*");
```

```
      myIntent.putExtra(Intent.EXTRA_STREAM, myUri);
      startActivity(myIntent);
    } catch (Exception e) {
      Toast.makeText(getApplicationContext(),
          e.getMessage().toString(), Toast.LENGTH_SHORT).show();
    } }
  @Override
  protected void onActivityResult(int myRequestCode,
              int resultCode, Intent myIntent) {      //获取选择的文件
    super.onActivityResult(myRequestCode, resultCode, myIntent);
    try {
      switch (myRequestCode) {
        case 1:
          if (resultCode == Activity.RESULT_OK) {
            myUri = myIntent.getData();
            String[] myPathColumns = {MediaStore.Audio.Media.DATA};
            Cursor myCursor = getContentResolver().query(myUri,
                    myPathColumns, null, null, null);
            myCursor.moveToFirst();
            int myIndex = myCursor.getColumnIndex(myPathColumns[0]);
            String myPath = myCursor.getString(myIndex);
            myEditText.setText(myPath);
          } } } catch (Exception e) { e.printStackTrace(); }
}
```

上面这段代码在 MyCode\MySample963\app\src\main\java\com\bin\luo\mysample\MainActivity.java 文件中。在这段代码中，myIntent.setType("image/*")表示文件类型是所有图像文件。myIntent.putExtra（Intent.EXTRA_STREAM，myUri）表示以流的形式传递图像文件。myComponentName = new ComponentName("com.tencent.mm","com.tencent.mm.ui.tools.AddFavoriteUI")用于指定图像文件的接收对象是微信的"我的收藏"。此实例的完整项目在 MyCode\MySample963 文件夹中。

226 使用 Intent 将视频发送到微信我的收藏

此实例主要通过在 Intent 的 setComponent()方法中使用"com.tencent.mm"和"com.tencent.mm.ui.tools.AddFavoriteUI"两个参数创建 ComponentName 对象，实现将本地视频文件通过 Intent 发送到微信的"我的收藏"。当实例运行之后，单击"选择视频文件"按钮，然后在弹出的窗口中选择本地视频文件并返回，该视频文件的路径信息将显示在"文件路径："输入框中，如图 226-1 的左图所示，单击"发送到微信收藏"按钮，该视频文件将立即出现在微信的"我的收藏"中，如图 226-1 的右图所示。注意：在测试此应用之前一定要安装微信 App。

主要代码如下：

```
public void onClickBtn1(View v) {          //响应单击"选择视频文件"按钮
  Intent myIntent = new Intent(Intent.ACTION_GET_CONTENT);
  myIntent.setType("video/*");
  myIntent.addCategory(Intent.CATEGORY_OPENABLE);
  try {
    startActivityForResult(Intent.createChooser(myIntent, null), 1);
  } catch (Exception ex) { }
}
```

第8章　Intent

图　226-1

```
public void onClickBtn2(View v) {    //响应单击"发送到微信收藏"按钮
  try {
    Intent myIntent = new Intent();
    ComponentName myComponentName = new ComponentName("com.tencent.mm",
                "com.tencent.mm.ui.tools.AddFavoriteUI");
    myIntent.setComponent(myComponentName);
    myIntent.setAction(Intent.ACTION_SEND);
    myIntent.setFlags(Intent.FLAG_ACTIVITY_NEW_TASK);
    myIntent.setType("*/*");
    myIntent.putExtra(Intent.EXTRA_STREAM, myUri);
    startActivity(myIntent);
  } catch (Exception e) {
    Toast.makeText(getApplicationContext(),
            e.getMessage().toString(), Toast.LENGTH_SHORT).show();
} }
@Override
protected void onActivityResult(int myRequestCode,
                int resultCode, Intent myIntent) { //获取选择的文件
  super.onActivityResult(myRequestCode, resultCode, myIntent);
  try {
    switch (myRequestCode) {
    case 1:
      if (resultCode == Activity.RESULT_OK) {
        myUri = myIntent.getData();
        String[] myPathColumns = {MediaStore.Audio.Media.DATA};
        Cursor myCursor = getContentResolver().query(myUri,
                            myPathColumns, null, null, null);
        myCursor.moveToFirst();
        int myIndex = myCursor.getColumnIndex(myPathColumns[0]);
        String myPath = myCursor.getString(myIndex);
        myEditText.setText(myPath);
    } } } catch (Exception e) { e.printStackTrace(); }
}
```

上面这段代码在 MyCode\MySample964\app\src\main\java\com\bin\luo\mysample\MainActivity.java 文件中。在这段代码中，myIntent.setType("video/*")表示文件类型是所有视频文件。myIntent.putExtra（Intent.EXTRA_STREAM，myUri）表示以流的形式传递视频文件。myComponentName = new ComponentName("com.tencent.mm","com.tencent.mm.ui.tools.AddFavoriteUI")用于指定视频文件的接收对象是微信的"我的收藏"。此实例的完整项目在 MyCode\MySample964 文件夹中。

227 使用 Intent 将本地视频分享给微信好友

此实例主要通过在 Intent 中指定微信包名"com.tencent.mm"，实现将本地视频文件仅分享到微信的好友对话中。当实例运行后，单击"选择视频文件"按钮，然后在弹出的窗口中选择视频文件并返回，该视频文件的路径信息将显示在"文件路径："输入框中，如图 227-1 的左图所示；单击"分享给微信好友"按钮，将弹出选择窗口，在选择窗口中选择微信好友，然后在弹出的发送窗口中单击"发送"按钮，如图 227-1 的右图所示，该视频将立即出现在微信的好友对话中。注意：在测试此应用之前一定要安装微信 App。

图 227-1

主要代码如下：

```
public void onClickBtn1(View v) {                    //响应单击"选择视频文件"按钮
    Intent myIntent = new Intent(Intent.ACTION_GET_CONTENT);
    myIntent.setType("video/mp4");
    startActivityForResult(myIntent, 0);
}
public void onClickBtn2(View v) {                    //响应单击"分享给微信好友"按钮
    try {
        Intent myIntent = new Intent(Intent.ACTION_SEND);
```

```
    myIntent.setType("video/mp4");
    myIntent.putExtra(Intent.EXTRA_STREAM, myUri);
    myIntent.setPackage("com.tencent.mm");        //微信应用包名
    startActivity(myIntent);
  } catch (Exception e) {
    Toast.makeText(getApplicationContext(),
            "请确认手机已经安装了微信!", Toast.LENGTH_SHORT).show();
  } }
//获取视频文件的路径
public static String getRealFilePath(final Context context, final Uri uri) {
  if (null == uri) return null;
  final String myScheme = uri.getScheme();
  String myData = null;
  if (myScheme == null)
    myData = uri.getPath();
  else if (ContentResolver.SCHEME_FILE.equals(myScheme)) {
    myData = uri.getPath();
  } else if (ContentResolver.SCHEME_CONTENT.equals(myScheme)) {
    Cursor myCursor = context.getContentResolver().query(uri,
            new String[]{MediaStore.Images.ImageColumns.DATA}, null, null, null);
    if (null != myCursor)
      if (myCursor.moveToFirst()) {
        int myIndex = myCursor.getColumnIndex(MediaStore.Images.ImageColumns.DATA);
        if (myIndex > -1) { myData = myCursor.getString(myIndex); }
      }
      myCursor.close();
  } }
  return myData;
}
@Override
protected void onActivityResult(int requestCode, int resultCode, Intent data) {
  if (resultCode == RESULT_OK) {
    myUri = data.getData();
    String myPath = getRealFilePath(MainActivity.this, myUri);
    myEditText.setText(myPath);
} }
```

上面这段代码在 MyCode \ MySample953 \ app \ src \ main \ java \ com \ bin \ luo \ mysample \ MainActivity.java 文件中。在这段代码中，myIntent.setPackage("com.tencent.mm")用于指定分享视频的应用是微信。myIntent.putExtra(Intent.EXTRA_STREAM，myUri)用于以流的形式传递视频文件。此实例的完整项目在 MyCode\MySample953 文件夹中。

228 使用 Intent 直接调启微信的扫一扫功能

此实例主要通过在 Intent 的 setComponent()方法中使用"com.tencent.mm"参数和"com. tencent.mm.ui.LauncherUI"参数创建 ComponentName 实例，同时在 Intent 的 putExtra()方法中使用"LauncherUI.From.Scaner.Shortcut"参数，实现从当前应用直接调启微信的扫一扫功能。当实例运行之后，单击"使用 Intent 直接调启微信的扫一扫功能"按钮，则将显示微信的扫一扫操作界面，效果分别如图 228-1 的左图和右图所示。注意：测试手机一定要安装微信 App。

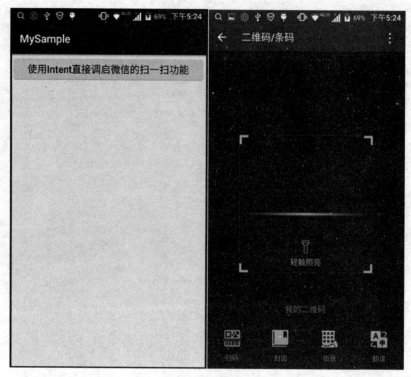

图　228-1

主要代码如下：

```
//响应单击"使用 Intent 直接调启微信的扫一扫功能"按钮
public void onClickBtn1(View v) {
  try {
    Intent myIntent = new Intent();
    myIntent.setComponent(new ComponentName("com.tencent.mm",
                                 "com.tencent.mm.ui.LauncherUI"));
    myIntent.putExtra("LauncherUI.From.Scaner.Shortcut", true);
    myIntent.setFlags(335544320);
    myIntent.setAction("android.intent.action.VIEW");
    startActivity(myIntent);
  } catch (Exception e) {
    Toast.makeText(MainActivity.this,
          e.getMessage().toString(), Toast.LENGTH_SHORT).show();
  } }
```

上面这段代码在 MyCode\MySampleX89\app\src\main\java\com\bin\luo\mysample\MainActivity.java 文件中。此实例的完整项目在 MyCode\MySampleX89 文件夹中。

229　使用 Intent 直接跳转到微信主操作界面

此实例主要通过根据应用包名"com.tencent.mm"创建 Intent，实现直接跳转到微信的主操作界面。当实例运行之后，单击"直接跳转到微信的主操作界面"按钮，将显示微信的主操作界面，效果分别如图 229-1 的左图和右图所示。注意：在测试此应用之前一定要安装微信 App。

图 229-1

主要代码如下：

```
//响应单击"直接跳转到微信的主操作界面"按钮
public void onClickmyBtn1(View v) {
  try {
    Intent myIntent =
            getPackageManager().getLaunchIntentForPackage("com.tencent.mm");
    startActivity(myIntent);
  } catch (Exception e) {
    Toast.makeText(this, "无法跳转到微信,请检查当前手机是否安装了微信!",
            Toast.LENGTH_SHORT).show();
  } }
```

上面这段代码在 MyCode\MySample377\app\src\main\java\com\bin\luo\mysample\MainActivity.java 文件中。在这段代码中，Intent myIntent = getPackageManager().getLaunchIntentForPackage("com.tencent.mm")用于根据微信包名创建 Intent。startActivity(myIntent)用于启动 Intent。此实例的完整项目在 MyCode\MySample377 文件夹中。

230 使用 Intent 根据号码启动 QQ 聊天界面

此实例主要通过使用 Intent 根据在 URI 中指定对方的 QQ 号码，实现直接启动与对方的 QQ 聊天界面。当实例运行之后，在"QQ 号码："输入框中输入 QQ 号码，如"2542640874"，如图 230-1 的左图所示；然后单击"打开 QQ 与对方聊天"按钮，将显示与对方的 QQ 聊天界面，如图 230-1 的右图所示。注意：在测试此应用之前一定要安装 QQ 手机版 App。

主要代码如下：

```
public void onClickmyBtn1(View v) {      //响应单击"打开 QQ 与对方聊天"按钮
  try{
```

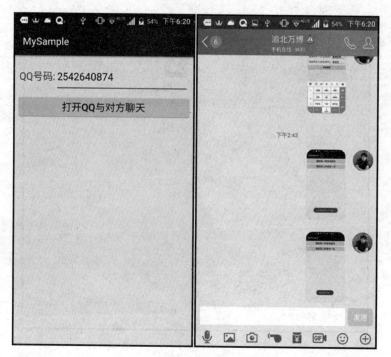

图 230-1

```
String myUrl = "mqqwpa://im/chat?chat_type = wpa&uin = "
        + myEditText.getText().toString();
startActivity(new Intent(Intent.ACTION_VIEW, Uri.parse(myUrl)));
}catch (Exception e){
    Toast.makeText(this, "测试时请确保手机中已经安装了移动QQ!",
            Toast.LENGTH_SHORT).show();
} }
```

上面这段代码在 MyCode\MySample376\app\src\main\java\com\bin\luo\mysample\MainActivity.java 文件中。在这段代码中,myUrl = "mqqwpa://im/chat? chat_type = wpa & uin = " + myEditText.getText().toString()中的 uin 参数指向的是对方的 QQ 号码,startActivity(new Intent(Intent.ACTION_VIEW, Uri.parse(myUrl)))则用于启动 myUrl 构造的 Intent。此实例的完整项目在 MyCode\MySample376 文件夹中。

231　使用 Intent 直接跳转到 QQ 主操作界面

此实例主要根据应用包名"com.tencent.mobileqq"创建 Intent,实现直接跳转到腾讯 QQ 的主操作界面。当实例运行之后,单击"直接跳转到腾讯 QQ 的主操作界面"按钮,将显示 QQ 的主操作界面,效果分别如图 231-1 的左图和右图所示。注意:在测试此应用之前一定要安装 QQ 手机版 App。

主要代码如下:

```
//响应单击"直接跳转到腾讯QQ的主操作界面"按钮
public void onClickmyBtn1(View v) {
  try {
    Intent myIntent =
        getPackageManager().getLaunchIntentForPackage("com.tencent.mobileqq");
```

图 231-1

```
    startActivity(myIntent);
} catch (Exception e) {
    Toast.makeText(MainActivity.this, "无法跳转到腾讯QQ,请检查当前手机是否安装了移动版的腾讯QQ!",
Toast.LENGTH_SHORT).show();
}}
```

上面这段代码在 MyCode\MySample378\app\src\main\java\com\bin\luo\mysample\MainActivity.java 文件中。在这段代码中，Intent myIntent = getPackageManager().getLaunchIntentForPackage("com.tencent.mobileqq")用于根据移动 QQ 包名创建 Intent。startActivity(myIntent)用于启动此 Intent。此实例的完整项目在 MyCode\MySample378 文件夹中。

232 使用 Intent 根据组件名称启动 QQ

此实例主要通过在 Intent 的 setComponent()方法中使用"com.tencent.mobileqq"和"com.tencent.mobileqq.activity.SplashActivity"两个参数创建 ComponentName，实现通过 Intent 启动 QQ。当实例运行之后，单击"启动 QQ"按钮，如图 232-1 的左图所示，将弹出 QQ 的启动界面，如图 232-1 的右图所示。注意：在测试此应用之前一定要安装 QQ 手机版 App。

主要代码如下：

```
public void onClickBtn1(View v) {        //响应单击"启动 QQ"按钮
    try {
    Intent myIntent = new Intent();
    ComponentName myComponentName = new ComponentName("com.tencent.mobileqq",
                    "com.tencent.mobileqq.activity.SplashActivity");
    myIntent.setComponent(myComponentName);
```

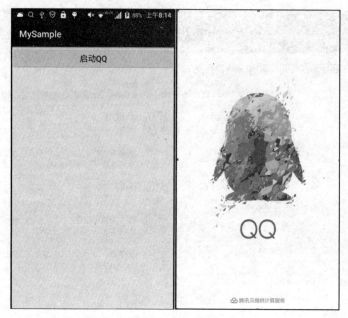

图 232-1

```
myIntent.setFlags(Intent.FLAG_ACTIVITY_NEW_TASK);
startActivity(myIntent);
} catch (ActivityNotFoundException e) {
Toast.makeText(MainActivity.this, e.getMessage().toString(),
                Toast.LENGTH_SHORT).show();
} }
```

上面这段代码在 MyCode\MySample895\app\src\main\java\com\bin\luo\mysample\MainActivity.java 文件中。此实例的完整项目在 MyCode\MySample895 文件夹中。

233 使用 Intent 直接跳转到 QQ 的我的电脑

此实例主要通过使用"com.tencent.mobileqq"和"com.tencent.mobileqq.activity.qfileJumpActivity"参数创建 ComponentName 的实例,实现通过 Intent 从当前应用直接跳转到 QQ 的我的电脑。当实例运行之后,单击"直接跳转到 QQ 的我的电脑"按钮,将弹出 QQ 的我的电脑,效果分别如图 233-1 的左图和右图所示。注意:在测试此应用之前一定要安装 QQ 手机版 App。

主要代码如下:

```
public void onClickBtn1(View v) {    //响应单击"直接跳转到 QQ 的我的电脑"按钮
Intent myIntent = new Intent(Intent.ACTION_SEND);
    ComponentName myComponentName = new ComponentName("com.tencent.mobileqq",
                "com.tencent.mobileqq.activity.qfileJumpActivity");
    myIntent.setComponent(myComponentName);
    startActivity(myIntent);
}
```

上面这段代码在 MyCode\MySample957\app\src\main\java\com\bin\luo\mysample\MainActivity.java 文件中。此实例的完整项目在 MyCode\MySample957 文件夹中。

图 233-1

234 使用 Intent 将本地图像发送到 QQ 的我的电脑

此实例主要通过在 Intent 的 setClassName()方法中使用"com.tencent.mobileqq"和"com.tencent.mobileqq.activity.qfileJumpActivity"作为参数,实现将本地图像通过 Intent 发送到 QQ 的我的电脑。当实例运行之后,将在 ImageView 控件中显示本地图像资源,如图 234-1 的左图所示,单击"发送到 QQ 的我的电脑"按钮,该图像将立即出现在 QQ 的我的电脑中,如图 234-1 的右图所示。注意:在测试此应用之前一定要安装 QQ 手机版 App。

图 234-1

主要代码如下：

```java
public void onClickBtn1(View v) {        //响应单击"发送到QQ的我的电脑"按钮
  try {
    Bitmap myBitmap =
            BitmapFactory.decodeResource(getResources(), R.mipmap.myimage1);
    String filePrefix = null;
    try {
      filePrefix = saveImageToGallery(this,myBitmap);
    } catch (IOException e) { e.printStackTrace(); }
    Intent myIntent = new Intent();
    myIntent.setAction(Intent.ACTION_SEND);
    myIntent.setClassName("com.tencent.mobileqq",
                    "com.tencent.mobileqq.activity.qfileJumpActivity");
    myIntent.setFlags(Intent.FLAG_ACTIVITY_NEW_TASK);
    myIntent.setType("image/jpg");
    Uri myUri = Uri.parse(filePrefix);
    myIntent.putExtra(Intent.EXTRA_STREAM,myUri);
    startActivity(myIntent);
  } catch (Exception e) {
    Toast.makeText(getApplicationContext(),
                e.getMessage().toString(), Toast.LENGTH_SHORT).show();
} }
public static String saveImageToGallery(Context context,
                                        Bitmap bmp) throws IOException {
  File myFile = new File(Environment.getExternalStorageDirectory(),
                        "luoGallery");      //保存图像文件
  if (!myFile.exists()) { myFile.mkdir(); }
  String myFileName = System.currentTimeMillis() + ".jpg";
  String myNameTemp;
  File myNewFile = new File(myFile, myFileName);
  myNewFile.createNewFile();
  try {
    FileOutputStream myStream = new FileOutputStream(myNewFile);
    bmp.compress(Bitmap.CompressFormat.JPEG, 100, myStream);
    myStream.flush();
    myStream.close();
  } catch (FileNotFoundException e) { e.printStackTrace(); }
  catch (IOException e) { e.printStackTrace();}
  //把图像文件插入到系统图库
  try {
    myNameTemp =
       MediaStore.Images.Media.insertImage(context.getContentResolver(),
       myNewFile.getAbsolutePath(), myFileName, null);
  } catch (FileNotFoundException e) {
    e.printStackTrace();
    myNameTemp = "";
  }
  //通知图库更新
  context.sendBroadcast(new Intent(Intent.ACTION_MEDIA_SCANNER_SCAN_FILE,
        Uri.parse(Environment.getExternalStorageDirectory().toString() +
        "/luoGallery")));
  return myNameTemp;
}
```

上面这段代码在 MyCode\MySample959\app\src\main\java\com\bin\luo\mysample\MainActivity.java 文件中。在这段代码中，myIntent.setClassName("com.tencent.mobileqq"，"com.tencent.mobileqq.activity.qfileJumpActivity")用于指定将图像仅发送到 QQ 的我的电脑。myIntent.putExtra(Intent.EXTRA_STREAM,myUri)表示以流的形式传递图像资源。自定义方法 saveImageToGallery()用于在发送图像前,首先将图像另存在存储卡上。此外,访问存储卡需要在 AndroidManifest.xml 文件中添加< uses-permission android:name="android.permission.WRITE_EXTERNAL_STORAGE"/>权限。此实例的完整项目在 MyCode\MySample959 文件夹中。

235 使用 Intent 将多首歌曲发送到 QQ 的我的电脑

此实例主要通过在 Intent 的 setComponent()方法中使用"com.tencent.mobileqq"和"com.tencent.mobileqq.activity.qfileJumpActivity"参数创建 ComponentName 对象,实现将本地的多首歌曲发送到 QQ 的我的电脑。当实例运行之后,单击"选择音乐文件"按钮,然后在弹出的窗口中选择一首音乐文件并返回,该音乐文件的路径信息将显示在"文件路径:"输入框中；再次单击"选择音乐文件"按钮,然后在弹出的窗口中选择一首音乐文件并返回,该音乐文件的路径信息也将显示在"文件路径:"输入框中,以此类推,可以选择多首音乐文件,如图 235-1 的左图所示；单击"发送到 QQ 我的电脑"按钮,这些音乐文件将会出现在 QQ 的我的电脑中,如图 235-1 的右图所示。注意：在测试此应用之前一定要安装 QQ 手机版 App。

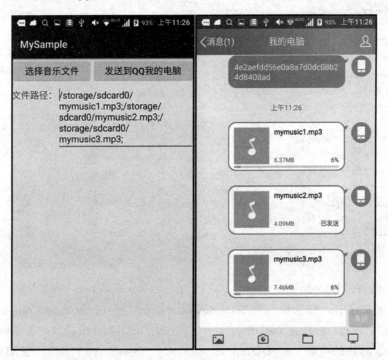

图 235-1

主要代码如下：

```
public void onClickBtn1(View v) {        //响应单击"选择音乐文件"按钮
    Intent myIntent = new Intent(Intent.ACTION_PICK, null);
    //设置"选择音乐曲目"窗口
```

```
      myIntent.setDataAndType(MediaStore.Audio.Media.EXTERNAL_CONTENT_URI, null);
      startActivityForResult(myIntent, 1);
    }
    public void onClickBtn2(View v) {  //响应单击"发送到 QQ 我的电脑"按钮
      try {
        Intent myQQIntent = new Intent();
        myQQIntent.setComponent(new ComponentName("com.tencent.mobileqq",
                        "com.tencent.mobileqq.activity.qfileJumpActivity"));
        if (myUriList.size() == 0) return;
        myQQIntent.setAction(Intent.ACTION_SEND_MULTIPLE);
        myQQIntent.setType("audio/*");
        myQQIntent.putExtra(Intent.EXTRA_STREAM, myUriList);
        startActivity(myQQIntent);
        myEditText.setText("");
      } catch (Exception e) {
        Toast.makeText(getApplicationContext(),
                        e.getMessage().toString(), Toast.LENGTH_SHORT).show();
      } }
    @Override
    protected void onActivityResult(int myRequestCode,
                        int resultCode, Intent myIntent) {  //获取选择的文件
      super.onActivityResult(myRequestCode, resultCode, myIntent);
      try {
        switch (myRequestCode) {
          case 1:
            if (resultCode == Activity.RESULT_OK) {
              myUri = myIntent.getData();
              myUriList.add(myUri);
              String[] myPathColumns = {MediaStore.Audio.Media.DATA};
              Cursor myCursor = getContentResolver().query(myUri,
                                myPathColumns, null, null, null);
              myCursor.moveToFirst();
              int myIndex = myCursor.getColumnIndex(myPathColumns[0]);
              String myPath = myCursor.getString(myIndex);
              String myContent = myEditText.getText().toString();
              myContent += myPath + ";";
              myEditText.setText(myContent);
        } } } catch (Exception e) {  e.printStackTrace(); }
    }
```

　　上面这段代码在 MyCode\MySample973\app\src\main\java\com\bin\luo\mysample\MainActivity.java 文件中。在这段代码中，myQQIntent.setComponent(new ComponentName("com.tencent.mobileqq","com.tencent.mobileqq.activity.qfileJumpActivity"))用于实现将选择的多首音乐文件发送到 QQ 的我的电脑。myQQIntent.setAction(Intent.ACTION_SEND_MULTIPLE)表示发送多个文件。myQQIntent.setType("audio/*")表示文件类型是音乐文件。myQQIntent.putExtra(Intent.EXTRA_STREAM，myUriList)表示以流的形式传递音乐文件。此外，读取在 SD 卡上的音乐文件需要在 AndroidManifest.xml 文件中添加<uses-permission android:name="android.permission.READ_EXTERNAL_STORAGE"/>权限。此实例的完整项目在 MyCode\MySample973 文件夹中。

236 使用 Intent 将音乐文件分享到 QQ 好友

此实例主要通过在 Intent 的 putExtra()方法中指定音乐文件的 URI,实现通过 Intent 将音乐文件分享到 QQ 好友。当实例运行之后,在"音乐文件:"输入框中输入在 SD 卡上存储的音乐文件名称,然后单击"开始播放音乐"按钮,将播放指定的音乐文件。单击"分享到移动 QQ"按钮,将弹出分享窗口,单击"仅此一次"按钮,如图 236-1 的左图所示;然后选择接收音乐文件的好友,如"渝北万博",如图 236-1 的右图所示,立即执行发送操作,如图 236-2 的左图所示,发送成功之后的效果如图 236-2 的右图所示。注意:在测试此应用之前一定要安装 QQ 手机版 App。

图　236-1

图　236-2

主要代码如下:

```java
public void onClickmyBtnPlay(View v) {           //响应单击"开始播放音乐"按钮
  try {
    MediaPlayer myMediaPlayer = new MediaPlayer();
    myMediaPlayer.setDataSource(myEditText.getText().toString());
    myMediaPlayer.prepare();
    myMediaPlayer.start();
  } catch (Exception e) {
    Toast.makeText(getApplicationContext(),
        e.getMessage().toString(),Toast.LENGTH_SHORT).show();
} }
public void onClickmyBtnShare(View v) {          //响应单击"分享到移动QQ"按钮
  try {
    Uri myUri = Uri.parse(myEditText.getText().toString());
    Intent myIntent = new Intent(Intent.ACTION_SEND);
    myIntent.setType("audio/mp3");
    myIntent.putExtra(Intent.EXTRA_STREAM, myUri);
    //指定移动QQ响应分享的音乐文件
    myIntent.setPackage("com.tencent.mobileqq");
     //执行分享操作
    startActivity(myIntent);
  } catch (Exception e) {
    Toast.makeText(getApplicationContext(),
        e.getMessage().toString(),Toast.LENGTH_SHORT).show();
} }
```

上面这段代码在 MyCode\MySample452\app\src\main\java\com\bin\luo\mysample\MainActivity.java 文件中。在这段代码中,myUri=Uri.parse(myEditText.getText().toString())用于将字符串形式的音乐文件路径转化为 URI 对象。myIntent = new Intent(Intent.ACTION_SEND)用于构造分享功能的 Intent。myIntent.putExtra(Intent.EXTRA_STREAM,myUri)用于设置此 Intent 分享的音乐文件。myIntent.setPackage("com.tencent.mobileqq")用于指定移动 QQ 响应分享的音乐文件。此外,读取在 SD 卡上的音乐文件需要在 AndroidManifest.xml 文件中添加<uses-permission android:name="android.permission.READ_EXTERNAL_STORAGE"/>权限。此实例的完整项目在 MyCode\MySample452 文件夹中。

237 使用 Intent 将多幅图像发送到 QQ 好友

此实例主要通过创建 Intent(Intent.ACTION_SEND_MULTIPLE)的实例,并使用 Intent 的 putParcelableArrayListExtra()方法,实现将存储卡上的多幅图像发送到 QQ 好友。当实例运行之后,单击"分享到我的移动 QQ"按钮,将弹出分享窗口,单击"仅此一次"按钮,如图 237-1 的左图所示;然后选择接收图像的好友,如"渝北万博",如图 237-1 的右图所示,立即执行发送操作,如图 237-2 的左图所示,发送成功之后的效果如图 237-2 的右图所示。注意:在测试此应用之前一定要安装 QQ 手机版 App。

第8章 Intent

图 237-1

图 237-2

主要代码如下：

```
public void onClickmyBtn1(View v) {        //响应单击"分享到我的移动 QQ"按钮
  try{
    ArrayList < Uri > myUris = new ArrayList <>();
    myUris.add(Uri.parse(Environment.getExternalStorageDirectory()
            + "/myimage1.jpg"));
    myUris.add(Uri.parse(Environment.getExternalStorageDirectory()
            + "/myimage2.jpg"));
    myUris.add(Uri.parse(Environment.getExternalStorageDirectory()
            + "/myimage3.jpg"));
    myUris.add(Uri.parse(Environment.getExternalStorageDirectory()
            + "/myimage4.jpg"));
```

```
        Intent myIntent = new Intent(Intent.ACTION_SEND_MULTIPLE);
        //附加多幅图像
        myIntent.putParcelableArrayListExtra(Intent.EXTRA_STREAM, myUris);
        //指定移动 QQ 响应分享的图像
        myIntent.setPackage("com.tencent.mobileqq");
        myIntent.setType("image/jpeg");
        startActivity(myIntent);
    }catch (Exception e){
        Toast.makeText(getApplicationContext(),
                "请确认手机已经安装了移动 QQ!", Toast.LENGTH_SHORT).show();
    }
}
```

上面这段代码在 MyCode \ MySample451 \ app \ src \ main \ java \ com \ bin \ luo \ mysample \ MainActivity.java 文件中。在这段代码中，myIntent = new Intent(Intent.ACTION_SEND_MULTIPLE)用于创建分享多文件的 Intent。myIntent.setPackage("com.tencent.mobileqq")表示使用移动 QQ 分享多个图像文件。myIntent.putParcelableArrayListExtra(Intent.EXTRA_STREAM, myUris)用于以流的形式传递多个图像文件。此实例的完整项目在 MyCode \ MySample451 文件夹中。

238　使用 Intent 实现截取屏幕部分区域

此实例主要通过使用 Intent("com.android.camera.action.CROP")创建裁剪 Intent，实现截取屏幕的部分区域。当实例运行之后，单击"选择图像文件"按钮，则弹出"选择照片"窗口，在该窗口中任意选择一个图像文件，该图像文件内容将显示在下面的 ImageView 控件中，如图 238-1 的左图所示；然后单击"截取部分屏幕"按钮，将弹出裁剪窗口，在裁剪窗口中可以拖动裁剪控制线改变裁剪区域大小，或者移动裁剪控制线到屏幕任意位置，如图 238-1 的右图所示；确认裁剪（修改）之后即可单击左上角的"保存"按钮将裁剪结果保存到存储卡的根目录中，文件名为"myClipImage.jpg"，同时显示在 ImageView 控件中。

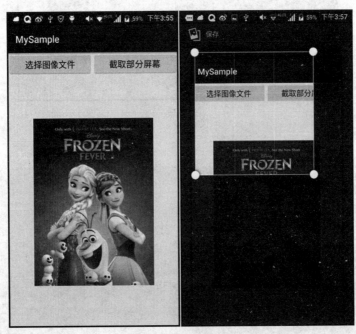

图　238-1

主要代码如下:

```java
@Override
public void onClick(View view) {
    if (view.getId() == R.id.myBtnClip) {           //响应单击"截取部分屏幕"按钮
        View myDecorView = getWindow().getDecorView();
        myDecorView.setDrawingCacheEnabled(true);
        myDecorView.buildDrawingCache();
        Bitmap myBitmap = myDecorView.getDrawingCache();
        SaveBitmap(myBitmap);
        Intent myIntent = new Intent("com.android.camera.action.CROP");
        myIntent.setDataAndType(Uri.parse(MediaStore.Images.Media.insertImage(
                getContentResolver(), myBitmap, null, null)), "image/*");
        myIntent.putExtra("crop", "true");
        //aspectX 和 aspectY 表示裁剪宽高的比例
        myIntent.putExtra("aspectX", 1);
        myIntent.putExtra("aspectY", 1);
        myIntent.putExtra("outputX", 340);
        myIntent.putExtra("outputY", 340);
        myIntent.putExtra("return-data", true);
        startActivityForResult(myIntent, 666);
    } else if (view.getId() == R.id.myBtnFile) {    //响应单击"选择图像文件"按钮
        Intent myIntent = new Intent(Intent.ACTION_PICK);
        myIntent.setType("image/*");
        startActivityForResult(myIntent, 3);
    }
}
public void SaveBitmap(Bitmap bitmap) {             //将裁剪图像保存为文件
    try {
        myStream = new FileOutputStream(myFile);
        bitmap.compress(Bitmap.CompressFormat.JPEG, 100, myStream);
        myStream.flush();
        myStream.close();
    } catch (Exception e) {   e.printStackTrace(); }
}
@Override
protected void onActivityResult(int requestCode, int resultCode, Intent data){
    if (requestCode == 3) {
        if (resultCode == RESULT_OK) {              //响应 Intent 选择图像文件
            myImageView.setImageURI(data.getData());
        }
    } else if (requestCode == 666) {                //如果同名文件已经存在,则删除该文件
        if (myFile.exists()) { myFile.delete(); }
        Bundle myExtras = data.getExtras();
        Bitmap myClipBitmap = myExtras.getParcelable("data");
        myImageView.setImageBitmap(myClipBitmap);
        SaveBitmap(myClipBitmap);
        Toast.makeText(this, "截取的图像保存成功!", Toast.LENGTH_SHORT).show();
    }
}
```

上面这段代码在 MyCode\MySample727\app\src\main\java\com\bin\luo\mysample\MainActivity.java 文件中。此外,访问存储卡需要在 AndroidManifest.xml 文件中添加< uses-permission android:name="android.permission.READ_EXTERNAL_STORAGE"/>权限和< uses-permission android:name="android.permission.WRITE_EXTERNAL_STORAGE"/>权限。此实

例的完整项目在 MyCode\MySample727 文件夹中。

239　使用 Intent 调用照相机拍照并裁剪头像

此实例主要通过创建 Intent（MediaStore.ACTION_IMAGE_CAPTURE）和 Intent（"com.android.camera.action.CROP"），实现使用相机拍摄照片并裁剪头像。当实例运行之后，单击"照相并裁剪头像"按钮，将打开照相机，拍摄照片并确认，如图 239-1 的左图所示，然后将显示裁剪窗口；在裁剪窗口中移动网格线剪裁图像，如图 239-1 的右图所示，单击左上角的"保存"，将在主窗口中显示裁剪之后的头像，如图 239-2 所示。

图　239-1

图　239-2

主要代码如下：

```java
public class MainActivity extends Activity {
    //照片文件
    private static final String IMAGE_FILE_NAME = "temp_head_image.jpg";
    //请求识别码
    private static final int CODE_CAMERA_REQUEST = 0xa1;
    private static final int CODE_RESULT_REQUEST = 0xa2;
    //裁剪图像的宽(X)和高(Y), 480 * 480 的正方形
    private static int output_X = 480;
    private static int output_Y = 480;
    private ImageView myImageView = null;
    @Override
    protected void onCreate(Bundle savedInstanceState) {
        super.onCreate(savedInstanceState);
        setContentView(R.layout.activity_main);
        myImageView = (ImageView) findViewById(R.id.myImageView);
    }
```

```java
public void onClickmyBtn1(View v) {              //响应单击"照相并裁剪头像"按钮
  Intent myIntent = new Intent(MediaStore.ACTION_IMAGE_CAPTURE);
  if (hasSdcard()) {                             //判断存储卡是否可用
   myIntent.putExtra(MediaStore.EXTRA_OUTPUT,Uri.fromFile(new File(
           Environment.getExternalStorageDirectory(), IMAGE_FILE_NAME)));
  }
  startActivityForResult(myIntent, CODE_CAMERA_REQUEST);
}
@Override
protected void onActivityResult(int requestCode,
                                             int resultCode,Intent intent) {
 //如果没有进行有效的设置操作,则直接返回
 if (resultCode == RESULT_CANCELED) {
  Toast.makeText(getApplication(), "取消", Toast.LENGTH_LONG).show();
  return;
 }
 switch (requestCode) {
  case CODE_CAMERA_REQUEST:
   if (hasSdcard()) {
    File myFile =
         new File(Environment.getExternalStorageDirectory(),IMAGE_FILE_NAME);
    cropRawPhoto(Uri.fromFile(myFile));
   } else { Toast.makeText(this, "没有 SDCard!", Toast.LENGTH_LONG).show(); }
   break;
  case CODE_RESULT_REQUEST:
   if (intent != null) { setImageToHeadView(intent); }
   break;
 }
 super.onActivityResult(requestCode, resultCode, intent);
}
public void cropRawPhoto(Uri uri) {              //裁剪头像
  Intent myIntentCrop = new Intent("com.android.camera.action.CROP");
  myIntentCrop.setDataAndType(uri, "image/*");
  myIntentCrop.putExtra("crop", "true");
  myIntentCrop.putExtra("aspectX", 1);
  myIntentCrop.putExtra("aspectY", 1);
  myIntentCrop.putExtra("outputX", output_X);
  myIntentCrop.putExtra("outputY", output_Y);
  myIntentCrop.putExtra("return-data", true);
  startActivityForResult(myIntentCrop, CODE_RESULT_REQUEST);
}
//提取保存裁剪之后的图像,并设置头像
private void setImageToHeadView(Intent intent) {
  Bundle extras = intent.getExtras();
  if (extras != null) {
   Bitmap myBmp = extras.getParcelable("data");
   myImageView.setImageBitmap(myBmp);
  }
}
//检查设备是否存在 SDCard 的方法
public static boolean hasSdcard() {
  String state = Environment.getExternalStorageState();
  if (state.equals(Environment.MEDIA_MOUNTED)) {
   return true;                                  //SD 卡存在
```

```
    } else {
     return false;              //SD卡不存在
} } }
```

上面这段代码在 MyCode\MySample187\app\src\main\java\com\bin\luo\mysample\MainActivity.java 文件中。在这段代码中，myIntent = new Intent(MediaStore.ACTION_IMAGE_CAPTURE)用于创建照相机 Intent，myIntent.putExtra(MediaStore.EXTRA_OUTPUT, Uri.fromFile(new File(Environment.getExternalStorageDirectory(), IMAGE_FILE_NAME)))用于指定照片保存位置及文件名称，startActivityForResult(myIntent, CODE_CAMERA_REQUEST)则用于打开照相机。myIntentCrop = new Intent("com.android.camera.action.CROP")用于创建裁剪 Intent，startActivityForResult(myIntent, CODE_CAMERA_REQUEST)用于启动裁剪窗口。照相和裁剪的结果将由 onActivityResult(int requestCode, int resultCode, Intent intent)方法根据它们不同的 requestCode 进行处理。此实例的完整项目在 MyCode\MySample187 文件夹中。

240　使用 Intent 实现允许或禁止按键截屏

此实例主要通过使用 Intent 传递参数，并使用 addFlags()和 clearFlags()方法，实现允许或禁止使用音量键和电源键组合进行截屏。当实例运行之后，单击"禁止截屏"按钮，使用音量键和电源键组合进行截屏时将在通知栏显示"无法抓取屏幕截图"，即截屏失败。单击"允许截屏"按钮，则可以使用音量键和电源键组合进行正常截屏，如图 240-1 所示。

图　240-1

主要代码如下：

```
public class MainActivity extends Activity {
    @Override
```

```java
protected void onCreate(Bundle savedInstanceState) {
  super.onCreate(savedInstanceState);
  boolean isScreenShotEnabled =
          getIntent().getBooleanExtra("isScreenShotEnabled", true);
  if(!isScreenShotEnabled){
   getWindow().addFlags(WindowManager.LayoutParams.FLAG_SECURE);
  }else{
   getWindow().clearFlags(WindowManager.LayoutParams.FLAG_SECURE);
  }
  setContentView(R.layout.activity_main);
}
public void onClickmyBtn1(View v) {             //响应单击"禁止截屏"按钮
  Toast.makeText(this, "现在已经禁止截屏!", Toast.LENGTH_SHORT).show();
  Intent myIntent = getIntent();
  myIntent.putExtra("isScreenShotEnabled",false);
  finish();
  startActivity(myIntent);
}
public void onClickmyBtn2(View v) {             //响应单击"允许截屏"按钮
  Toast.makeText(this, "现在已经允许截屏!", Toast.LENGTH_SHORT).show();
  Intent myIntent = getIntent();
  myIntent.putExtra("isScreenShotEnabled",true);
  finish();
  startActivity(myIntent);
} }
```

上面这段代码在 MyCode\MySample248\app\src\main\java\com\bin\luo\mysample\MainActivity.java 文件中。在这段代码中,getWindow().addFlags(WindowManager.LayoutParams.FLAG_SECURE)用于禁止截屏,getWindow().clearFlags(WindowManager.LayoutParams.FLAG_SECURE)用于恢复正常截屏。此实例的完整项目在 MyCode\MySample248 文件夹中。

241 使用 Intent 在应用市场中查找包名详情

此实例主要通过使用应用包名创建 Intent(Intent.ACTION_VIEW, myUri),实现在应用市场中根据包名查询该应用详情。当实例运行之后,在"应用包名:"输入框中输入应用包名,如 QQ 音乐的应用包名"com.tencent.qqmusic",如图 241-1 的左图所示,单击"查询详情"按钮,将显示该应用在应用市场中的详细信息,如图 241-1 的右图所示。

主要代码如下:

```java
public void onClickmyBtn1(View v) {             //响应单击"查询详情"按钮
    //爱奇艺包名: com.qiyi.video
    //优酷包名: com.youku.phone
    //新浪微博包名: com.sina.weibo
  Uri myUri = Uri.parse("market://details?id=" + myEditText.getText().toString());
  Intent myIntent = new Intent(Intent.ACTION_VIEW, myUri);
  startActivity(myIntent);
}
```

图 241-1

上面这段代码在 MyCode\MySample211\app\src\main\java\com\bin\luo\mysample\MainActivity.java 文件中。在这段代码中，myUri = Uri.parse("market://details?id=" + myEditText.getText().toString())用于创建一个在应用市场上查询指定应用详情的URI。myIntent = new Intent(Intent.ACTION_VIEW，myUri)用于根据指定的Uri创建一个Intent，用以显示Uri指定的内容。此实例的完整项目在 MyCode\MySample211 文件夹中。

242　使用 Intent 根据包名卸载手机应用

此实例主要通过创建 Intent(Intent.ACTION_DELETE，myUri)的实例，实现根据应用包名称卸载手机中已安装应用。当实例运行之后，在"应用包名："输入框中输入手机中已经安装的应用包名称，如实例自身"com.bin.luo.mysample"，如图242-1的左图所示，单击"卸载应用"按钮，将显示该应用的卸载界面，如图242-1的右图所示。单击"确定"按钮即可卸载。

主要代码如下：

```java
public void onClickmyBtn1(View v) {        //响应单击"卸载应用"按钮
    String myPackageName = myEditText.getText().toString();
    Uri myUri = Uri.parse("package:" + myPackageName);
    Intent myIntent = new Intent(Intent.ACTION_DELETE, myUri);
    startActivity(myIntent);
}
```

上面这段代码在 MyCode\MySample197\app\src\main\java\com\bin\luo\mysample\MainActivity.java 文件中。在这段代码中，myIntent = new Intent(Intent.ACTION_DELETE，myUri)用于创建一个卸载应用 Intent，参数 myUri 必须是类似于"package：com.bin.luo.mysample"这种格式。此实例的完整项目在 MyCode\MySample197 文件夹中。

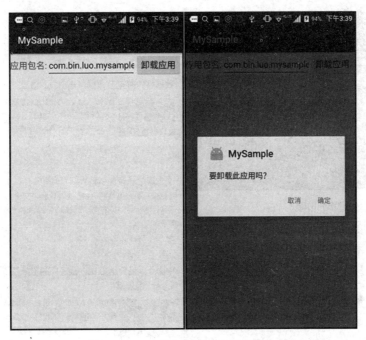

图 242-1

243 使用Intent根据内容跳转到搜索工具

此实例主要通过创建 Intent(Intent.ACTION_WEB_SEARCH)的实例,实现根据搜索内容跳转到手机上安装的搜索工具。当实例运行之后,在"搜索内容:"输入框中输入内容,然后单击"开始搜索"按钮,将显示系统已经安装的搜索工具(如百度等),如图243-1的左图所示;选择百度,将在手机百度中显示搜索结果,如图243-1的右图所示。注意:如果手机没有安装搜索工具,在测试实例时可能会报错。

主要代码如下:

```
public void onClickmyBtn1(View v) {              //响应单击"开始搜索"按钮
    Intent myIntent = new Intent(Intent.ACTION_WEB_SEARCH);
    myIntent.putExtra(SearchManager.QUERY, myEditText.getText().toString());
    startActivity(myIntent);
}
```

上面这段代码在 MyCode\MySample204\app\src\main\java\com\bin\luo\mysample\MainActivity.java 文件中。在这段代码中,myIntent = new Intent(Intent.ACTION_WEB_SEARCH)用于创建搜索 Intent,myIntent.putExtra(SearchManager.QUERY,myEditText.getText().toString())用于在 Intent 中预置搜索内容。此实例的完整项目在 MyCode\MySample204 文件夹中。

244 使用Intent指定应用打开PDF文件

此实例主要通过使用 Intent 的 setDataAndType(myUri,"application/pdf")方法,实现选择指定的应用打开 PDF 格式的文件。当实例运行之后,在"文件路径:"输入框中输入在 SD 卡上存储的

图 243-1

PDF 文件路径,然后单击"打开文件"按钮,将在底部滑出支持 PDF 格式的应用选择窗口;在该窗口中选择将要打开文件的应用,如"使用 PDF 查看器打开",然后单击"仅此一次"按钮,将使用该应用打开 PDF 文件,效果分别如图 244-1 的左图和右图所示。

图 244-1

主要代码如下：

```java
public void onClickmyBtn1(View v) {              //响应单击"打开文件"按钮
    Intent myIntent = new Intent("android.intent.action.VIEW");
    myIntent.addCategory("android.intent.category.DEFAULT");
    myIntent.addFlags(Intent.FLAG_ACTIVITY_NEW_TASK);
    Uri myUri = Uri.fromFile(new File(myEditText.getText().toString()));
    myIntent.setDataAndType(myUri, "application/pdf");
    this.startActivity(myIntent);
}
```

上面这段代码在 MyCode\MySample498\app\src\main\java\com\bin\luo\mysample\MainActivity.java 文件中。在这段代码中，myIntent.setDataAndType(myUri, "application/pdf")用于设置 Intent 的数据和类型，该方法的两个参数通常需要配对，常用的文件后缀名与类型的配对关系如下：{".3gp", "video/3gpp"}、{".apk", "application/vnd.android.package-archive"}、{".asf", "video/x-ms-asf"}、{".avi", "video/x-msvideo"}、{".bin", "application/octet-stream"}、{".bmp", "image/bmp"}、{".c", "text/plain"}、{".class", "application/octet-stream"}、{".conf", "text/plain"}、{".cpp", "text/plain"}、{".doc", "application/msword"}、{".docx", "application/vnd.openxmlformats-officedocument.wordprocessingml.document"}、{".xls", "application/vnd.ms-excel"}、{".xlsx", "application/vnd.openxmlformats-officedocument.spreadsheetml.sheet"}、{".exe", "application/octet-stream"}、{".gif", "image/gif"}、{".gtar", "application/x-gtar"}、{".gz", "application/x-gzip"}、{".h", "text/plain"}、{".htm", "text/html"}、{".html", "text/html"}、{".jar", "application/java-archive"}、{".java", "text/plain"}、{".jpeg", "image/jpeg"}、{".jpg", "image/jpeg"}、{".js", "application/x-javascript"}、{".log", "text/plain"}、{".m3u", "audio/x-mpegurl"}、{".m4a", "audio/mp4a-latm"}、{".m4b", "audio/mp4a-latm"}、{".m4p", "audio/mp4a-latm"}、{".m4u", "video/vnd.mpegurl"}、{".m4v", "video/x-m4v"}、{".mov", "video/quicktime"}、{".mp2", "audio/x-mpeg"}、{".mp3", "audio/x-mpeg"}、{".mp4", "video/mp4"}、{".mpc", "application/vnd.mpohun.certificate"}、{".mpe", "video/mpeg"}、{".mpeg", "video/mpeg"}、{".mpg", "video/mpeg"}、{".mpg4", "video/mp4"}、{".mpga", "audio/mpeg"}、{".msg", "application/vnd.ms-outlook"}、{".ogg", "audio/ogg"}、{".pdf", "application/pdf"}、{".png", "image/png"}、{".pps", "application/vnd.ms-powerpoint"}、{".ppt", "application/vnd.ms-powerpoint"}、{".pptx", "application/vnd.openxmlformats-officedocument.presentationml.presentation"}、{".prop", "text/plain"}、{".rc", "text/plain"}、{".rmvb", "audio/x-pn-realaudio"}、{".rtf", "application/rtf"}、{".sh", "text/plain"}、{".tar", "application/x-tar"}、{".tgz", "application/x-compressed"}、{".txt", "text/plain"}、{".wav", "audio/x-wav"}、{".wma", "audio/x-ms-wma"}、{".wmv", "audio/x-ms-wmv"}、{".wps", "application/vnd.ms-works"}、{".xml", "text/plain"}、{".z", "application/x-compress"}、{".zip", "application/x-zip-compressed"}、{"", "*/*"}。

此外，读取 SD 卡文件需要在 AndroidManifest.xml 文件中添加< uses-permission android:name="android.permission.READ_EXTERNAL_STORAGE"/>权限。此实例的完整项目在 MyCode\MySample498 文件夹中。

245　使用 Intent 启动应用打开文本文件

此实例主要通过使用 Intent 的 setDataAndType(myUri,"text/plain")方法,实现指定应用打开选择的文本文件。当实例运行后,单击"选择文本文件"按钮,然后在弹出的窗口中选择一个文本文件并返回,该文本文件的路径信息将显示在"文件路径:"输入框中,如图 245-1 的左图所示;然后单击"查看文本文件"按钮,将在底部滑出支持查看文本文件内容的应用选择窗口,在该窗口中选择将要打开文本文件的应用,将显示该文本文件的内容,如图 245-1 的右图所示。

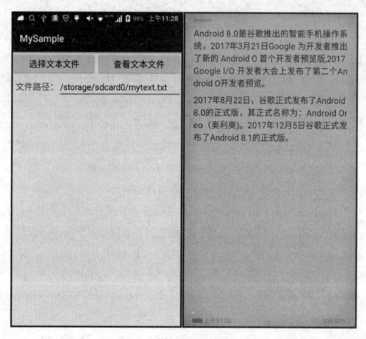

图　245-1

主要代码如下:

```java
public void onClickBtn1(View v) {                    //响应单击"选择文本文件"按钮
    Intent myIntent = new Intent(Intent.ACTION_GET_CONTENT);
    myIntent.setType("text/plain");
    myIntent.addCategory(Intent.CATEGORY_OPENABLE);
    try {
        startActivityForResult(Intent.createChooser(myIntent, null), 1);
    } catch (Exception ex) { }
}
public void onClickBtn2(View v) {                    //响应单击"查看文本文件"按钮
    try {
    Intent myIntent = new Intent("android.intent.action.VIEW");
        myIntent.addCategory("android.intent.category.DEFAULT");
        myIntent.addFlags(Intent.FLAG_ACTIVITY_NEW_TASK);
        Uri myUri = Uri.fromFile(new File(myEditText.getText().toString()));
        myIntent.setDataAndType(myUri, "text/plain");
        this.startActivity(myIntent);
    } catch (Exception e) {
    Toast.makeText(getApplicationContext(),
            e.getMessage().toString(), Toast.LENGTH_SHORT).show();
```

```
} }
@Override
protected void onActivityResult(int myRequestCode,
                int resultCode, Intent myIntent) {                    //获取选择的文件
  super.onActivityResult(myRequestCode, resultCode, myIntent);
  try {
    switch (myRequestCode) {
      case 1:
        if (resultCode == Activity.RESULT_OK) {
          myUri = myIntent.getData();
          String[] myPathColumns = {MediaStore.Files.FileColumns.DATA};
          Cursor myCursor = getContentResolver().query(myUri,
                                    myPathColumns, null, null, null);
          myCursor.moveToFirst();
          int myIndex = myCursor.getColumnIndex(myPathColumns[0]);
          String myPath = myCursor.getString(myIndex);
          myEditText.setText(myPath);
        } } } catch (Exception e) {   e.printStackTrace(); }
}
```

上面这段代码在 MyCode \ MySample980 \ app \ src \ main \ java \ com \ bin \ luo \ mysample \ MainActivity.java 文件中。此实例的完整项目在 MyCode\MySample980 文件夹中。

246　使用 Intent 启动应用打开 Excel 文件

此实例主要通过使用 Intent 的 setDataAndType(myUri, "application/vnd.ms-excel")方法，实现指定应用打开选择的 Excel 格式文件。当实例运行之后，单击"选择 Excel 格式文件"按钮，然后在弹出的窗口中选择一个 Excel 文件并返回，该 Excel 文件的路径信息将显示在"文件路径："输入框中，如图 246-1 的左图所示；然后单击"查看 Excel 格式文件"按钮，将在底部滑出支持查看 Excel 文件内容的应用选择窗口，在该窗口中选择将要打开 Excel 文件的应用，将显示该 Excel 文件的内容，如图 246-1 的右图所示。

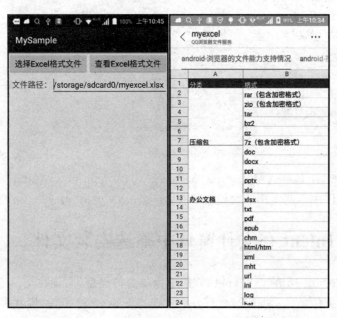

图　246-1

主要代码如下：

```java
public void onClickBtn1(View v) {            //响应单击"选择 Excel 格式文件"按钮
    Intent myIntent = new Intent(Intent.ACTION_GET_CONTENT);
    myIntent.setType("application/*");
    myIntent.addCategory(Intent.CATEGORY_OPENABLE);
    try {
        startActivityForResult(Intent.createChooser(myIntent, null), 1);
    } catch (Exception ex) { }
}
public void onClickBtn2(View v) {            //响应单击"查看 Excel 格式文件"按钮
    try {
        Intent myIntent = new Intent("android.intent.action.VIEW");
        myIntent.addCategory("android.intent.category.DEFAULT");
        myIntent.addFlags(Intent.FLAG_ACTIVITY_NEW_TASK);
        Uri myUri = Uri.fromFile(new File(myEditText.getText().toString()));
        myIntent.setDataAndType(myUri, "application/vnd.ms-excel");
        //myIntent.setDataAndType(myUri, "application/msexcel");
        //myIntent.setDataAndType(myUri, "application/excel");
        //myIntent.setDataAndType(myUri, "application/luoexcel");
        startActivity(myIntent);
    } catch (Exception e) {
        Toast.makeText(getApplicationContext(),
                e.getMessage().toString(), Toast.LENGTH_SHORT).show();
    }
}
@Override
protected void onActivityResult(int myRequestCode,
                  int resultCode, Intent myIntent) {    //获取选择的文件
    super.onActivityResult(myRequestCode, resultCode, myIntent);
    try {
        switch (myRequestCode) {
            case 1:
                if (resultCode == Activity.RESULT_OK) {
                    myUri = myIntent.getData();
                    String[] myPathColumns = {MediaStore.Files.FileColumns.DATA};
                    Cursor myCursor = getContentResolver().query(myUri,
                                        myPathColumns, null, null, null);
                    myCursor.moveToFirst();
                    int myIndex = myCursor.getColumnIndex(myPathColumns[0]);
                    String myPath = myCursor.getString(myIndex);
                    myEditText.setText(myPath);
    } } } catch (Exception e) { e.printStackTrace(); }
}
```

上面这段代码在 MyCode\MySample983\app\src\main\java\com\bin\luo\mysample\MainActivity.java 文件中。此实例的完整项目在 MyCode\MySample983 文件夹中。

247　使用 Intent 在文件窗口中筛选安装文件

此实例主要通过在 Intent 的 setType()方法中指定文件类型，实现在文件选择（内部存储空间）窗口中筛选指定类型的文件。当实例运行之后，单击"选择安装文件"按钮，则在弹出的"内部存储空间"窗口中只有 APK 类型的安装文件可以选择，单击该文件，则执行安装操作，效果分别如图 247-1 的左

图和右图所示。

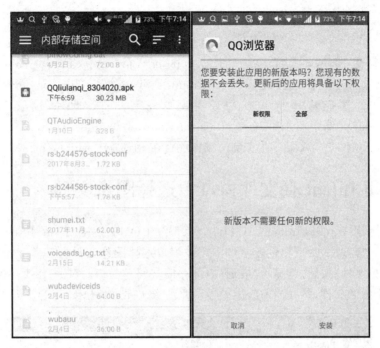

图 247-1

主要代码如下：

```
public void onClickBtn1(View v) {                    //响应单击"选择安装文件"按钮
    Intent myIntent = new Intent(Intent.ACTION_GET_CONTENT);
    //筛选文件后缀名是.apk的安装文件
    myIntent.setType("application/vnd.android.package-archive");
    //myIntent.setType("image/jpeg");                //筛选JPG格式的图像文件
    myIntent.addCategory(Intent.CATEGORY_OPENABLE);
    startActivityForResult(myIntent,1);              //跳转至文件选择界面
}
@Override
protected void onActivityResult(int requestCode, int resultCode, Intent data){
    if(resultCode == RESULT_OK){                     //返回用户所选APK文件信息
        Uri myUri = data.getData();
        String myPath = getRealPathFromURI(myUri);   //将URI转换为绝对路径
        Intent myAPKIntent = new Intent(Intent.ACTION_VIEW);
        myAPKIntent.setDataAndType(Uri.fromFile(new File(myPath)),
                        "application/vnd.android.package-archive");
        startActivity(myAPKIntent);                  //开始安装所选APK安装文件
}}
public String getRealPathFromURI(Uri uri){
    String myPath = null;
    String[] myProjection = {MediaStore.Images.Media.DATA};
    Cursor myCursor = getContentResolver().query(uri,myProjection,null,null,null);
    //在数据库中查询指定URI,并返回符合条件的文件路径
    if(myCursor.moveToFirst()){
        int myIndex = myCursor.getColumnIndexOrThrow(MediaStore.Images.Media.DATA);
        myPath = myCursor.getString(myIndex);
    }
```

```
    myCursor.close();
    return myPath;
}
```

上面这段代码在 MyCode\MySample872\app\src\main\java\com\bin\luo\mysample\MainActivity.java 文件中。在这段代码中,myIntent.setType("application/vnd.android.package-archive")用于在文件选择窗口中筛选 APK 类型的安装文件。此外,读取 SD 卡文件需要在 AndroidManifest.xml 文件中添加< uses-permission android:name="android.permission.READ_EXTERNAL_STORAGE"/>权限。此实例的完整项目在 MyCode\MySample872 文件夹中。

248　使用 Intent 在文件窗口中选择图像文件

此实例主要使用 Intent(Intent.ACTION_GET_CONTENT)构造 Intent 的实例,并调用 setType("image/*")方法,实现在最近窗口中允许用户选择图像文件并显示。当实例运行之后,单击"选择并显示图像文件"按钮,将显示"最近(内部存储空间)"窗口,如图 248-1 的左图所示。在该窗口中任意选择一个图像文件,将在当前应用中显示此图像文件,如图 248-1 的右图所示。

图　248-1

主要代码如下:

```
public void onClickmyBtn1(View v) {                     //响应单击"选择并显示图像文件"按钮
    myTextView.setText("显示图像文件全路径");
    Intent myIntent = new Intent(Intent.ACTION_GET_CONTENT);
    myIntent.addCategory(Intent.CATEGORY_OPENABLE);
    myIntent.setType("image/*");
    startActivityForResult(myIntent, 1);
}
```

```
@Override
protected void onActivityResult(int myRequestCode,
                  int resultCode, Intent myIntent) {      //显示选择的图像文件
  super.onActivityResult(myRequestCode, resultCode, myIntent);
  try {
   switch (myRequestCode) {
   case 1:
    if (resultCode == Activity.RESULT_OK) {
     Uri myUri = myIntent.getData();
     String[] myPathColumns = {MediaStore.Images.Media.DATA};
     Cursor myCursor = getContentResolver().query(myUri,
                         myPathColumns, null, null, null);
     myCursor.moveToFirst();
     int myIndex = myCursor.getColumnIndex(myPathColumns[0]);
     String myPath = myCursor.getString(myIndex);       //获取选择的图像文件
     Bitmap myBmp = BitmapFactory.decodeFile(myPath);
     myImageView.setImageBitmap(myBmp);
     myTextView.setText(myPath);
    }
    break;
  } } catch (Exception e) { e.printStackTrace(); }
}
```

上面这段代码在 MyCode\MySample418\app\src\main\java\com\bin\luo\mysample\MainActivity.java 文件中。在这段代码中，myIntent = new Intent(Intent.ACTION_GET_CONTENT)用于创建可进行选择的 Intent。myIntent.setType("image/*")表示此 Intent 的数据类型为图像文件。onActivityResult()用于处理在选择图像文件之后返回的信息，如图像文件全路径等。此外，读取在 SD 卡上的文件需要在 AndroidManifest.xml 文件中添加<uses-permission android:name="android.permission.READ_EXTERNAL_STORAGE"/>权限。此实例的完整项目在 MyCode\MySample418 文件夹中。

249　使用 Intent 查询支持多个图像分享包名

此实例主要通过使用 PackageManager 的 queryIntentActivities()方法，实现查询在当前手机中支持多个图像文件分享的应用包名。当实例运行之后，单击"查询当前手机支持多个图像分享的包名"按钮，将在"包名列表："中显示查询结果，如图 249-1 所示。由于在测试手机中安装了微信和 QQ，因此查询结果表明两者均支持多个图像文件的分享。

主要代码如下：

```
//响应单击"查询当前手机支持多个图像分享的包名"按钮
public void onClickBtn1(View v) {
  try {
   myEditText.setText("");
   Intent myIntent = new Intent(Intent.ACTION_SEND_MULTIPLE);
   myIntent.addCategory(Intent.CATEGORY_DEFAULT);
   myIntent.setType("image/*");         //分享图像文件
   List<ResolveInfo> myResolveInfos =
         getPackageManager().queryIntentActivities(myIntent,
```

图 249-1

```
            PackageManager.COMPONENT_ENABLED_STATE_DEFAULT);
    for (ResolveInfo info : myResolveInfos) {
     ActivityInfo myActivityInfo = info.activityInfo;
     String myText = myEditText.getText().toString();
     if(myActivityInfo.packageName.length()> 0)
      myText += myActivityInfo.packageName + "\n";
     myEditText.setText(myText);
    }
   } catch (Exception e) {
    Toast.makeText(getApplicationContext(),
        e.getMessage().toString(), Toast.LENGTH_SHORT).show();
} }
```

上面这段代码在 MyCode\MySample976\app\src\main\java\com\bin\luo\mysample\ MainActivity.java 文件中。在这段代码中，myResolveInfos = getPackageManager(). queryIntentActivities(myIntent, PackageManager.COMPONENT_ENABLED_STATE_ DEFAULT)用于查询当前手机中符合 myIntent 的所有 ResolveInfo 对象(本质上是 Activity)。myIntent = new Intent(Intent.ACTION_SEND_MULTIPLE)表示 myIntent 支持多个文件分享。myIntent.setType("image/*")表示 myIntent 分享的是图像文件。如果是 myIntent.setType("audio/*"),则表示 myIntent 分享的是音乐文件。此实例的完整项目在 MyCode\MySample976 文件夹中。

250　使用 Intent 启用默认网络文件下载器

此实例主要实现了调用 Intent 将下载链接通过浏览器打开,把下载任务交给浏览器,让浏览器调用下载器下载网络文件;下载成功后,下载文件通常存放在浏览器默认的下载文件夹中;使用这种下载方法完全把下载工作交给了浏览器,自己的应用不需要申请任何权限,方便简单快捷。当实例运行之后,在"下载网址:"输入框中输入下载地址,如"http://gdown.baidu.com/data/wisegame/0904344dee4a2d92/QQ_718.apk",如图250-1的左图所示;然后单击"开始下载"按钮,在弹出的下载应用中选择浏览器,浏览器立即接管下载任务执行下载操作,如图250-1的右图所示。

图 250-1

主要代码如下:

```
public void startDownload(View v){        //响应单击"开始下载"按钮
    Intent myIntent = new Intent();
    myIntent.setAction(Intent.ACTION_VIEW);
    myIntent.addCategory(Intent.CATEGORY_BROWSABLE);
    myIntent.setData(Uri.parse(myEditText.getText().toString()));
    startActivity(myIntent);
}
```

上面这段代码在 MyCode\MySample893\app\src\main\java\com\bin\luo\mysample\MainActivity.java 文件中。此实例的完整项目在 MyCode\MySample893 文件夹中。

251 使用 Intent 发送带附件的邮件

此实例主要通过在 Intent 的 putExtra() 方法中设置 Intent.EXTRA_STREAM 参数和 URI 参数,实现将邮件附件以数据流的方式进行发送。当实例运行之后,在"收件人:""附件:""主题:""内容:"等输入框中输入相关的内容,单击"发送邮件"按钮,将显示手机已经安装的邮件应用列表,选择"使用通过邮件发送分享",然后单击"仅此一次"按钮,如图251-1的左图所示,将启动手机指定的邮件发送应用发送邮件,如图251-1的右图所示。单击右上角的"发送"按钮即可成功发送邮件。

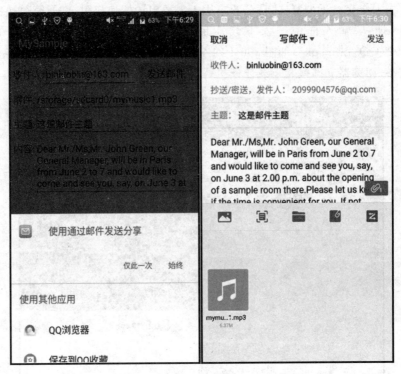

图 251-1

主要代码如下：

```
public void onClickmyBtn1(View v) {                          //响应单击"发送邮件"按钮
    Intent myIntent = new Intent(Intent.ACTION_SEND);
    String[] myTomails = {myEditTomail.getText().toString()};
    myIntent.putExtra(Intent.EXTRA_EMAIL, myTomails);        //设置收件邮箱
    myIntent.putExtra(Intent.EXTRA_SUBJECT,
            myEditSubject.getText().toString());             //设置邮件主题
    myIntent.putExtra(Intent.EXTRA_STREAM, Uri.parse("file://"
            + myEditAttachment.getText().toString()));       //设置发送附件
    myIntent.setType("application/octet - stream");          //设置以数据流形式上传
    myIntent.putExtra(Intent.EXTRA_TEXT,
            myEditBody.getText().toString());                //设置邮件内容
    myIntent.setType("text/plain");
    startActivity(myIntent);
}
```

上面这段代码在 MyCode\MySample499\app\src\main\java\com\bin\luo\mysample\MainActivity.java 文件中。此实例的完整项目在 MyCode\MySample499 文件夹中。

252 使用 Intent 跳转到系统无障碍设置界面

此实例主要通过创建 Intent(Settings. ACTION_ACCESSIBILITY_SETTINGS)的实例，实现从当前应用直接跳转到无障碍设置界面。当实例运行之后，单击"直接跳转到无障碍设置界面"按钮，将显示无障碍设置界面，效果分别如图 252-1 的左图和右图所示。

图 252-1

主要代码如下:

```
//响应单击"直接跳转到无障碍设置界面"按钮
public void onClickmyBtn1(View v) {
  try {
    Intent myIntent = new Intent(Settings.ACTION_ACCESSIBILITY_SETTINGS);
    startActivity(myIntent);
  } catch (Exception e) {
    Toast.makeText(MainActivity.this,
            e.getMessage().toString(), Toast.LENGTH_SHORT).show();
} }
```

上面这段代码在 MyCode\MySample388\app\src\main\java\com\bin\luo\mysample\MainActivity.java 文件中。在这段代码中，myIntent = new Intent(Settings.ACTION_ACCESSIBILITY_SETTINGS)用于创建管理无障碍设置界面的 Intent，startActivity(myIntent)则通过 Intent 显示无障碍设置界面。此实例的完整项目在 MyCode\MySample388 文件夹中。

第9章 第三方SDK开发

253　使用腾讯 SDK 将指定图像分享给 QQ 好友

此实例主要通过使用腾讯 SDK，实现将指定的图像分享给 QQ 好友。当实例运行之后，在各个输入框中输入对应的内容，如图 253-1 的左图所示，然后单击"分享给 QQ 好友"按钮，将启用 QQ 分享功能并分享给指定的好友，如图 253-1 的右图所示。

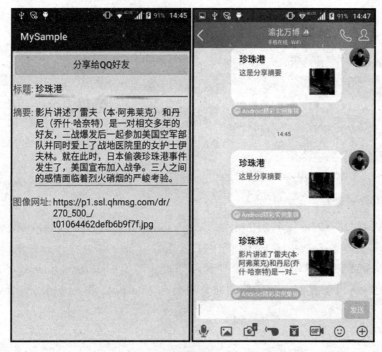

图　253-1

主要代码如下：

```
public class MainActivity extends Activity implements IUiListener {
    Tencent myTencent;
    EditText myEditTitle;
    EditText myEditSummary;
    EditText myEditAddress;
```

```java
@Override
protected void onCreate(Bundle savedInstanceState) {
 super.onCreate(savedInstanceState);
 //1106416651 是腾讯反馈给申请人的 ID,每个应用都不相同
 myTencent = Tencent.createInstance("1106416651", getApplicationContext());
 setContentView(R.layout.activity_main);
 myEditTitle = (EditText) findViewById(R.id.myEditTitle);
 myEditSummary = (EditText) findViewById(R.id.myEditSummary);
 myEditAddress = (EditText) findViewById(R.id.myEditAddress);
}
public void onClickBtn1(View v) {                                   //响应单击"分享给 QQ 好友"按钮
 Bundle myBundle = new Bundle();
 //设置分享类型为应用分享类型
 myBundle.putInt(QQShare.SHARE_TO_QQ_KEY_TYPE, QQShare.SHARE_TO_QQ_TYPE_APP);
 myBundle.putString(QQShare.SHARE_TO_QQ_TITLE,
                    myEditTitle.getText().toString());              //设置分享标题
 myBundle.putString(QQShare.SHARE_TO_QQ_SUMMARY,
                    myEditSummary.getText().toString());            //设置分享摘要
 myBundle.putString(QQShare.SHARE_TO_QQ_IMAGE_URL,
                    myEditAddress.getText().toString());            //设置分享图像网址
 myTencent.shareToQQ(MainActivity.this, myBundle, this);
}
@Override
public void onComplete(Object o) {
 Toast.makeText(MainActivity.this,
                "应用分享成功!", Toast.LENGTH_SHORT).show();
}
@Override
public void onError(UiError uiError) { }
@Override
public void onCancel() { }
@Override
//处理回调操作
protected void onActivityResult(int requestCode, int resultCode, Intent data) {
 Tencent.onActivityResultData(requestCode, resultCode, data, this);
} }
```

上面这段代码在 MyCode\MySample941\app\src\main\java\com\bin\luo\mysample\MainActivity.java 文件中。需要说明的是,此实例需要在 MyCode\MySample941\app\libs 文件夹中添加库文件 open_sdk_r5886_lite.jar,该库文件的下载地址是 http://wiki.connect.qq.com/sdk%E4%B8%8B%E8%BD%BD。然后在 MyCode\MySample941\app\build.gradle 文件的根节点下添加如下代码:

```
repositories { flatDir { dirs 'libs' } }
```

并在该文件的 dependencies 节点下添加依赖项:

```
compile files('libs/open_sdk_r5886_lite.jar')
compile 'com.android.support:support-v4:26.0.0-alpha1'
```

同时还要在 MyCode\MySample941\app\src\main\AndroidManifest.xml 文件中按照下列粗体

字所示的内容进行修改：

```xml
<?xml version = "1.0" encoding = "utf-8"?>
<manifest xmlns:android = "http://schemas.android.com/apk/res/android"
    package = "com.bin.luo.mysample">
    <application android:allowBackup = "true"
        android:icon = "@mipmap/ic_launcher"
        android:label = "@string/app_name"
        android:supportsRtl = "true"
        android:theme = "@style/AppTheme">
        <activity android:name = ".MainActivity">
            <intent-filter>
                <action android:name = "android.intent.action.MAIN"/>
                <category android:name = "android.intent.category.LAUNCHER"/>
            </intent-filter>
        </activity>
        <activity android:name = "com.tencent.tauth.AuthActivity"
            android:launchMode = "singleTask"
            android:noHistory = "true">
            <intent-filter>
                <action android:name = "android.intent.action.VIEW"/>
                <category android:name = "android.intent.category.DEFAULT"/>
                <category android:name = "android.intent.category.BROWSABLE"/>
                <data android:scheme = "tencent1106416651"/><!-- 填入你的 APP ID -->
            </intent-filter>
        </activity>
        <activity android:name = "com.tencent.connect.common.AssistActivity"
            android:configChanges = "orientation|keyboardHidden|screenSize"/>
    </application>
    <uses-permission android:name = "android.permission.INTERNET"/>
    <uses-permission android:name = "android.permission.ACCESS_NETWORK_STATE"/>
</manifest>
```

此外，使用腾讯 SDK 服务需要在腾讯开放平台上创建应用并获取 AppID，步骤如下。

(1) 登录腾讯开放平台 (http://open.qq.com)，进入管理中心页面。

(2) 单击右侧的"创建应用"开始创建操作，选择"移动应用安卓"选项，然后单击"创建应用"按钮进行下一步操作。在弹出的对话框中选择"软件"选项，并单击"确定"按钮进行下一步操作。

(3) 在表单中填写应用的相关信息，填写完成后单击底部的"提交审核"按钮即可完成应用创建操作。应用审核一般会在 24 小时内完成并反馈 ID 号码 (如此实例的"1106416651")。

此实例的完整项目在 MyCode\MySample941 文件夹中。

254 使用腾讯 SDK 将指定链接分享到 QQ 空间

此实例主要通过使用腾讯 SDK 实现将指定的链接分享到 QQ 空间。当实例运行后，在各个输入框中输入对应的内容，如图 254-1 的左图所示，然后单击"分享到 QQ 空间"按钮，将跳转到 QQ 空间的发表窗口；在发表窗口中单击右上角的"发表"按钮，如图 254-1 的右图所示，分享链接将会出现在 QQ 空间中，并在当前应用中提示"分享成功！"。

图 254-1

主要代码如下：

```java
public void onClickBtn1(View v) {                              //响应单击"分享到QQ空间"按钮
    Bundle myBundle = new Bundle();
    //设置在QQ空间分享链接
    myBundle.putInt(QQShare.SHARE_TO_QQ_EXT_INT,
                    QQShare.SHARE_TO_QQ_FLAG_QZONE_AUTO_OPEN);
    myBundle.putString(QQShare.SHARE_TO_QQ_TITLE,
                    this.myEditTitle.getText().toString());    //设置分享标题
    //设置该分享在单击时要跳转的URL链接
    myBundle.putString(QQShare.SHARE_TO_QQ_TARGET_URL,
                    myEditURL.getText().toString());
    myTencent.shareToQQ(this,myBundle,this);
}
@Override                                                       //分享回调处理
protected void onActivityResult(int requestCode, int resultCode, Intent data) {
    Tencent.onActivityResultData(requestCode, resultCode, data, this);
}
@Override
public void onComplete(Object o) {                              //分享成功时的回调函数
    Toast.makeText(MainActivity.this, "分享成功!", Toast.LENGTH_SHORT).show();
}
@Override
public void onError(UiError uiError) {                          //分享失败时的回调函数
    Toast.makeText(MainActivity.this, "分享失败!", Toast.LENGTH_SHORT).show();
}
@Override
public void onCancel() {                                        //取消分享操作时的回调函数
    Toast.makeText(MainActivity.this, "取消分享操作!",
                    Toast.LENGTH_SHORT).show();
}
```

上面这段代码在 MyCode\MySample944\app\src\main\java\com\bin\luo\mysample\MainActivity.java 文件中。需要说明的是，此实例需要添加腾讯 SDK 库文件，及修改 build.gradle 文件和 AndroidManifest.xml 文件，相关操作请参考实例 253。此实例的完整项目在 MyCode\MySample944 文件夹中。

255 使用腾讯 SDK 将本地视频发布到 QQ 空间

此实例主要通过使用腾讯 SDK 实现将本地视频发布到 QQ 空间。当实例运行之后，单击"选择视频文件"按钮，然后在弹出的窗口中选择视频文件，视频文件路径将显示在输入框中，如图 255-1 的左图所示；然后单击"发布到 QQ 空间"按钮，将跳转到上传窗口；在上传窗口中单击"上传"按钮，该视频将会出现在 QQ 空间中，如图 255-1 的右图所示。

图 255-1

主要代码如下：

```java
public void onClickBtn1(View v) {           //响应单击"选择视频文件"按钮
    Intent myIntent = new Intent(Intent.ACTION_GET_CONTENT);
    myIntent.setType("video/mp4");
    startActivityForResult(myIntent, 0);
}
public void onClickBtn2(View v) {           //响应单击"发布到 QQ 空间"按钮
    final Bundle myBundle = new Bundle();
    myBundle.putInt(QzonePublish.PUBLISH_TO_QZONE_KEY_TYPE,
                    QzonePublish.PUBLISH_TO_QZONE_TYPE_PUBLISHVIDEO);
    myBundle.putString(QzonePublish.PUBLISH_TO_QZONE_VIDEO_PATH,
                        myEditText.getText().toString());
    myTencent.publishToQzone(MainActivity.this, myBundle, null);
}
//获取视频文件的路径
public static String getRealFilePath(final Context context, final Uri uri){
```

```
    if (null == uri) return null;
    final String scheme = uri.getScheme();
    String data = null;
    if (scheme == null)
     data = uri.getPath();
    else if (ContentResolver.SCHEME_FILE.equals(scheme)) {
     data = uri.getPath();
    } else if (ContentResolver.SCHEME_CONTENT.equals(scheme)) {
     Cursor cursor = context.getContentResolver().query(uri,
         new String[]{MediaStore.Images.ImageColumns.DATA}, null, null, null);
     if (null != cursor) {
      if (cursor.moveToFirst()) {
       int index = cursor.getColumnIndex(MediaStore.Images.ImageColumns.DATA);
       if (index > -1) { data = cursor.getString(index); }
      }
      cursor.close();
     } }
    return data;
  }
  @Override
  protected void onActivityResult(int requestCode, int resultCode, Intent data) {
    if (resultCode == RESULT_OK) {
     Uri myUri = data.getData();
     String myPath = getRealFilePath(MainActivity.this, myUri);
     myEditText.setText(myPath);
    }else{
    Tencent.onActivityResultData(requestCode, resultCode, data, null);
    }
  }
```

上面这段代码在 MyCode\MySample950\app\src\main\java\com\bin\luo\mysample\MainActivity.java 文件中。需要说明的是，此实例需要添加腾讯 SDK 库文件，及修改 build.gradle 文件和 AndroidManifest.xml 文件，相关操作请参考实例 253。此实例的完整项目在 MyCode\MySample950 文件夹中。

256 使用微信 SDK 将本地图像分享到朋友圈

此实例主要通过使用微信 SDK 的 WXMediaMessage 和 SendMessageToWX，实现将选择的本地图像分享到微信的朋友圈。当实例运行之后，单击"选择本地图像"按钮，然后在弹出的窗口中选择一幅本地图像，该图像将显示在下面的 ImageView 控件中，如图 256-1 的左图所示；然后单击"分享至微信朋友圈"按钮，该图像将出现在微信朋友圈中，如图 256-1 的右图所示。

主要代码如下：

```
public class MainActivity extends Activity{
  IWXAPI myWeChat;
  ImageView myImageView;
  @Override
  protected void onCreate(Bundle savedInstanceState){
   super.onCreate(savedInstanceState);
//创建微信接口对象
   myWeChat = WXAPIFactory.createWXAPI(this,"wx9fbf8c966226923a",true);
```

图 256-1

```java
//将应用的 AppID 注册至微信
myWeChat.registerApp("wx9fbf8c966226923a");
setContentView(R.layout.activity_main);
myImageView = (ImageView)findViewById(R.id.myImageView);
}
public void onClickBtn2(View v){        //响应单击"分享至微信朋友圈"按钮
Bitmap myBitmap = ((BitmapDrawable)myImageView.getDrawable()).getBitmap();
    WXImageObject myImageObject = new WXImageObject(myBitmap);
    WXMediaMessage myWXMediaMessage = new WXMediaMessage();
    myWXMediaMessage.mediaObject = myImageObject;
    Bitmap myThumbnailBitmap = ThumbnailUtils.extractThumbnail(myBitmap,100,100);
    ByteArrayOutputStream myBitmapStream = new ByteArrayOutputStream();
    myThumbnailBitmap.compress(Bitmap.CompressFormat.JPEG,100,myBitmapStream);
    //将图像转换为 byte 数组
    myWXMediaMessage.thumbData = myBitmapStream.toByteArray();
    SendMessageToWX.Req myRequest = new SendMessageToWX.Req();
    //设置唯一标识符
    myRequest.transaction = String.valueOf(System.currentTimeMillis());
    myRequest.message = myWXMediaMessage;
    myRequest.scene = SendMessageToWX.Req.WXSceneTimeline;   //微信朋友圈
    myWcChat.sendReq(myRequest);          //请求分享操作
}
    public void onClickBtn1(View v){    //响应单击"选择本地图像"按钮
    Intent myIntent = new Intent(Intent.ACTION_GET_CONTENT);
    myIntent.setType("image/*");
    startActivityForResult(myIntent,0);
}
@Override
protected void onActivityResult(int requestCode,int resultCode,Intent data){
    //将选择的图像显示在 ImageView 控件上
    myImageView.setImageURI(data.getData());
}}
```

上面这段代码在 MyCode\MySample925\app\src\main\java\com\bin\luo\mysample\MainActivity.java 文件中。此外,此实例需要在 MyCode\MySample925\app\build.gradle 文件中添加 compile 'com.tencent.mm.opensdk:wechat-sdk-android-without-mta:+' 依赖项。同时还要修改 MyCode\MySample925\app\src\main\AndroidManifest.xml 文件的内容,如下面的粗体字所示:

```xml
<?xml version="1.0" encoding="utf-8"?>
<manifest xmlns:android="http://schemas.android.com/apk/res/android"
    package="com.bin.luo.mysample">
    <application
        android:allowBackup="true"
        android:icon="@mipmap/ic_launcher"
        android:label="@string/app_name"
        android:supportsRtl="true"
        android:theme="@style/AppTheme">
        <activity android:name=".MainActivity">
            <intent-filter>
                <action android:name="android.intent.action.MAIN"/>
                <category android:name="android.intent.category.LAUNCHER"/>
            </intent-filter>
        </activity>
    </application>
    <uses-permission android:name="android.permission.INTERNET"/>
    <uses-permission android:name="android.permission.ACCESS_NETWORK_STATE"/>
    <uses-permission android:name="android.permission.ACCESS_WIFI_STATE"/>
    <uses-permission android:name="android.permission.READ_PHONE_STATE"/>
    <uses-permission android:name="android.permission.WRITE_EXTERNAL_STORAGE"/>
</manifest>
```

使用微信 SDK 需要在微信开放平台上创建应用并获取 AppID,操作步骤如下。

(1) 登录微信开放平台(https://open.weixin.qq.com),进入"管理中心",单击"创建移动应用"按钮创建新应用。

(2) 在表单内填写应用的相关信息,完成后单击"下一步"按钮。

(3) 在应用平台选项中选择"Android 应用",并在其下方表单内填入相关信息,完成后单击"提交审核"按钮即可完成应用创建操作。一般应用审核需要 7 个工作日左右时间,其中应用签名需要通过签名生成工具来获取,该工具可以在微信开放平台的资源中心下载。

(4) 应用创建审核通过后,会自动生成一个对应的 AppID 值(如"wx9fbf8c966226923a")。用户需要将该 AppID 值传入微信接口对象构造器中以调用微信 SDK 相关功能。

此实例的完整项目在 MyCode\MySample925 文件夹中。

257 使用微信 SDK 将本地图像分享至微信好友

此实例主要通过使用微信 SDK 的 WXMediaMessage 和 SendMessageToWX,实现将选择的本地图像分享至指定的微信好友(对话)。当实例运行之后,单击"选择本地图像"按钮,然后在弹出的窗口中选择一幅本地图像,该图像将显示在下面的 ImageView 控件中,如图 257-1 的左图所示;然后单击"分享至微信好友"按钮,该图像将出现在微信的好友对话中,如图 257-1 的右图所示。

图 257-1

主要代码如下：

```java
public void onClickBtn1(View v){        //响应单击"分享至微信好友"按钮
  Bitmap myBitmap = ((BitmapDrawable)myImageView.getDrawable()).getBitmap();
  WXImageObject myImageObject = new WXImageObject(myBitmap);
  WXMediaMessage myWXMediaMessage = new WXMediaMessage();
  myWXMediaMessage.mediaObject = myImageObject;
  //创建缩略图
  Bitmap myThumbnailBitmap = ThumbnailUtils.extractThumbnail(myBitmap,100,100);
  ByteArrayOutputStream myBitmapStream = new ByteArrayOutputStream();
  myThumbnailBitmap.compress(Bitmap.CompressFormat.JPEG,100,myBitmapStream);
  //将图像转换为 byte 数组
  myWXMediaMessage.thumbData = myBitmapStream.toByteArray();
  SendMessageToWX.Req myRequest = new SendMessageToWX.Req();
  //设置唯一标识符
  myRequest.transaction = String.valueOf(System.currentTimeMillis());
  myRequest.message = myWXMediaMessage;
  //发送到微信好友对话中
  myRequest.scene = SendMessageToWX.Req.WXSceneSession;
  //请求分享操作
  myWeChat.sendReq(myRequest);
}
public void onClickBtn2(View v){        //响应单击"选择本地图像"按钮
  Intent myIntent = new Intent(Intent.ACTION_GET_CONTENT);
  myIntent.setType("image/*");
  startActivityForResult(myIntent,0);
}
@Override
protected void onActivityResult(int requestCode,int resultCode,Intent data){
  //将选择的图像显示在 ImageView 控件上
  myImageView.setImageURI(data.getData());
}
```

上面这段代码在 MyCode\MySampleX85\app\src\main\java\com\bin\luo\mysample\MainActivity.java 文件中。此外，此实例需要修改 AndroidManifest.xml 和 build.gradle 等文件，以及获取每个应用唯一的 AppID，相关操作请参考实例 256。此实例的完整项目在 MyCode\MySampleX85 文件夹中。

258 使用微信 SDK 将音乐链接分享至微信好友

此实例主要通过使用微信 SDK 的 WXMediaMessage 和 SendMessageToWX，实现将指定的音乐链接分享到指定的微信好友对话中。当实例运行后，在"音乐文件链接："输入框中输入音乐链接，如 "http://staff2.ustc.edu.cn/~wdw/softdown/index.asp/0042515_05.ANDY.mp3"，如图 258-1 的左图所示；然后单击"将下面的音乐链接分享至微信好友"按钮，该内容将出现在指定的微信好友对话中，如图 258-1 的右图所示。

图 258-1

主要代码如下：

```
public void onClickBtn1(View v) {                       //响应单击"将下面的音乐链接分享至微信好友"按钮
    WXMusicObject myMusicObject = new WXMusicObject();
    //设置音乐分享的 URL 链接
    myMusicObject.musicUrl = myEditText.getText().toString();
    WXMediaMessage myMessage = new WXMediaMessage(myMusicObject);
    myMessage.mediaObject = myMusicObject;
    myMessage.title = "音乐分享测试";                    //设置音乐链接分享标题
    myMessage.setThumbImage(BitmapFactory.decodeResource(
            getResources(),R.mipmap.ic_launcher));       //设置音乐链接分享缩略图样式
    SendMessageToWX.Req myRequest = new SendMessageToWX.Req();
    //设置该请求对象所对应的唯一标识符
    myRequest.transaction = String.valueOf(System.currentTimeMillis());
```

```
myRequest.message = myMessage;
myRequest.scene = SendMessageToWX.Req.WXSceneSession;    //发送到微信好友对话
myWeChat.sendReq(myRequest);                              //发送分享请求
}
```

上面这段代码在 MyCode\MySampleX86\app\src\main\java\com\bin\luo\mysample\MainActivity.java 文件中。此外，此实例需要修改 AndroidManifest.xml 和 build.gradle 等文件，以及获取每个应用唯一的 AppID，相关操作请参考实例 256。此实例的完整项目在 MyCode\MySampleX86 文件夹中。

259 使用微信 SDK 将视频链接分享到朋友圈

此实例主要通过使用微信 SDK 的 WXMediaMessage 和 SendMessageToWX，实现将指定的视频文件链接分享到微信朋友圈。当实例运行之后，在"视频文件链接："输入框中输入视频文件链接，如"http://v.youku.com/v_show/id_XMzA5MDk5OTcxMg==.html?spm=a2hww.20027244.m_250166.5~5!2~5~5~5~5~A"，如图 259-1 的左图所示；然后单击"将下面的视频链接分享至微信朋友圈"按钮，该视频链接内容将出现在微信朋友圈中，如图 259-1 的右图所示。

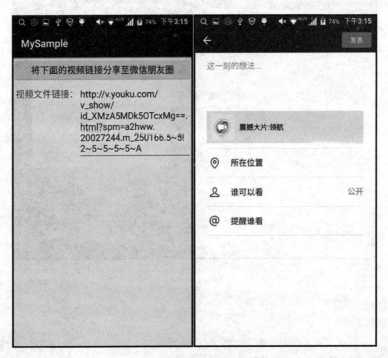

图 259-1

主要代码如下：

```
//响应单击"将下面的视频链接分享至微信朋友圈"按钮
public void onClickBtn1(View v) {
    WXVideoObject myVideoObject = new WXVideoObject();
    //设置分享视频 URL 链接
    myVideoObject.videoUrl = myEditText.getText().toString();
    WXMediaMessage myWXMediaMessage = new WXMediaMessage(myVideoObject);
```

```
    myWXMediaMessage.title = "震撼大片:领航";              //设置分享视频标题
    SendMessageToWX.Req myRequest = new SendMessageToWX.Req();
    //设置请求唯一标识符
    myRequest.transaction = String.valueOf(System.currentTimeMillis());
    myRequest.message = myWXMediaMessage;
    myRequest.scene = SendMessageToWX.Req.WXSceneTimeline;    //发送到微信朋友圈
    myWeChat.sendReq(myRequest);                              //发起分享请求
}
```

上面这段代码在 MyCode \ MySample928 \ app \ src \ main \ java \ com \ bin \ luo \ mysample \ MainActivity.java 文件中。此外，此实例需要修改 AndroidManifest.xml 和 build.gradle 等文件，以及获取每个应用唯一的 AppID，相关操作请参考实例 256。此实例的完整项目在 MyCode \ MySample928 文件夹中。

260　使用新浪 SDK 将文本分享到当前微博

此实例主要通过使用新浪微博 SDK，实现将指定的文本内容分享到当前的微博账号。当实例运行之后，在"分享内容："输入框中输入分享文本，如"这是我分享到新浪微博的文本内容"，如图 260-1 的左图所示；然后单击"分享至手机当前登录的新浪微博账号"按钮，则该分享内容将出现在新浪微博中，如图 260-1 的右图所示。

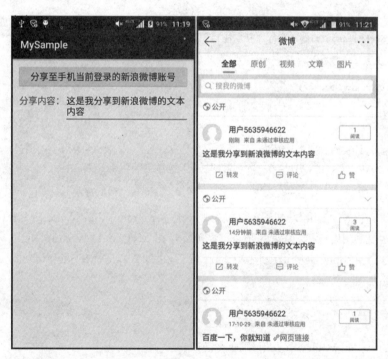

图　260-1

主要代码如下：

```
public class MainActivity extends Activity implements WbShareCallback {
    WbShareHandler myHandler;
    EditText myEditText;
    @Override
```

```java
protected void onCreate(Bundle savedInstanceState) {
    super.onCreate(savedInstanceState);
    WbSdk.install(this, new AuthInfo(this, "639620816",
                        "http://www.sina.com", null));        //初始化微博 SDK
    myHandler = new WbShareHandler(this);
    myHandler.registerApp();                                  //注册应用
    setContentView(R.layout.activity_main);
    myEditText = (EditText) findViewById(R.id.myEditText);
}
//响应单击"分享至手机当前登录的新浪微博账号"按钮
public void onClickBtn1(View v) {
    WeiboMultiMessage myWeiboMultiMessage = new WeiboMultiMessage();
    TextObject myTextObject = new TextObject();
    //获取并存储将要分享的文本内容
    myTextObject.text = myEditText.getText().toString();
    myWeiboMultiMessage.textObject = myTextObject;
    myHandler.shareMessage(myWeiboMultiMessage, false);       //进行分享操作
}
@Override
protected void onNewIntent(Intent intent) {
    super.onNewIntent(intent);
    myHandler.doResultIntent(intent, this);
}
@Override
public void onWbShareSuccess() {                              //分享成功时的回调函数
    Toast.makeText(MainActivity.this, "分享成功!", Toast.LENGTH_SHORT).show();
}
@Override
public void onWbShareCancel() {                               //主动取消分享操作的回调函数
    Toast.makeText(MainActivity.this, "取消分享!", Toast.LENGTH_SHORT).show();
}
@Override
public void onWbShareFail() {                                 //分享失败时的回调函数
    Toast.makeText(MainActivity.this, "分享失败!", Toast.LENGTH_SHORT).show();
}}
```

上面这段代码在 MyCode\MySample929\app\src\main\java\com\bin\luo\mysample\MainActivity.java 文件中。需要说明的是,此实例需要在 MyCode\MySample929\app\libs 文件夹中添加库文件 core-4.1.0-openDefaultRelease.aar,然后在 MyCode\MySample929\app\build.gradle 文件的根节点下添加如下代码:

```
repositories { flatDir { dirs 'libs' } }
```

并在该文件的 dependencies 节点下添加依赖项:

```
compile 'com.sina.weibo.sdk:core:4.1.0:openDefaultRelease@aar'
```

同时在 MyCode\MySample929\app\src\main\AndroidManifest.xml 文件中添加下列权限:

```xml
<uses-permission android:name="android.permission.INTERNET" />
<uses-permission android:name="android.permission.ACCESS_WIFI_STATE" />
<uses-permission android:name="android.permission.ACCESS_NETWORK_STATE" />
```

此外，使用新浪微博 SDK 需要在微博开放平台上创建应用并获取 AppKey，应用创建步骤如下：

（1）登录新浪微博开放平台（http://open.weibo.com/development/mobile），单击"移动应用"下方的"立即接入"按钮，并在弹出的对话框中单击"继续创建"按钮。

（2）在弹出的页面表单中填入移动应用相关信息，完成后单击"创建"按钮即可完成应用创建操作。

（3）此外，还需完善一部分应用信息，即需要在表单内输入应用包名、应用签名、应用下载地址、Android 下载地址、应用简介、应用介绍等必需信息。

（4）应用创建审核通过后，将会收到 AppKey（如 "639620816"），在使用 WbSdk.install() 进行初始化时将会使用此 AppKey。

此实例的完整项目在 MyCode\MySample929 文件夹中。

261　使用新浪 SDK 实现获取最新发布的微博

此实例主要通过使用新浪微博 SDK，实现获取最新发布的微博信息。当实例运行之后，单击"获取最新发布微博"按钮，则最新发布的新浪微博信息将出现在微博列表中，效果分别如图 261-1 的左图和右图所示。

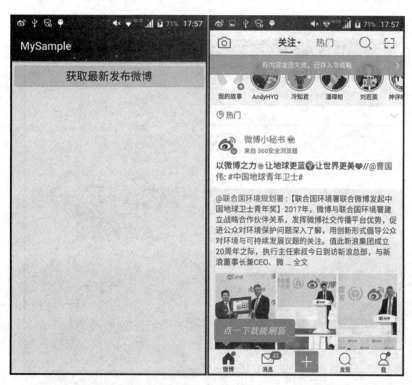

图　261-1

主要代码如下：

```
public class MainActivity extends Activity {
 AuthInfo myAuthInfo;
 @Override
 protected void onCreate(Bundle savedInstanceState) {
  super.onCreate(savedInstanceState);
```

```
myAuthInfo = new AuthInfo(this, "639620816", "http://www.sina.com", null);
WbSdk.install(this, myAuthInfo);                    //初始化微博 SDK
setContentView(R.layout.activity_main);
}
public void onClickBtn1(View v) {                   //响应单击"获取最新发布微博"按钮
    //调用浏览器或新浪微博应用来浏览最新微博
    WeiboPageUtils.getInstance(MainActivity.this,
                    myAuthInfo).gotoMyHomePage(false);
}}
```

上面这段代码在 MyCode\MySample931\app\src\main\java\com\bin\luo\mysample\MainActivity.java 文件中。需要说明的是，此实例需要添加新浪微博库文件，以及修改权限等，相关操作请参考实例 260。此实例的完整项目在 MyCode\MySample931 文件夹中。

262 使用新浪 SDK 实现第三方登录微博账号

此实例主要通过使用新浪微博 SDK，实现第三方登录新浪微博账号。当实例运行之后，单击"登录新浪微博账号"按钮，将弹出登录窗口自动登录新浪微博账号，效果分别如图 262-1 的左图和右图所示。

图 262-1

主要代码如下：

```
public void onClickBtn1(View v) {                   //响应单击"登录新浪微博账号"按钮
    myHandler = new SsoHandler(this);
    //设置授权登录回调监听器
    myHandler.authorizeClientSso(new WbAuthListener() {
        @Override
```

```java
    public void onSuccess(Oauth2AccessToken token) {          //登录成功时的回调函数
      AccessTokenKeeper.writeAccessToken(MainActivity.this, token);
      Toast.makeText(MainActivity.this, "登录成功!", Toast.LENGTH_SHORT).show();
    }
    @Override
    public void cancel() {                                     //取消登录时的回调函数
   Toast.makeText(MainActivity.this,
                "用户取消登录操作!", Toast.LENGTH_SHORT).show();
    }
    @Override
    //登录失败时的回调函数
    public void onFailure(WbConnectErrorMessage wbConnectErrorMessage) {
      Toast.makeText(MainActivity.this, "登录失败!", Toast.LENGTH_SHORT).show();
    } });
}
@Override
protected void onActivityResult(int requestCode, int resultCode, Intent data) {
    super.onActivityResult(requestCode, resultCode, data);
    //通过 SsoHandler 对象来处理回调结果
    if (myHandler != null)
       myHandler.authorizeCallBack(requestCode, resultCode, data);
}
```

上面这段代码在 MyCode\MySample932\app\src\main\java\com\bin\luo\mysample\MainActivity.java 文件中。需要说明的是，此实例需要添加新浪微博库文件，以及修改权限等，相关操作请参考实例 260。此实例的完整项目在 MyCode\MySample932 文件夹中。

263 使用新浪 SDK 实现分享链接地址至微博

此实例主要通过使用新浪微博 SDK，实现将链接地址分享到新浪微博。当实例运行之后，在"链接地址："输入框中输入分享链接，如"http://www.baidu.com"，如图 263-1 的左图所示。单击"分享至新浪微博"按钮，则该链接地址将被分享到新浪微博中，如图 263-1 的右图所示。

图 263-1

主要代码如下:

```java
public void onClickBtn1(View v) {                          //响应单击"分享至新浪微博"按钮
    WebpageObject myWebpageObject = new WebpageObject();
    myWebpageObject.title = "百度搜索";                      //设置链接标题
    myWebpageObject.description = "百度一下,你就知道";        //设置链接简介
    Bitmap myBitmap =
        BitmapFactory.decodeResource(getResources(),R.mipmap.ic_launcher);
    myWebpageObject.setThumbImage(myBitmap);               //设置将要显示的链接缩略图
    //设置分享的 URL 链接
    myWebpageObject.actionUrl = myEditText.getText().toString();
    WeiboMultiMessage myWeiboMultiMessage = new WeiboMultiMessage();
    myWeiboMultiMessage.mediaObject = myWebpageObject;
    //分享指定 URL 链接至微博
    myHandler.shareMessage(myWeiboMultiMessage, false);
}
```

上面这段代码在 MyCode\MySample934\app\src\main\java\com\bin\luo\mysample\MainActivity.java 文件中。需要说明的是,此实例需要添加新浪微博库文件,以及修改权限等,相关操作请参考实例 260。此实例的完整项目在 MyCode\MySample934 文件夹中。

264 使用新浪 SDK 实现跳转到微博账户简介

此实例主要通过使用新浪微博 SDK,实现从当前应用直接跳转到新浪微博的个人账户的简介页面的功能。当实例运行之后,单击"跳转到新浪微博个人账户简介页面"按钮,将启动新浪微博应用并跳转到个人账户的简介页面,效果分别如图 264-1 的左图和右图所示。

图 264-1

主要代码如下：

```
//响应单击"跳转到新浪微博个人账户简介页面"按钮
public void onClickBtn1(View v) {
    WeiboPageUtils.getInstance(MainActivity.this,myAuthInfo).gotoMyProfile();
}
```

上面这段代码在 MyCode\MySample936\app\src\main\java\com\bin\luo\mysample\MainActivity.java 文件中。需要说明的是，此实例需要添加新浪微博库文件，以及修改权限等，相关操作请参考实例 260。此实例的完整项目在 MyCode\MySample936 文件夹中。

265 使用百度 SDK 获取当前手机的经纬度值

此实例主要通过使用百度地图 SDK，实现获取当前手机的经纬度值和地址信息。当实例运行之后，单击"获取当前手机位置信息"按钮，将在下面显示手机的当前经度、纬度和位置信息，效果如图 265-1 所示。

图　265-1

主要代码如下：

```
public class MainActivity extends Activity {
    private LocationClient myClient = null;
    Handler myHandler = new Handler() {
        @Override
        public void handleMessage(Message msg) {
            myTextView.setText((String) msg.obj);        //显示定位结果
            myClient.stop();                              //停止定位
        }
    };
```

```java
TextView myTextView;
@Override
protected void onCreate(Bundle savedInstanceState) {
    super.onCreate(savedInstanceState);
    setContentView(R.layout.activity_main);
    myTextView = (TextView) findViewById(R.id.myTextView1);
    myClient = new LocationClient(MainActivity.this);
    LocationClientOption myOption = new LocationClientOption();
    myOption.setIsNeedAddress(true);              //获取并显示当前位置信息
    myClient.setLocOption(myOption);
    myClient.registerLocationListener(new BDLocationListener() {
        @Override
        public void onReceiveLocation(final BDLocation location) {
            new Thread() {
                @Override
                public void run() {
                    Message myMessage = new Message();
                    myMessage.obj = "经度:" + location.getLongitude() + "\n纬度:" + location.
                         getLatitude() + "\n您所在的位置:" + location.getAddrStr();
                    myHandler.sendMessage(myMessage);      //在主线程显示定位结果
                } }.start();
        } });
}
public void onClickmyBtn1(View v) {           //响应单击"获取当前手机位置信息"按钮
    myTextView.setText("正在获取位置信息...");
    if(myClient.isStarted()){ myClient.stop();}
    myClient.start();                         //开始定位
} }
```

上面这段代码在 MyCode\MySample752\app\src\main\java\com\bin\luo\mysample\MainActivity.java 文件中。此外，此实例需要引入百度地图库文件 BaiduLBS_Android.jar，具体操作步骤如下：首先将 BaiduLBS_Android.jar 文件及其关联的目录复制到 libs 目录中，右击 BaiduLBS_Android.jar 文件，在弹出的右键菜单中选择 Add As Library 菜单，则弹出 Create Library 窗口。在"Create Library"窗口中单击 OK 按钮，则 BaiduLBS_Android.jar 文件就可以像普通库那样进行调用了。然后修改 MyCode\MySample752\app\src\main\AndroidManifest.xml 文件，如下面的粗体字所示：

```xml
<?xml version="1.0" encoding="utf-8"?>
<manifest xmlns:android="http://schemas.android.com/apk/res/android"
          package="com.bin.luo.mysample">
  <application
      android:allowBackup="true"
      android:icon="@mipmap/ic_launcher"
      android:label="@string/app_name"
      android:supportsRtl="true"
      android:theme="@style/AppTheme">
    <meta-data android:name="com.baidu.lbsapi.API_KEY"
               android:value="f6GHYRd9fqudNt7654TWdlrzzyX9VqGX"/>
    <activity android:name=".MainActivity">
      <intent-filter>
        <action android:name="android.intent.action.MAIN"/>
```

```xml
        <category android:name = "android.intent.category.LAUNCHER"/>
      </intent-filter>
    </activity>
    <service
        android:name = "com.baidu.location.f"
        android:enabled = "true"
        android:permission = "android.permission.BAIDU_LOCATION_SERVICE"
        android:process = ":remote">
      <intent-filter>
        <action android:name = "com.baidu.location.service_v2.6"></action>
      </intent-filter>
    </service>
  </application>
  <uses-permission android:name = "android.permission.ACCESS_COARSE_LOCATION"/>
<uses-permission android:name = "android.permission.ACCESS_FINE_LOCATION"/>
<uses-permission android:name = "android.permission.ACCESS_WIFI_STATE"/>
<uses-permission android:name = "android.permission.ACCESS_NETWORK_STATE"/>
<uses-permission android:name = "android.permission.CHANGE_WIFI_STATE"/>
<uses-permission android:name = "android.permission.READ_PHONE_STATE"/>
<uses-permission android:name = "android.permission.WRITE_EXTERNAL_STORAGE"/>
<uses-permission android:name = "android.permission.INTERNET"/>
<uses-permission android:name = "android.permission.MOUNT_UNMOUNT_FILESYSTEMS"/>
</manifest>
```

关于如何获取 API_KEY，请参考官方网址（http://lbsyun.baidu.com/sdk）。接下来，按照下面粗体字所示的内容修改 MyCode\MySample752\app\build.gradle 文件：

```gradle
apply plugin: 'com.android.application'
android {
    compileSdkVersion 24
    buildToolsVersion "25.0.2"
    defaultConfig {
        applicationId "com.bin.luo.mysample"
        minSdkVersion 22
        targetSdkVersion 24
        versionCode 1
     versionName "1.0"
     testInstrumentationRunner "android.support.test.runner.AndroidJUnitRunner"
    }
    buildTypes {
      release {
        minifyEnabled false
        proguardFiles getDefaultProguardFile('proguard-android.txt'),
'proguard-rules.pro'
      }  }
    sourceSets { main { jniLibs.srcDirs = ['libs'] } }
}
dependencies {
  compile fileTree(include: ['*.jar'], dir: 'libs')
 androidTestCompile('com.android.support.test.espresso:espresso-core:2.2.2',
  {exclude group: 'com.android.support', module: 'support-annotations'})
    testCompile 'junit:junit:4.12'
    compile files('libs/BaiduLBS_Android.jar')
}
```

此实例的完整项目在 MyCode\MySample752 文件夹中。

266 使用百度 SDK 在地图中定位指定的地名

此实例主要通过使用百度地图 SDK，实现在百度地图中以指定的地址为中心显示地图。当实例运行后，在"地址："输入框中输入地址，如"重庆市沙坪坝区沙正街 174 号"，然后单击"搜索"按钮，将在百度地图中显示该地址（即重庆大学），如图 266-1 的左图所示。如果在"地址："输入框中输入地址，如"重庆市金开大道 2009 号"，然后单击"搜索"按钮，将在地图中显示该地址（即长福路与回兴之间），如图 266-1 的右图所示。注意，由于该实例是通过网络访问百度服务器获取的数据，因此需要等待一定的时间。

图　266-1

主要代码如下：

```
public class MainActivity extends Activity {
  BaiduMap myBaiduMap;
  @Override
  protected void onCreate(Bundle savedInstanceState) {
    super.onCreate(savedInstanceState);
    SDKInitializer.initialize(getApplicationContext());
    setContentView(R.layout.activity_main);
    MapView myMapView = (MapView) findViewById(R.id.myMapView);
    myBaiduMap = myMapView.getMap();
  }
  public void onClickBtn1(View v) {            //响应单击"搜索"按钮
    EditText myEditText = (EditText) findViewById(R.id.myEditText);
    GeoCoder myGeoCoder = GeoCoder.newInstance();
    myGeoCoder.setOnGetGeoCodeResultListener(new OnGetGeoCoderResultListener(){
      @Override
      //将搜索地址显示在地图上
```

```java
    public void onGetGeoCodeResult(GeoCodeResult result) {
      MapStatus myMapStatus = new MapStatus.Builder().target(
            result.getLocation()).zoom(15).build();
      myBaiduMap.setMapStatus(MapStatusUpdateFactory.newMapStatus(myMapStatus));
    }
    @Override
    public void onGetReverseGeoCodeResult(
            ReverseGeoCodeResult reverseGeoCodeResult) { }
  });
  String myAddress = myEditText.getText().toString();
  //对输入地址进行解析操作
  myGeoCoder.geocode(new GeoCodeOption().city(myAddress.substring(0, 2)).
            address(myAddress.substring(2, myAddress.length() - 1)));
} }
```

上面这段代码在 MyCode\MySample912\app\src\main\java\com\bin\luo\mysample\MainActivity.java 文件中。显示位置的百度地图 MapView 控件的布局如下：

```xml
<com.baidu.mapapi.map.MapView
    android:id = "@+id/myMapView"
    android:layout_width = "match_parent"
    android:layout_height = "match_parent"
    android:clickable = "true"/>
```

上面这段代码在 MyCode\MySample912\app\src\main\res\layout\activity_main.xml 文件中。此外，此实例需要引入 MyCode\MySample912\app\libs\BaiduLBS_Android.jar 文件及 MyCode\MySample912\app\libs 下的所有文件和子目录，有了这些文件之后，就不再需要安装百度地图 App，即使在纯净的模拟器环境中也能正常测试大部分应用。然后修改 MyCode\MySample912\app\src\main\AndroidManifest.xml 文件，如下面的粗体字所示：

```xml
<?xml version = "1.0" encoding = "utf-8"?>
<manifest xmlns:android = "http://schemas.android.com/apk/res/android"
          package = "com.bin.luo.mysample">
  <application android:allowBackup = "true"
            android:icon = "@mipmap/ic_launcher"
            android:label = "@string/app_name"
            android:supportsRtl = "true"
            android:theme = "@style/AppTheme">
    <meta-data android:name = "com.baidu.lbsapi.API_KEY"
            android:value = "21xq8w6f8yS6kmvzLxM8MeB0c1k0TXIX"/>
    <activity android:name = ".MainActivity">
      <intent-filter>
        <action android:name = "android.intent.action.MAIN"/>
        <category android:name = "android.intent.category.LAUNCHER"/>
      </intent-filter>
    </activity>
  </application>
  <uses-permission android:name = "android.permission.ACCESS_NETWORK_STATE"/>
  <uses-permission android:name = "android.permission.INTERNET"/>
  <uses-permission android:name = "com.android.launcher.permission.READ_SETTINGS"/>
  <uses-permission android:name = "android.permission.WAKE_LOCK"/>
```

```xml
<uses-permission android:name="android.permission.CHANGE_WIFI_STATE"/>
<uses-permission android:name="android.permission.ACCESS_WIFI_STATE"/>
<uses-permission android:name="android.permission.GET_TASKS"/>
<uses-permission android:name="android.permission.WRITE_EXTERNAL_STORAGE"/>
<uses-permission android:name="android.permission.WRITE_SETTINGS"/>
</manifest>
```

接下来，按照下面粗体字所示修改 MyCode\MySample912\app\build.gradle 文件：

```gradle
apply plugin: 'com.android.application'
android {
    compileSdkVersion 24
    buildToolsVersion "25.0.2"
    defaultConfig {
        applicationId "com.bin.luo.mysample"
        minSdkVersion 22
        targetSdkVersion 24
        versionCode 1
        versionName "1.0"
        testInstrumentationRunner "android.support.test.runner.AndroidJUnitRunner"
    }
    buildTypes {
        release {
            minifyEnabled false
            proguardFiles getDefaultProguardFile('proguard-android.txt'),
            'proguard-rules.pro'
        }
    }
    sourceSets { main { jniLibs.srcDirs = ['libs'] } }
}
dependencies {
    compile fileTree(include: ['*.jar'], dir: 'libs')
    androidTestCompile('com.android.support.test.espresso:espresso-core:2.2.2', {
        exclude group: 'com.android.support', module: 'support-annotations'
    })
    testCompile 'junit:junit:4.12'
    compile files('libs/BaiduLBS_Android.jar')
}
```

关于百度地图 SDK 开发说明请参考官方网址（http://lbsyun.baidu.com/sdk），在这里详细介绍了如何获取 API_KEY，每个应用必须是独立的 API_KEY，在 AndroidManifest.xml 文件中需要登记此 API_KEY，申请 API_KEY 的操作步骤如下。

（1）登录百度地图开放平台官网（http://lbsyun.baidu.com/sdk），进入自己账号的 API 控制台，并单击"创建应用"按钮创建应用（需要注册成为百度开发者）。

（2）在弹出的窗口中填入应用相关信息，将应用类型选择为 Android SDK，然后单击"提交"按钮即可完成应用的创建。

（3）创建成功之后，将在应用列表中显示，其中的 AK 值就是在应用中填写的 API_KEY。

此实例的完整项目在 MyCode\MySample912 文件夹中。

267 使用百度SDK查询指定地点的卫星图

此实例主要通过使用百度地图SDK,实现查询指定经度和纬度位置的卫星图。当实例运行之后,在"纬度:"输入框中输入"29.716470",在"经度:"输入框中输入"106.645030",然后单击"显示指定经纬度位置的卫星图"按钮,将在百度地图中显示重庆江北国际机场的卫星图,效果如图267-1的左图所示;如果在"纬度:"输入框中输入"31.239690",在"经度:"输入框中输入"121.499720",然后单击"显示指定经纬度位置的卫星图"按钮,将在百度地图中显示上海东方明珠电视塔的卫星图,效果如图267-1的右图所示。

图 267-1

主要代码如下:

```java
public void onClickBtn1(View v) {        //响应单击"显示指定经纬度位置的卫星图"按钮
    myBaiduMap = myMapView.getMap();
    LatLng myLocation = new LatLng(
            Double.parseDouble(myEditLatitude.getText().toString()),
            Double.parseDouble(myEditLongitude.getText().toString()));
    MapStatus myMapStatus =
            new MapStatus.Builder().target(myLocation).build();
    myBaiduMap.setMapStatus(MapStatusUpdateFactory.newMapStatus(myMapStatus));
    myBaiduMap.setMapType(BaiduMap.MAP_TYPE_SATELLITE); //设置卫星地图模式
    MapStatusUpdate myMapStatusUpdate = MapStatusUpdateFactory.zoomTo(16);
    myBaiduMap.animateMapStatus(myMapStatusUpdate);    //设置地图缩放级别16
}
```

上面这段代码在 MyCode\MySample917\app\src\main\java\com\bin\luo\mysample\MainActivity.java 文件中。此外,此实例需要引入百度地图SDK库文件,并修改 build.gradle 和 AndroidManifest.xml 等文件,相关操作请参考实例266。此实例的完整项目在 MyCode\MySample917 文件夹中。

268 使用百度 SDK 在地图上自定义热力图

此实例主要通过使用百度地图 SDK 中的 HeatMap,实现在百度地图指定的地理范围内添加自定义热力图。当实例运行之后,单击"在指定的地理范围添加自定义热力图"按钮,将在指定的矩形范围(重庆北碚的经纬度值表示矩形左上角坐标,重庆南岸的经纬度值表示矩形右下角坐标)内显示用多个随机渐变色构造的热力图,效果分别如图 268-1 的左图和右图所示。

图 268-1

主要代码如下:

```
//响应单击"在指定的地理范围添加自定义热力图"按钮
public void onClickBtn1(View v) {
    myBaiduMap.clear();
    int[] DEFAULT_GRADIENT_COLORS = {Color.rgb(255, 0, 0),
            Color.rgb(0, 255, 0), Color.rgb(0, 0, 255) }; //设置渐变颜色值
    //设置每个颜色的渐变起始值(百分比值)
    float[] DEFAULT_GRADIENT_START_POINTS = { 0.2f, 0.5f, 1f };
    //构造颜色渐变对象
    Gradient myGradient = new Gradient(DEFAULT_GRADIENT_COLORS,
                                       DEFAULT_GRADIENT_START_POINTS);
    List<LatLng> myPoints = new ArrayList<LatLng>();
    Random myRandom = new Random();
    for (int i = 0; i < 150; i++) {
        //热力图的矩形覆盖范围
        //左上角是北碚(106.402277,29.811138),右下角是南岸(106.65078,29.506913)
        //106.650780 - 106.402277 = 0.248503,29.506913 - 29.811138 = - 0.304225
        int myLat = myRandom.nextInt(304225);
        int myLng = myRandom.nextInt(248503);
```

```
    int myLatitude = 29506913 + myLat;
    int myLongitude = 106402277 + myLng;
    LatLng myPoint = new LatLng(myLatitude / 1E6, myLongitude / 1E6);
    myPoints.add(myPoint);
}
HeatMap myHeatmap = new HeatMap.Builder().data(myPoints).
    gradient(myGradient).build();               //根据位置数据和渐变色创建自定义热力图
myBaiduMap.addHeatMap(myHeatmap);               //在百度地图上添加自定义热力图
}
```

上面这段代码在 MyCode\MySampleX40\app\src\main\java\com\bin\luo\mysample\MainActivity.java 文件中。此外，此实例需要引入百度地图 SDK 库文件，并修改 build.gradle 和 AndroidManifest.xml 等文件，相关操作请参考实例 266。此实例的完整项目在 MyCode\MySampleX40 文件夹中。

269　使用百度 SDK 实现计算指定范围的面积

此实例主要通过使用百度地图 SDK 的 AreaUtil，实现根据右上角和左下角的经纬度值计算该地理范围内的矩形区域面积。当实例运行之后，在"右上角纬度："输入框中输入重庆中央公园右上角的纬度值，如"29.732368"，在"右上角经度："输入框中输入重庆中央公园右上角的经度值，如"106.594239"；在"左下角纬度："输入框中输入重庆中央公园左下角的纬度值，如"29.711416"，在"左下角经度："输入框中输入重庆中央公园左下角的经度值，如"106.585974"，然后单击"计算该地理范围的面积"按钮，将在弹出的 Toast 中显示重庆中央公园的面积，如图 269-1 的左图所示。如果在"右上角纬度："输入框中输入重庆外环右上角的纬度值，如"29.756608"，在"右上角经度："输入框中输入重庆外环右上角的经度值，如"106.70915"；在"左下角纬度："输入框中输入重庆外环左下角的纬度值，如"29.318189"，在"左下角经度："输入框中输入重庆外环左下角的经度值，如"106.306709"，然后单击"计算该地理范围的面积"按钮，将在弹出的 Toast 中显示重庆外环的面积，如图 269-1 的右图所示。

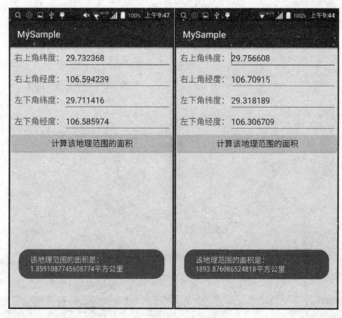

图 269-1

主要代码如下:

```
public void onClickBtn1(View v) {        //响应单击"计算该地理范围的面积"按钮
    //重庆中央公园的右上角经纬度(106.594239,29.732368);
    //重庆中央公园的左下角经纬度(106.585974,29.711416);
    //重庆外环的右上角经纬度(106.70915,29.756608);
    //重庆外环的左下角经纬度(106.306709,29.318189);
    LatLng myPoint1 =
            new LatLng(Double.valueOf(myEditPoint1A.getText().toString()),
                Double.valueOf(myEditPoint1B.getText().toString()));
    LatLng myPoint2 =
            new LatLng(Double.valueOf(myEditPoint2A.getText().toString()),
                Double.valueOf(myEditPoint2B.getText().toString()));
    double myDistance = AreaUtil.calculateArea(myPoint1, myPoint2)/1000/1000;
    Toast.makeText(MainActivity.this,"该地理范围的面积是:"
            + myDistance +"平方公里", Toast.LENGTH_SHORT).show();
}
```

上面这段代码在 MyCode\MySampleX48\app\src\main\java\com\bin\luo\mysample\MainActivity.java 文件中。此外,此实例需要引入百度地图 SDK 库文件,并修改 build.gradle 和 AndroidManifest.xml 等文件,相关操作请参考实例 266。此实例的完整项目在 MyCode\MySampleX48 文件夹中。

270 使用百度 SDK 在地图上叠加圆点覆盖物

此实例主要通过使用百度地图 SDK 的 DotOptions,实现在指定位置叠加圆点覆盖物。当实例运行之后,将显示以武汉为中心的百度地图,如图 270-1 的左图所示。单击"以武汉为中心叠加圆点覆盖物"按钮,将以武汉为中心,以 150px 为半径绘制一个半透明的圆形,如图 270-1 的右图所示。

图 270-1

主要代码如下:

```
//响应单击"以武汉为中心叠加圆点覆盖物"按钮
public void onClickBtn1(View v) {
    myBaiduMap.clear();
    DotOptions myDotOptions = new DotOptions();
    myDotOptions.center(new LatLng(30.53553, 114.324234));   //设置圆心坐标
    myDotOptions.color(0XCCBEFFFF);                          //半透明青色
    myDotOptions.radius(150);                                //设置半径
    myBaiduMap.addOverlay(myDotOptions);
}
```

上面这段代码在 MyCode \ MySampleX80 \ app \ src \ main \ java \ com \ bin \ luo \ mysample \ MainActivity.java 文件中。此外,此实例需要引入百度地图 SDK 库文件,并修改 AndroidManifest. xml 和 build.gradle 等文件,相关操作请参考实例 266。此实例的完整项目在 MyCode \ MySampleX80 文件夹中。

271 使用百度 SDK 在地图上添加半透明椭圆

此实例主要通过使用百度地图 SDK 中的 GroundOverlayOptions,实现在百度地图指定的地理范围内添加半透明的椭圆。当实例运行之后,单击"在指定的地理范围添加半透明椭圆"按钮,将在指定的矩形范围(重庆两江新区的左上角和右下角的纬度经度值限定的区域)内显示一个半透明的椭圆及文字"重庆两江新区",效果分别如图 271-1 的左图和右图所示。

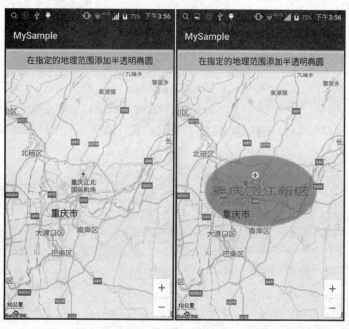

图　271-1

主要代码如下:

```
//响应单击"在指定的地理范围添加半透明椭圆"按钮
public void onClickBtn1(View v) {
```

```java
myBaiduMap.clear();
//定义椭圆的地理范围
LatLng myPoint1 = new LatLng(29.810734,106.405711);      //左上角纬度经度
LatLng myPoint2 = new LatLng(29.517988,106.949006);      //右下角纬度经度
LatLngBounds myBounds = new LatLngBounds.Builder()
        .include(myPoint1).include(myPoint2).build();
TextView myTextView = new TextView(getApplicationContext());
myTextView.setWidth(250);
myTextView.setHeight(250);
myTextView.setText("重庆两江新区");
myTextView.setGravity(Gravity.CENTER);                    //设置文字居中对齐
myTextView.setTextSize(36);                               //设置文字大小
myTextView.setTextColor(Color.BLUE);                      //设置文字颜色
//在 TextView 上设置椭圆背景
myTextView.setBackground(getDrawable(R.drawable.myshape));
//创建线性布局管理器
LinearLayout myLinearLayout = new LinearLayout(this);
//以垂直方式布局界面控件
myLinearLayout.setOrientation(LinearLayout.VERTICAL);
//在线性布局管理器中添加 myTextView 控件
myLinearLayout.addView(myTextView);
//将线性布局转换为 BitmapDescriptor
BitmapDescriptor myShape = BitmapDescriptorFactory.fromView(myLinearLayout);
//根据地理范围和图形创建 GroundOverlayOptions
GroundOverlayOptions myGroundOverlayOptions = new GroundOverlayOptions()
        .positionFromBounds(myBounds)                     //设置地理范围
        .image(myShape)                                   //设置覆盖图形
        .transparency(0.5f);                              //设置图像透明度
myBaiduMap.addOverlay(myGroundOverlayOptions);
}
```

上面这段代码在 MyCode\MySampleX35\app\src\main\java\com\bin\luo\mysample\MainActivity.java 文件中。在这段代码中，myTextView.setBackground(getDrawable(R.drawable.myshape))表示使用 XML 文件 myshape 配置的椭圆设置 TextView 控件的背景，myshape 文件的主要内容如下：

```xml
<?xml version="1.0" encoding="utf-8"?>
<shape xmlns:android="http://schemas.android.com/apk/res/android"
       android:shape="oval">
    <size android:width="250dp" android:height="250dp"/>
    <solid android:color="#BA55D3"/>
    <stroke android:width="5dp"
            android:color="#0f0"
            android:dashGap="5sp"
            android:dashWidth="5dp"/>
</shape>
```

上面这段代码在 MyCode\MySampleX35\app\src\main\res\drawable\myshape.xml 文件中。此外，此实例需要引入百度地图 SDK 库文件，并修改 AndroidManifest.xml 和 build.gradle 等文件，相关操作请参考实例 266。此实例的完整项目在 MyCode\MySampleX35 文件夹中。

272 使用百度 SDK 在地图的指定位置添加标记

此实例主要通过使用百度地图 SDK，实现在百度地图的指定经纬度位置显示自定义标记（PNG 图像）。当实例运行之后，在"纬度："输入框中输入纬度值，如"29.85724"，在"经度："输入框中输入经度值，如"107.074115"，然后单击"在指定经纬度位置添加自定义标记"按钮，将在长寿古镇上添加自定义标记，如图 272-1 的左图所示。如果在"纬度："输入框中输入"29.722571"，在"经度："输入框中输入"106.590326"，然后单击"在指定经纬度位置添加自定义标记"按钮，将在重庆中央公园上添加自定义标记，如图 272-1 的右图所示。

图 272-1

主要代码如下：

```
public void onClickBtn1(View v) {    //响应单击"在指定经纬度位置添加自定义标记"按钮
    myBaiduMap = myMapView.getMap();
    //长寿古镇经纬度(107.074115,29.85724)
    //重庆中央公园经纬度(106.590326,29.722571)
    //定义 Marker 坐标点
    LatLng myLocation = new LatLng(Double.parseDouble(myEditLatitude.getText().toString()),Double.parseDouble(myEditLongitude.getText().toString()));
    //构建 Marker 图标
    BitmapDescriptor myBitmap =
                BitmapDescriptorFactory.fromResource(R.mipmap.myimage);
    //构建 MarkerOptions，用于在地图上添加 Marker
    OverlayOptions myOverlayOptions =
                new MarkerOptions().position(myLocation).icon(myBitmap);
    //在地图上添加 Marker 并显示
    myBaiduMap.addOverlay(myOverlayOptions);
    MapStatus myMapStatus = new MapStatus.Builder().target(myLocation).build();
```

```
        myBaiduMap.setMapStatus(MapStatusUpdateFactory.newMapStatus(myMapStatus));
        //设置缩放比例16
        MapStatusUpdate myMapStatusUpdate = MapStatusUpdateFactory.zoomTo(16);
        myBaiduMap.animateMapStatus(myMapStatusUpdate);
    }
```

上面这段代码在 MyCode\MySampleX09\app\src\main\java\com\bin\luo\mysample\MainActivity.java 文件中。此外,此实例需要引入百度地图 SDK 库文件,并修改 build.gradle 和 AndroidManifest.xml 等文件,相关操作请参考实例 266。此实例的完整项目在 MyCode\MySampleX09 文件夹中。

273　使用百度 SDK 实现在地图上添加图像按钮

此实例主要通过使用百度地图 SDK 中的 InfoWindow,实现在百度地图的指定位置添加图像按钮。当实例运行之后,单击"在指定的位置添加图像按钮"按钮,将在指定楼盘的位置添加一个图像按钮(带箭头的气泡图像),单击该图像按钮,将在下面弹出一个 Toast 显示"新都汇楼盘全体员工欢迎您的光临!",效果分别如图 273-1 的左图和右图所示。

图　273-1

主要代码如下:

```
public void onClickBtn1(View v) {                //响应单击"在指定的位置添加图像按钮"按钮
    //定义图像按钮的坐标点(纬度、经度)
    LatLng myPoint = new LatLng(29.694177,106.609398);
    Button myButton = new Button(getApplicationContext());        //创建图像按钮
    myButton.setOnClickListener(new View.OnClickListener() {
        @Override
        public void onClick(View v) {            //单击图像按钮时显示 Toast
```

```
        Toast.makeText(MainActivity.this,
                "新都汇楼盘全体员工欢迎您的光临!", Toast.LENGTH_SHORT).show();
    } });
    myButton.setBackgroundResource(R.mipmap.myimage1);
    //在 InfoWindow 中加载图像按钮
    InfoWindow myInfoWindow = new InfoWindow(myButton, myPoint, -10);
    myBaiduMap.showInfoWindow(myInfoWindow);        //显示 InfoWindow
}
```

上面这段代码在 MyCode\MySampleX19\app\src\main\java\com\bin\luo\mysample\MainActivity.java 文件中。此外，此实例需要引入百度地图 SDK 库文件，并修改 build.gradle 和 AndroidManifest.xml 等文件，相关操作请参考实例 266。此实例的完整项目在 MyCode\MySampleX19 文件夹中。

274 使用百度 SDK 在地图的城市之间绘制虚线

此实例主要通过使用百度地图 SDK，实现在百度地图中使用虚线连接两个城市。当实例运行之后，单击"显示四点连线使用虚线效果"按钮，将使用虚线将武汉、长沙、南昌、九江连接起来，形成一个封闭的四边形，效果分别如图 274-1 的左图和右图所示。当缩放百度地图时，四边形将同步缩放。

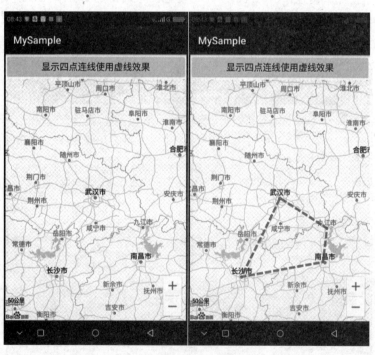

图 274-1

主要代码如下：

```
public void onClickBtn1(View v) {        //响应单击"显示四点连线使用虚线效果"按钮
    LatLng myPoint1 = new LatLng(30.604435,114.311586);     //武汉纬度、经度
    LatLng myPoint2 = new LatLng(28.241,112.94547);         //长沙纬度、经度
    LatLng myPoint3 = new LatLng(28.695539,115.859986);     //南昌纬度、经度
    LatLng myPoint4 = new LatLng(29.713348,116.002934);     //九江纬度、经度
```

```
    myPoints = new ArrayList<LatLng>();
    myPoints.add(myPoint1);
    myPoints.add(myPoint2);
    myPoints.add(myPoint3);
    myPoints.add(myPoint4);
    myPoints.add(myPoint1);
    OverlayOptions myOverlayOptions = new PolylineOptions().width(10).
          color(0xAAFF0000).points(myPoints);       //绘制折线(连线)
    Polyline myPolyline = (Polyline) myBaiduMap.addOverlay(myOverlayOptions);
    myPolyline.setDottedLine(true);                 //允许连线使用虚线
}
```

上面这段代码在 MyCode\MySampleX11\app\src\main\java\com\bin\luo\mysample\MainActivity.java 文件中。此外,此实例需要引入百度地图 SDK 库文件,并修改 build.gradle 和 AndroidManifest.xml 等文件,相关操作请参考实例 266。此实例的完整项目在 MyCode\MySampleX11 文件夹中。

275 使用百度 SDK 实现在地图上绘制多边形

此实例主要通过使用百度地图 SDK,实现在百度地图中以指定的多个位置绘制半透明的多边形。当实例运行之后,单击"以指定的多个位置绘制半透明多边形"按钮,将以武汉、长沙、南昌、九江为顶点绘制一个半透明的多边形,效果分别如图 275-1 的左图和右图所示。

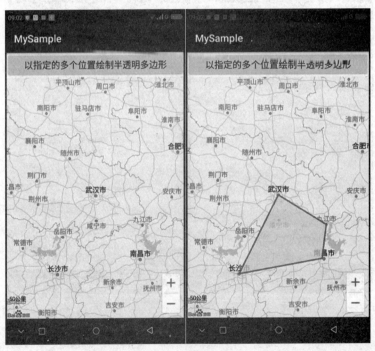

图 275-1

主要代码如下:

```
//响应单击"以指定的多个位置绘制半透明多边形"按钮
public void onClickBtn1(View v) {
```

```
LatLng myPoint1 = new LatLng(30.604435,114.311586);    //武汉纬度、经度
LatLng myPoint2 = new LatLng(28.241,112.94547);        //长沙纬度、经度
LatLng myPoint3 = new LatLng(28.695539,115.859986);    //南昌纬度、经度
LatLng myPoint4 = new LatLng(29.713348,116.002934);    //九江纬度、经度
List<LatLng>  myPoints = new ArrayList<LatLng>();
myPoints.add(myPoint1);
myPoints.add(myPoint2);
myPoints.add(myPoint3);
myPoints.add(myPoint4);
myPoints.add(myPoint5);
//根据4个顶点绘制四边形
//0xCCBFEFFF 的 CC 表示透明度值,BFEFFF 是多边形的填充颜色
//5 表示线宽,Color.RED 表示线条是红色
OverlayOptions myPolygonOptions = new PolygonOptions()
    .points(myPoints).fillColor(0xCCBFEFFF).stroke(new Stroke(5, Color.RED));
myBaiduMap.addOverlay(myPolygonOptions);
}
```

上面这段代码在 MyCode\MySampleX17\app\src\main\java\com\bin\luo\mysample\MainActivity.java 文件中。此外,此实例需要引入百度地图 SDK 库文件,并修改 build.gradle 和 AndroidManifest.xml 等文件,相关操作请参考实例 266。此实例的完整项目在 MyCode\MySampleX17 文件夹中。

276 使用百度 SDK 在地图的三点位置绘制弧线

此实例主要通过使用百度地图 SDK 的 ArcOptions,实现在百度地图中根据三点的经纬度值绘制弧线。当实例运行之后,单击"显示根据三点位置绘制的弧线"按钮,将把武汉、长沙、南昌三点连接在同一条弧线上,效果分别如图 276-1 的左图和右图所示。

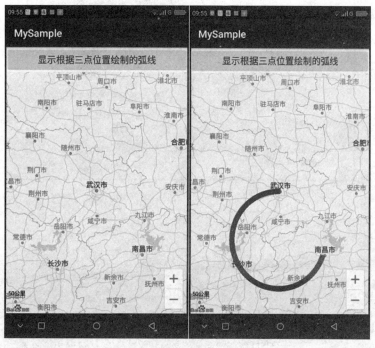

图 276-1

主要代码如下:

```
public void onClickBtn1(View v) {        //响应单击"显示根据三点位置绘制的弧线"按钮
    LatLng myPoint1 = new LatLng(30.604435,114.311586);   //武汉纬度、经度
    LatLng myPoint2 = new LatLng(28.241,112.94547);       //长沙纬度、经度
    LatLng myPoint3 = new LatLng(28.695539,115.859986);   //南昌纬度、经度
    OverlayOptions ooArc = new ArcOptions().color(Color.BLUE).
        width(20).points(myPoint1, myPoint2, myPoint3);   //根据3点坐标绘制弧线
    myBaiduMap.addOverlay(ooArc);
}
```

上面这段代码在 MyCode\MySampleX15\app\src\main\java\com\bin\luo\mysample\MainActivity.java 文件中。此外，此实例需要引入百度地图 SDK 库文件，并修改 build.gradle 和 AndroidManifest.xml 等文件，相关操作请参考实例 266。此实例的完整项目在 MyCode\MySampleX15 文件夹中。

277 使用百度 SDK 在地图上添加生长型动画

此实例主要通过使用百度地图 SDK 中的 MarkerOptions，实现在百度地图的指定位置添加生长型动画。当实例运行之后，单击"在指定的位置添加生长型动画"按钮，将在指定的位置（重庆鳄鱼中心）从小到大放大一幅魔兽图像，效果分别如图 277-1 的左图和右图所示。

图 277-1

主要代码如下:

```
public void onClickBtn1(View v) {//响应单击"在指定的位置添加生长型动画"按钮
    myBaiduMap.clear();
    //定义动画图像(重庆鳄鱼中心)坐标点(纬度、经度)
    LatLng myPoint = new LatLng(29.650778,106.599107);
    BitmapDescriptor myBitmap = BitmapDescriptorFactory.
                    fromResource(R.mipmap.myimage1);    //获取图像资源
```

```
MarkerOptions myMarkerOptions =
            new MarkerOptions().position(myPoint).icon(myBitmap)
            .zIndex(0).period(10);              //根据图像创建 MarkerOptions
//在 MarkerOptions 中设置生长型动画 MarkerAnimateType.grow
myMarkerOptions.animateType(MarkerOptions.MarkerAnimateType.grow);
myBaiduMap.addOverlay(myMarkerOptions);         //在百度地图中添加生长型动画
}
```

上面这段代码在 MyCode\MySampleX31\app\src\main\java\com\bin\luo\mysample\MainActivity.java 文件中。此外，此实例需要引入百度地图 SDK 库文件，并修改 build.gradle 和 AndroidManifest.xml 等文件，相关操作请参考实例 266。此实例的完整项目在 MyCode\MySampleX31 文件夹中。

278 使用百度 SDK 在地图上添加降落型动画

此实例主要通过使用百度地图 SDK 中的 MarkerOptions，实现在百度地图的指定位置添加降落型动画。当实例运行之后，单击"在指定的位置添加降落型动画"按钮，一只小鸟将从屏幕顶端降落在指定的位置（重庆鳄鱼中心），效果分别如图 278-1 的左图和右图所示。

图 278-1

主要代码如下：

```
public void onClickBtn1(View v) {    //响应单击"在指定的位置添加降落型动画"按钮
    myBaiduMap.clear();
    //定义动画图像(重庆鳄鱼中心)坐标点(纬度、经度)
    LatLng myPoint = new LatLng(29.650778,106.599107);
    BitmapDescriptor myBitmap = BitmapDescriptorFactory.
                    fromResource(R.mipmap.myimage1);       //获取图像资源
    MarkerOptions myMarkerOptions =
                new MarkerOptions().position(myPoint).icon(myBitmap)
                .zIndex(0).period(10);         //根据图像创建 MarkerOptions
```

```
        //在MarkerOptions中设置降落型动画MarkerAnimateType.drop
        myMarkerOptions.animateType(MarkerOptions.MarkerAnimateType.drop);
        myBaiduMap.addOverlay(myMarkerOptions);        //在百度地图中添加降落型动画
    }
```

上面这段代码在 MyCode\MySampleX32\app\src\main\java\com\bin\luo\mysample\MainActivity.java 文件中。此外，此实例需要引入百度地图 SDK 库文件，并修改 build.gradle 和 AndroidManifest.xml 等文件，相关操作请参考实例 266。此实例的完整项目在 MyCode\MySampleX32 文件夹中。

279 使用百度 SDK 在地图上添加淡入放大动画

此实例主要通过使用百度地图 SDK 中的 InfoWindow，实现在百度地图的指定位置添加淡入放大动画。当实例运行之后，单击"在指定的位置添加淡入放大动画"按钮，将在指定的位置（重庆江北嘴金融城）浮出一个由透明变为不透明、同时由小变大的建筑物图像，效果分别如图 279-1 的左图和右图所示。

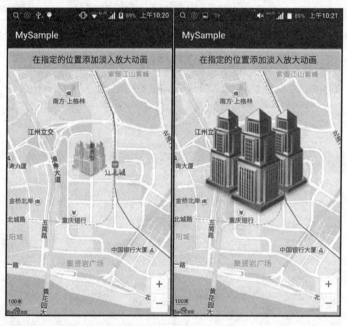

图 279-1

主要代码如下：

```
public void onClickBtn1(View v) {                //响应单击"在指定的位置添加淡入放大动画"按钮
    //重庆江北嘴金融城的纬度、经度
    LatLng myPoint = new LatLng(29.580759,106.576481);
    ImageView myImageView = new ImageView(getApplicationContext());
    myImageView.setImageResource(R.mipmap.myimage1);
    //创建淡入图像动画(0f 表示完全透明,1.0f 表示完全不透明)
    AlphaAnimation myAnimation1 = new AlphaAnimation(0f, 1.0f);
    myAnimation1.setDuration(15000);              //设置动画持续时间 15s
    //创建以自身为中心，由小变大地放大动画
    ScaleAnimation myAnimation2 = new ScaleAnimation(0f, 1.0f, 0f, 1.0f,
            Animation.RELATIVE_TO_SELF, 0.5f, Animation.RELATIVE_TO_SELF, 0.5f);
```

```
myAnimation2.setDuration(15000);                          //设置动画持续时间 15s
AnimationSet myAnimationSet = new AnimationSet(true);     //创建动画集合对象
myAnimationSet.addAnimation(myAnimation1);                //在动画集合对象中添加淡入动画
myAnimationSet.addAnimation(myAnimation2);                //在动画集合对象中添加放大动画
myAnimationSet.setFillAfter(true);                        //保留动画最后的状态
myAnimationSet.setFillEnabled(true);
//在 myImageView 上添加淡入动画和放大动画
myImageView.startAnimation(myAnimationSet);
//创建线性布局管理器
LinearLayout myLinearLayout = new LinearLayout(this);
//以垂直方式布局界面控件
myLinearLayout.setOrientation(LinearLayout.VERTICAL);
//在线性布局管理器中添加 myImageView 控件
myLinearLayout.addView(myImageView);
//在 InfoWindow 中加载 myLinearLayout,实现淡入放大图像动画
InfoWindow myInfoWindow = new InfoWindow(myLinearLayout, myPoint,160);
myBaiduMap.showInfoWindow(myInfoWindow);                  //显示 InfoWindow
}
```

上面这段代码在 MyCode \ MySampleX23 \ app \ src \ main \ java \ com \ bin \ luo \ mysample \ MainActivity.java 文件中。此外,此实例需要引入百度地图 SDK 库文件,并修改 build.gradle 和 AndroidManifest.xml 等文件,相关操作请参考实例 266。此实例的完整项目在 MyCode \ MySampleX23 文件夹中。

280 使用百度 SDK 在地图上添加水平展开动画

此实例主要通过使用百度地图 SDK 中的 InfoWindow,实现在百度地图指定的位置添加水平展开图像的动画。当实例运行之后,单击"在指定的位置添加水平展开动画"按钮,将在指定的位置(重庆人民大礼堂)在水平方向上从中心向左右两端展开图像,效果分别如图 280-1 的左图和右图所示。

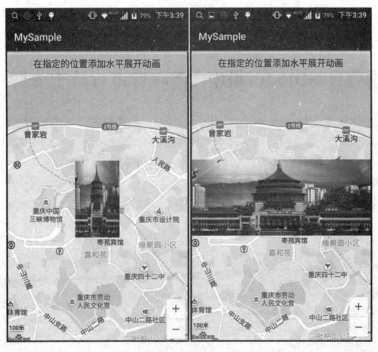

图 280-1

主要代码如下：

```java
public void onClickBtn1(View v) {           //响应单击"在指定的位置添加水平展开动画"按钮
    //定义图像(重庆人民大礼堂)坐标点(纬度、经度)
    LatLng myPoint = new LatLng(29.567848,106.560459);
    ImageView myImageView = new ImageView(getApplicationContext());
    myImageView.setImageResource(R.mipmap.myimage1);
    //创建水平展开图像的动画
    //Animation.RELATIVE_TO_SELF, 0.5f 表示以自身中心为展开中心
    //0f, 1.0f 表示在水平方向上从最小展开到最大
    //1f, 1f 表示在垂直方向上大小始终保持不变
    ScaleAnimation myAnimation = new ScaleAnimation(0f, 1.0f, 1f, 1f,
            Animation.RELATIVE_TO_SELF, 0.5f, Animation.RELATIVE_TO_SELF, 0.5f);
    //设置动画持续时间 5s
    myAnimation.setDuration(5000);
    //在 myImageView 上添加水平展开动画 myAnimation
    myImageView.startAnimation(myAnimation);
    //创建线性布局管理器
    LinearLayout myLinearLayout = new LinearLayout(this);
    //以垂直方式布局界面控件
    myLinearLayout.setOrientation(LinearLayout.VERTICAL);
    //在线性布局管理器中添加 myImageView 控件
    myLinearLayout.addView(myImageView);
    //在 InfoWindow 中加载 myLinearLayout,实现水平展开图像的动画
    InfoWindow myInfoWindow = new InfoWindow(myLinearLayout, myPoint,160);
    //显示 InfoWindow
    myBaiduMap.showInfoWindow(myInfoWindow);
}
```

上面这段代码在 MyCode\MySampleX30\app\src\main\java\com\bin\luo\mysample\MainActivity.java 文件中。此外，此实例需要引入百度地图 SDK 库文件，并修改 build.gradle 和 AndroidManifest.xml 等文件，相关操作请参考实例 266。此实例的完整项目在 MyCode\MySampleX30 文件夹中。

281 使用百度 SDK 在地图上查询省市行政中心

此实例主要通过使用百度地图 SDK 的 DistrictSearch，实现查询指定省市县的行政中心。当实例运行之后，在"省市县名称："输入框中输入省名、市名或县名，如"武汉"，然后单击"开始查询"按钮，将在百度地图中使用半透明的颜色标注武汉市的行政区域并在武汉市的上方显示一个灯泡，如图 281-1 的左图所示。如果在"省市县名称："输入框中输入"湖北"，然后单击"开始查询"按钮，将在百度地图中使用半透明的颜色标注湖北的行政区域并在武汉市的上方显示一个灯泡，如图 281-1 的右图所示。

主要代码如下：

```java
public void onClickBtn1(View v) {           //响应单击"开始查询"按钮
    myBaiduMap.clear();
    //建立查询条件
    DistrictSearchOption myDistrictSearchOption =
        new DistrictSearchOption().cityName(myEditText.getText().toString());
    //设置查询监听事件
```

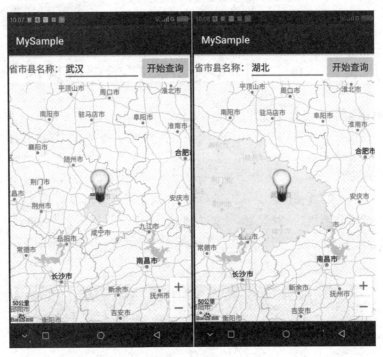

图 281-1

```
myDistrictSearch.setOnDistrictSearchListener(
    new OnGetDistricSearchResultListener() {
    @Override
    public void onGetDistrictResult(DistrictResult districtResult) {
     if (districtResult.error == SearchResult.ERRORNO.NO_ERROR) {
      List<List<LatLng>> pointsList = districtResult.getPolylines();
      if (pointsList == null) return;
      for (List<LatLng> polyline : pointsList) {
       //绘制行政区域多边形
       OverlayOptions myPolygonOptions =
               new PolygonOptions().points(polyline)
               .stroke(new Stroke(2, 0xAA00FF88)).fillColor(0xCCBEFFFF);
       myBaiduMap.addOverlay(myPolygonOptions);//添加 Overlay
      }
      //获取行政中心的经纬度
      LatLng myPoint = districtResult.getCenterPt();
      ImageView myImageView = new ImageView(getApplicationContext());
      myImageView.setBackgroundResource(R.mipmap.myimage1);
      //在 InfoWindow 中加载行政中心图像
      InfoWindow myInfoWindow = new InfoWindow(myImageView, myPoint, 0);
      //在百度地图中使用图像指示行政中心
      myBaiduMap.showInfoWindow(myInfoWindow);
     } } });
    //执行行政区域的查询
    myDistrictSearch.searchDistrict(myDistrictSearchOption);
}
```

上面这段代码在 MyCode\MySampleX49\app\src\main\java\com\bin\luo\mysample\MainActivity.java 文件中。此外,此实例需要引入百度地图 SDK 库文件,并修改 build.gradle 和

AndroidManifest.xml 等文件，相关操作请参考实例 266。此实例的完整项目在 MyCode\MySampleX49 文件夹中。

282 使用百度 SDK 判断某地是否在指定区域内

此实例主要通过使用百度地图 SDK 的 SpatialRelationUtil、DistrictSearch 和 GeoCoder，从而判断指定地点是否在指定行政区域（多边形范围）内。当实例运行之后，在"楼盘名称："输入框中输入楼盘名称，如"苏州吴江绿地香奈"，然后单击"查询楼盘的经纬度值"按钮，将在弹出的 Toast 中显示该楼盘所在位置的经纬度值，如图 282-1 的左图所示。在"省市县名称："输入框中输入中国境内无重名的省市县名称，如"江苏"或"苏州"或"吴江"，然后单击"判断两者的所属关系"按钮，将在弹出的 Toast 中提示前者在后者的管辖范围内，如图 282-1 的右图所示。如果在"省市县名称："输入框中输入"上海"或"广西"或"重庆"等，在单击"判断两者的所属关系"按钮后，将在弹出的 Toast 中提示无法判断两者的关系。

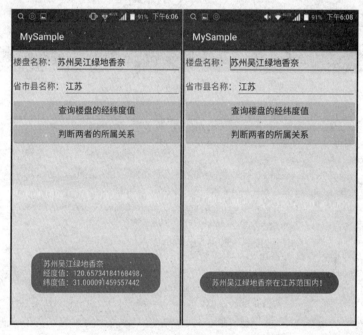

图 282-1

主要代码如下：

```
public void onClickBtn1(View v) {            //响应单击"查询楼盘的经纬度值"按钮
    GeoCoder myGeoCoder = GeoCoder.newInstance();
    myGeoCoder.setOnGetGeoCodeResultListener(new OnGetGeoCoderResultListener(){
        @Override
        public void onGetGeoCodeResult(GeoCodeResult result){
            if (result == null || result.error != SearchResult.ERRORNO.NO_ERROR){
                Toast.makeText(MainActivity.this,
                        "暂时未搜索到此地的经纬度值", Toast.LENGTH_SHORT).show();
                return;
            }
            myPoint = new LatLng(result.getLocation().latitude,
                                 result.getLocation().longitude);
```

```
        Toast.makeText(MainActivity.this, myEditPlace.getText().toString()
                + "\n经度值:" + result.getLocation().longitude + ",\n纬度值:"
                + result.getLocation().latitude, Toast.LENGTH_SHORT).show();
    }
    @Override
    public void onGetReverseGeoCodeResult(
            ReverseGeoCodeResult reverseGeoCodeResult) {
    }});
    String myAddress = myEditPlace.getText().toString();
    //对输入楼盘地址进行解析操作
    myGeoCoder.geocode(new GeoCodeOption().city(myAddress.substring(0, 2)).
            address(myAddress.substring(2, myAddress.length() - 1)));
}
public void onClickBtn2(View v) {            //响应单击"判断两者的所属关系"按钮
    if (myPoint == null) return;
    DistrictSearchOption myDistrictSearchOption =
        new DistrictSearchOption().cityName(myEditDistrict.getText().toString());
    //设置查询区域监听事件
    myDistrictSearch.setOnDistrictSearchListener(
            new OnGetDistricSearchResultListener(){
                @Override
                public void onGetDistrictResult(DistrictResult districtResult){
                    if (districtResult.error == SearchResult.ERRORNO.NO_ERROR){
                        List<List<LatLng>> pointsList = districtResult.getPolylines();
                        if (pointsList == null) return;
                        boolean myResult = false;
                        for (List<LatLng> polyline : pointsList){
                            if (SpatialRelationUtil.isPolygonContainsPoint(polyline, myPoint)){
                                myResult = true;
                            }}
                        if (myResult) {
                            Toast.makeText(MainActivity.this, myEditPlace.getText().toString()
                                    + "在" + myEditDistrict.getText().toString() + "范围内!",
                                Toast.LENGTH_SHORT).show();
                        } else {
                            Toast.makeText(MainActivity.this, "无法确认两者的关系!",
                                    Toast.LENGTH_SHORT).show();
                        }}}});
    //执行行政区域的查询
    myDistrictSearch.searchDistrict(myDistrictSearchOption);
}
```

上面这段代码在 MyCode\MySampleX50\app\src\main\java\com\bin\luo\mysample\MainActivity.java 文件中。此外，此实例需要引入百度地图 SDK 库文件，并修改 build.gradle 和 AndroidManifest.xml 等文件，相关操作请参考实例 266。此实例的完整项目在 MyCode\MySampleX50 文件夹中。

283 使用百度 SDK 在地图上自定义行政区颜色

此实例主要通过使用百度地图 SDK 的 DistrictSearch，实现使用自定义颜色显示指定行政区。当实例运行之后，在"重庆市的区县名称:"输入框中输入区县名称，如"九龙坡"，然后单击"开始查询"按钮,将在百度地图中使用半透明的颜色标注九龙坡区的行政区域,如图 283-1 的左图所示。如果在"重庆市的区县名称:"输入框中输入"沙坪坝",然后单击"开始查询"按钮,将在百度地图中使用半透明的颜色标注沙坪坝区的行政区域,如图 283-1 的右图所示。

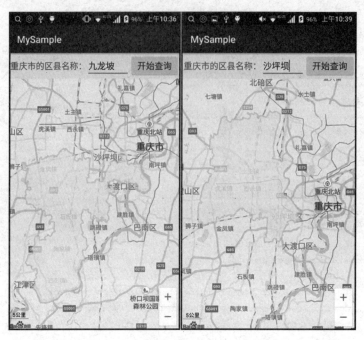

图 283-1

主要代码如下:

```java
public void onClickBtn1(View v) {                        //响应单击"开始查询"按钮
    myBaiduMap.clear();
    //建立查询条件
    DistrictSearchOption myDistrictSearchOption = new DistrictSearchOption().
            cityName("重庆").districtName(myEditText.getText().toString());
    //设置查询监听事件
    myDistrictSearch.setOnDistrictSearchListener(
            new OnGetDistricSearchResultListener() {
                @Override
                public void onGetDistrictResult(DistrictResult districtResult) {
                    if (districtResult.error == SearchResult.ERRORNO.NO_ERROR) {
                        List<List<LatLng>> pointsList = districtResult.getPolylines();
                        if (pointsList == null) return;
                        for (List<LatLng> polyline : pointsList) {
                            //绘制区域多边形(0xCCBEFFFF 表示半透明的青色)
                            OverlayOptions myPolygonOptions =
                                new PolygonOptions().points(polyline)
                                    .stroke(new Stroke(5, 0xAA00FF88)).fillColor(0xCCBEFFFF);
                            myBaiduMap.addOverlay(myPolygonOptions);       //添加 Overlay
                        } } });
    //执行行政区域的查询
    myDistrictSearch.searchDistrict(myDistrictSearchOption);
}
```

上面这段代码在 MyCode\MySampleX45\app\src\main\java\com\bin\luo\mysample\MainActivity.java 文件中。此外,此实例需要引入百度地图 SDK 库文件,并修改 build.gradle 和

AndroidManifest.xml 等文件，相关操作请参考实例 266。此实例的完整项目在 MyCode\MySampleX45 文件夹中。

284　使用百度 SDK 查询城市兴趣点并显示街景

此实例主要通过使用百度地图 SDK 的 BaiduMapPoiSearch，从而实现在百度地图中查询指定城市的兴趣点（POI）信息并显示街景。当实例运行之后，在"城市："输入框中输入城市名称，如"渝北"，在"兴趣点："输入框中输入"轻轨站"，然后单击"搜索兴趣点并查看街（全）景"按钮，将在百度地图中显示该地区的轻轨站分布图，如图 284-1 的左图所示。单击其中悬浮的任意一个图标（轻轨站点），将在弹出的 Toast 中显示该轻轨站的名称，如果该轻轨站有街景，将启动百度地图 App 显示对应的街景，图 284-1 右图所示的街景是在单击 4 号图标之后显示的重庆轻轨 3 号线回兴站的街景。

图　284-1

主要代码如下：

```java
public void onClickBtn1(View v){                    //响应单击"搜索兴趣点并查看街(全)景"按钮
  myPoiSearch = PoiSearch.newInstance();
  myPoiSearch.setOnGetPoiSearchResultListener(
                    new OnGetPoiSearchResultListener(){
   @Override
   public void onGetPoiResult(PoiResult result) {
    if (result.error == SearchResult.ERRORNO.NO_ERROR) {
     myBaiduMap.clear();                    //重置地图
     MarkerOverlay myOverlay = new MarkerOverlay(myBaiduMap);
     myBaiduMap.setOnMarkerClickListener(myOverlay);
     myOverlay.setData(result);             //传入目标兴趣点相关数据
     myOverlay.addToMap();                  //在地图上显示
     myOverlay.zoomToSpan();                //自适应缩放
    } else {
     Toast.makeText(MainActivity.this,
```

```
                    "暂时未搜索到相关" + myEditTarget, Toast.LENGTH_SHORT).show();
        } }
        @Override
        public void onGetPoiDetailResult(PoiDetailResult poiDetailResult){ }
        @Override
        public void onGetPoiIndoorResult(PoiIndoorResult poiIndoorResult){ }
    });
    //在指定城市搜索兴趣点
    myPoiSearch.searchInCity((new PoiCitySearchOption())
            .city(myEditCity.getText().toString())
            .keyword(myEditTarget.getText().toString())
            .pageNum(10));
}
class MarkerOverlay extends PoiOverlay {
    public MarkerOverlay(BaiduMap baiduMap) { super(baiduMap); }
    @Override
    public boolean onPoiClick(int i) {                          //自定义标记单击监听事件
        //获取单击位置对应的信息
        PoiInfo myPoiInfo = getPoiResult().getAllPoi().get(i);
        Toast.makeText(MainActivity.this,
                    myPoiInfo.name, Toast.LENGTH_SHORT).show();    //显示名称
        //根据 PoiInfo.uid 显示街景,显示街景需要安装百度地图 App;
        //如果此地没有街景,则不启动百度地图 App 显示街景
        BaiduMapPoiSearch.openBaiduMapPanoShow(myPoiInfo.uid, MainActivity.this);
        return true;
    } }
```

上面这段代码在 MyCode\MySampleX59\app\src\main\java\com\bin\luo\mysample\MainActivity.java 文件中。此外,此实例需要引入百度地图 SDK 库文件,并修改 AndroidManifest.xml 和 build.gradle 等文件,相关操作请参考实例 266。此实例的完整项目在 MyCode\MySampleX59 文件夹中。

285 使用百度 SDK 查询指定位置附近的兴趣点

此实例主要通过使用百度地图 SDK,实现查询指定位置附近的宾馆超市等分类信息。当实例运行之后,在"纬度:"输入框中输入"29.6876",在"经度:"输入框中输入"106.5979",在"搜索半径:"输入框中输入"5000",在"搜索目标:"输入框中输入"宾馆",然后单击"开始搜索"按钮,将在百度地图中显示以指定位置(29.6876,106.5979)为中心,半径在 5000 米内的宾馆,红色的(打点)图标表示该宾馆的所在位置,单击该图标,将在弹出的 Toast 中显示宾馆名称,效果分别如图 285-1 的左图和右图所示。

主要代码如下:

```
public void onClickBtn1(View v) {                     //响应单击"开始搜索"按钮
    myPoiSearch = PoiSearch.newInstance();
    myPoiSearch.setOnGetPoiSearchResultListener(
                        new OnGetPoiSearchResultListener(){
        @Override
        public void onGetPoiResult(PoiResult result) {
            if (result.error == SearchResult.ERRORNO.NO_ERROR) {
                myBaiduMap.clear();                    //重置地图
                MarkerOverlay myOverlay = new MarkerOverlay(myBaiduMap);
```

图 285-1

```
        myBaiduMap.setOnMarkerClickListener(myOverlay);
        myOverlay.setData(result);           //传入目标兴趣点相关数据
        myOverlay.addToMap();                //在地图上显示
        myOverlay.zoomToSpan();              //自适应缩放
    } else {
    Toast.makeText(MainActivity.this,
            "暂时未搜索到相关" + myEditTarget, Toast.LENGTH_SHORT).show();
    } }
    @Override
    public void onGetPoiDetailResult(PoiDetailResult poiDetailResult){ }
    @Override
    public void onGetPoiIndoorResult(PoiIndoorResult poiIndoorResult){ }
});
double myLatitude = Double.parseDouble(myEditLatitude.getText().toString());
double myLongitude = Double.parseDouble(myEditLongitude.getText().toString());
int myRadius = Integer.parseInt(myEditRadius.getText().toString());
//在指定坐标位置范围内搜索目标,最多显示 25 个,其中最大显示数可自定义
myPoiSearch.searchNearby((new PoiNearbySearchOption()).location(
        new LatLng(myLatitude, myLongitude)).radius(myRadius).keyword(
        myEditTarget.getText().toString()).pageNum(25));
}
class MarkerOverlay extends PoiOverlay {
    public MarkerOverlay(BaiduMap baiduMap) { super(baiduMap); }
    @Override
    public boolean onPoiClick(int i) {         //自定义标记单击监听事件
        //获取单击位置对应的信息
        PoiInfo myPoiInfo = getPoiResult().getAllPoi().get(i);
        Toast.makeText(MainActivity.this,
                            myPoiInfo.name, Toast.LENGTH_SHORT).show();
        return true;
    }}
```

上面这段代码在 MyCode\MySample919\app\src\main\java\com\bin\luo\mysample\MainActivity.java 文件中。此外，此实例需要引入百度地图 SDK 库文件，并修改 build.gradle 和 AndroidManifest.xml 等文件，相关操作请参考实例 266。此实例的完整项目在 MyCode\MySample919 文件夹中。

286　使用百度 SDK 查询在指定区域内的兴趣点

此实例主要通过使用百度地图 SDK，实现在百度地图中查询指定地理范围的宾馆、美食等兴趣点（POI）信息。当实例运行之后，在"左上角纬度："输入框中输入重庆照母山的纬度值，如"29.629167"，在"左上角经度："输入框中输入重庆照母山的经度值，如"106.505325"，在"右下角纬度："输入框中输入重庆北站的纬度值，如"29.622918"，在"右下角经度："输入框中输入重庆北站的经度值，如"106.556103"，在"兴趣点："输入框中输入"楼盘"，然后单击"开始搜索"按钮，将在百度地图中显示该区域内的楼盘分布，单击其中悬浮的任意一个图标，将在弹出的 Toast 中显示该楼盘的名称，如图 286-1 的左图所示。如果在"左上角纬度："输入框中输入重庆石桥铺的纬度值，如"29.536115"，在"左上角经度："输入框中输入重庆石桥铺的经度值，如"106.490393"，在"右下角纬度："输入框中输入重庆杨家坪的纬度值，如"29.513536"，在"右下角经度："输入框中输入重庆杨家坪的经度值，如"106.520423"，在"兴趣点："输入框中输入"火锅"，然后单击"开始搜索"按钮，将在百度地图中显示该区域内的火锅店分布，单击其中悬浮的任意一个图标，将在弹出的 Toast 中显示该火锅店的名称，如图 286-1 的右图所示。

图　286-1

主要代码如下：

```
public void onClickBtn1(View v) {                   //响应单击"开始搜索"按钮
    myPoiSearch = PoiSearch.newInstance();
```

```java
        myPoiSearch.setOnGetPoiSearchResultListener(
                            new OnGetPoiSearchResultListener() {
            @Override
            public void onGetPoiResult(PoiResult result) {
                if (result.error == SearchResult.ERRORNO.NO_ERROR) {
                    myBaiduMap.clear();                 //重置地图
                    MarkerOverlay myOverlay = new MarkerOverlay(myBaiduMap);
                    myBaiduMap.setOnMarkerClickListener(myOverlay);
                    myOverlay.setData(result);          //传入目标兴趣点相关数据
                    myOverlay.addToMap();               //在地图上显示
                    myOverlay.zoomToSpan();             //自适应缩放
                } else {
                    Toast.makeText(MainActivity.this,
                            "暂时未搜索到相关" + myEditTarget, Toast.LENGTH_SHORT).show();
                } }
            @Override
            public void onGetPoiDetailResult(PoiDetailResult poiDetailResult){ }
            @Override
            public void onGetPoiIndoorResult(PoiIndoorResult poiIndoorResult){ }
        });
        //左上(下)角是照母山(106.505325,29.629167),
        //右下(上)角是重庆北站龙头寺(106.556103,29.622918),楼盘
        //左上(下)角是石桥铺(106.490393,29.536115),
        //右下(上)角是杨家坪(106.520423,29.513536),火锅
        LatLng myPoint1 =
                new LatLng(Double.valueOf(myEditPoint1A.getText().toString()),
                Double.valueOf(myEditPoint1B.getText().toString()));
        LatLng myPoint2 =
                new LatLng(Double.valueOf(myEditPoint2A.getText().toString()),
                Double.valueOf(myEditPoint2B.getText().toString()));
        LatLngBounds myBounds = new LatLngBounds.Builder().include(myPoint1)
                .include(myPoint2).build();
        myPoiSearch.searchInBound(new PoiBoundSearchOption()
        .bound(myBounds).keyword(myEditTarget.getText().toString()).pageNum(10));
    }
    class MarkerOverlay extends PoiOverlay {
        public MarkerOverlay(BaiduMap baiduMap) { super(baiduMap);}
        @Override
        public boolean onPoiClick(int i) {        //自定义标记单击监听事件
        //获取单击位置对应的信息
            PoiInfo myPoiInfo = getPoiResult().getAllPoi().get(i);
            Toast.makeText(MainActivity.this,
                                myPoiInfo.name, Toast.LENGTH_SHORT).show();
            return true;
} } }
```

上面这段代码在 MyCode \ MySampleX42 \ app \ src \ main \ java \ com \ bin \ luo \ mysample \ MainActivity.java 文件中。此外,此实例需要引入百度地图 SDK 库文件,并修改 build.gradle 和 AndroidManifest.xml 等文件,相关操作请参考实例 266。此实例的完整项目在 MyCode \ MySampleX42 文件夹中。

287 使用百度 SDK 根据起止地点规划出行线路

此实例主要通过使用百度地图 SDK 实现根据起点和终点规划出行线路。当实例运行之后,在"城市:"输入框中输入城市名称,如"重庆",在"起点:"输入框中输入出行起点名称,如"朝天门",在"终点:"输入框中输入终点名称,如"江北机场",然后单击"规划线路"按钮,将在地图中显示起点和终点之间的出行线路;单击该线路上的中转站图标,将在弹出的 Toast 中显示在该中转站点的出行建议,效果分别如图 287-1 的左图和右图所示。

图 287-1

主要代码如下:

```
public void onClickBtn1(View v) {                    //响应单击"规划线路"按钮
    RoutePlanSearch myRoutePlanSearch = RoutePlanSearch.newInstance();
    myRoutePlanSearch.setOnGetRoutePlanResultListener(
                            new OnGetRoutePlanResultListener(){
        @Override
        public void onGetWalkingRouteResult(WalkingRouteResult walkingRouteResult){}
        @Override
        public void onGetTransitRouteResult(TransitRouteResult result) {
            if (result.getRouteLines() == null) {
                Toast.makeText(MainActivity.this,
                        "抱歉,未找到相关线路!", Toast.LENGTH_SHORT).show();
            } else {
                myBaiduMap.clear();                   //重置地图状态,防止叠加
                MyRouteOverlay myRouteOverlay = new MyRouteOverlay(myBaiduMap);
                //传入线路数据
                myRouteOverlay.setData(result.getRouteLines().get(0));
                myRouteOverlay.addToMap();            //将线路图添加至地图中
```

```
            myRouteOverlay.zoomToSpan();                    //自适应缩放
            //设置换乘站单击监听
            myBaiduMap.setOnMarkerClickListener(myRouteOverlay);
        } }
        @Override
        public void onGetMassTransitRouteResult(
                MassTransitRouteResult massTransitRouteResult) { }
        @Override
        public void onGetDrivingRouteResult(DrivingRouteResult drivingRouteResult){ }
        @Override
        public void onGetIndoorRouteResult(IndoorRouteResult indoorRouteResult){ }
        @Override
        public void onGetBikingRouteResult(BikingRouteResult bikingRouteResult){ }
    });
    PlanNode myStartNode =
        PlanNode.withCityNameAndPlaceName(myEditCity.getText().toString(), myEditBegin.getText().
            toString());                                    //设置起点
    PlanNode myEndNode =
        PlanNode.withCityNameAndPlaceName(myEditCity.getText().toString(),
            myEditEnd.getText().toString());                //设置终点
    //根据起点和终点进行线路规划,并将线路图显示在地图上
    myRoutePlanSearch.transitSearch((new TransitRoutePlanOption()).
        from(myStartNode).city(myEditCity.getText().toString()).to(myEndNode));
}
class MyRouteOverlay extends TransitRouteOverlay {
    public MyRouteOverlay(BaiduMap baiduMap) { super(baiduMap); }
    @Override
    public boolean onRouteNodeClick(int i) {
        //通过 Toast 显示换乘点建议
        Toast.makeText(MainActivity.this, "此处建议:" + mRouteLine.getAllStep().
                get(i).getInstructions(), Toast.LENGTH_SHORT).show();
        return true;
    }
  }
}
```

上面这段代码在 MyCode\MySample914\app\src\main\java\com\bin\luo\mysample\MainActivity.java 文件中。此外,此实例需要引入百度地图 SDK 库文件,并修改 build.gradle 和 AndroidManifest.xml 等文件,相关操作请参考实例 266。此实例的完整项目在 MyCode\MySample914 文件夹中。

288 使用百度 SDK 在地图中搜索指定公交线路

此实例主要通过使用百度地图 SDK 实现查询指定城市指定公交线路的路线及停靠站点等信息。当实例运行之后,在"城市:"输入框中输入城市名称,如"重庆",在"公交线路:"输入框中输入公交线路名称,如"877"路,然后单击"在百度地图中搜索"按钮,将在地图中显示重庆主城 877 路公交车行车线路(即蓝色公交车图标串联起来的线路);单击该线路下面的任意一个公交车图标,将在弹出的 Toast 中显示 877 路的停靠站点名称,如"北湖郡",如图 288-1 的左图所示。如果在"城市:"输入框中输入城市名称,如"重庆",在"公交线路:"输入框中输入公交线路名称,如"619"路,然后单击"在百度地图中搜索"按钮,将在地图中显示重庆主城 619 路公交车行车线路(即蓝色公交车图标串联起来的

线路）；单击该线路下面的任意一个公交车图标，将在弹出的 Toast 中显示 619 路的停靠站点名称，如"西政渝北校区南门"，如图 288-1 的右图所示。

图　288-1

主要代码如下：

```
public void onClickBtn1(View v) {                           //响应单击"在百度地图中搜索"按钮
    final PoiSearch myPoiSearch = PoiSearch.newInstance();
    myPoiSearch.setOnGetPoiSearchResultListener(
        new OnGetPoiSearchResultListener(){
            @Override
            public void onGetPoiResult(PoiResult poiResult) {
                for (PoiInfo poiInfo : poiResult.getAllPoi()) {
                    BusLineSearch myBusLineSearch = BusLineSearch.newInstance();
                    myBusLineSearch.setOnGetBusLineSearchResultListener(
                        new OnGetBusLineSearchResultListener() {
                            @Override
                            public void onGetBusLineResult(BusLineResult result) {
                                myBaiduMap.clear();           //重置地图状态，防止线路叠加
                                MyBuslineOverlay myBusLineOverlay =
                                        new MyBuslineOverlay(myBaiduMap);
                                myBusLineOverlay.setData(result); //获取公交线路信息
                                myBusLineOverlay.addToMap();     //将线路显示在地图上
                                myBusLineOverlay.zoomToSpan();   //自动缩放地图
                                //设置公交站单击监听
                                myBaiduMap.setOnMarkerClickListener(myBusLineOverlay);
                            } });
                    //根据 UID 查询指定公交线路
                    myBusLineSearch.searchBusLine((new BusLineSearchOption().city(
                            myEditCity.getText().toString()).uid(poiInfo.uid)));
                } }
            @Override
            public void onGetPoiDetailResult(PoiDetailResult poiDetailResult){}
```

```
            @Override
            public void onGetPoiIndoorResult(PoiIndoorResult poiIndoorResult){}
        });
    //通过POI搜索指定城市的指定公交线路
    myPoiSearch.searchInCity((new PoiCitySearchOption()).city(
            myEditCity.getText().toString()).keyword(
            myEditBus.getText().toString()));
}
class MyBuslineOverlay extends BusLineOverlay {
    public MyBuslineOverlay(BaiduMap baiduMap) { super(baiduMap); }
    @Override
    public boolean onBusStationClick(int index) {
      Toast.makeText(MainActivity.this, "刚才单击的公交站名为: " + mBusLineResult.getStations().get
(index).getTitle(),Toast.LENGTH_SHORT).show();
        return true;
} }
```

上面这段代码在 MyCode\MySample913\app\src\main\java\com\bin\luo\mysample\MainActivity.java 文件中。此外，此实例需要引入百度地图 SDK 库文件，并修改 build.gradle 和 AndroidManifest.xml 等文件，相关操作请参考实例 266。此实例的完整项目在 MyCode\MySample913 文件夹中。

289 使用百度 SDK 查询百度地图的公交线规划

此实例主要通过使用百度地图 SDK 的 BaiduMapRoutePlan，实现根据目的地和出发地规划两者之间的公交线路。当实例运行之后，在"出发地："输入框中输入"重庆北站"，在"目的地："输入框中输入"重庆渝北中央公园"，如图 289-1 的左图所示，然后单击"查询两地的公交线路规划"按钮，将启动百度地图显示这两地之间的公交线路规划，如图 289-1 的右图所示。

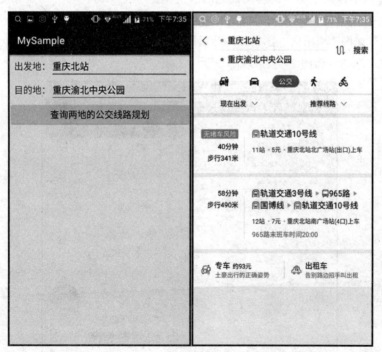

图 289-1

主要代码如下：

```
public void onClickBtn1(View v) {         //响应单击"查询两地的公交线路规划"按钮
    //构建搜索参数以及策略,起点和终点也可以使用 LatLng 构造
    RouteParaOption myRouteParaOption = new RouteParaOption()
            .startName(myEditStart.getText().toString())
            .endName(myEditEnd.getText().toString())
            .busStrategyType(RouteParaOption.EBusStrategyType.bus_recommend_way);
    try {
        //显示百度地图的公交线路规划
        BaiduMapRoutePlan.openBaiduMapTransitRoute(myRouteParaOption, this);
    } catch (Exception e) {
        Toast.makeText(MainActivity.this,
                    e.getMessage().toString(), Toast.LENGTH_SHORT).show();
    }
    //调用 finish()方法以释放相关资源
    BaiduMapRoutePlan.finish(this);
}
```

上面这段代码在 MyCode\MySampleX52\app\src\main\java\com\bin\luo\mysample\MainActivity.java 文件中。此外,此实例需要引入百度地图 SDK 库文件,并修改 build.gradle 和 AndroidManifest.xml 等文件,相关操作请参考实例 266。此实例的完整项目在 MyCode\MySampleX52 文件夹中。

290　使用百度 SDK 调用百度地图的步行导航

此实例主要通过使用百度地图 SDK 的 BaiduMapNavigation,实现根据起点和终点对步行线路进行导航。当实例运行之后,在"起点纬度:"输入框中输入重庆中央公园北门的纬度值"29.733082",在"起点经度:"输入框中输入重庆中央公园北门的经度值"106.590041",在"终点纬度:"输入框中输入重庆中央公园樱花园的纬度值"29.726484",在"终点经度:"输入框中输入重庆中央公园樱花园的经度值"106.591068",如图 290-1 的左图所示；然后单击"调用百度地图的步行导航功能"按钮,将启动百度地图的步行导航功能,如图 290-1 的右图所示。

图　290-1

主要代码如下：

```
public void onClickBtn1(View v) {        //响应单击"调用百度地图的步行导航功能"按钮
    //重庆中央公园北门的经度和纬度(106.590041,29.733082)
    //重庆中央公园樱花园的经度和纬度(106.591068,29.726484)
    //此实例需要安装百度地图 App,模拟器无法测试
    LatLng myPoint1 =
            new LatLng(Double.valueOf(myEditPoint1A.getText().toString()),
                    Double.valueOf(myEditPoint1B.getText().toString()));
    LatLng myPoint2 =
            new LatLng(Double.valueOf(myEditPoint2A.getText().toString()),
                    Double.valueOf(myEditPoint2B.getText().toString()));
    NaviParaOption myNaviParaOption = new NaviParaOption()
            .startPoint(myPoint1).endPoint(myPoint2)
            .startName("我的起点").endName("我的终点");
    try {
        //调用百度地图步行导航
        BaiduMapNavigation.openBaiduMapWalkNavi(myNaviParaOption, this);
    } catch (Exception e) {
        Toast.makeText(MainActivity.this,
                e.getMessage().toString(), Toast.LENGTH_SHORT).show();
    }
}
```

上面这段代码在 MyCode \ MySampleX53 \ app \ src \ main \ java \ com \ bin \ luo \ mysample \ MainActivity.java 文件中。此外，此实例需要引入百度地图 SDK 库文件，并修改 build.gradle 和 AndroidManifest.xml 等文件，相关操作请参考实例 266。此实例的完整项目在 MyCode \ MySampleX53 文件夹中。

291 使用百度 SDK 调用百度地图的骑行导航

此实例主要通过使用百度地图 SDK 的 BaiduMapNavigation，实现根据起点和终点对骑行(自行车)线路进行导航。当实例运行之后，在"起点纬度："输入框中输入重庆中央公园北门的纬度值"29.733082"，在"起点经度："输入框中输入重庆中央公园北门的经度值"106.590041"，在"终点纬度："输入框中输入重庆中央公园南入口的纬度值"29.711557"，在"终点经度："输入框中输入重庆中央公园南入口的经度值"106.589102"，如图 291-1 的左图所示；然后单击"调用百度地图的骑行导航功能"按钮，将启动百度地图的骑行导航功能，如图 291-1 的右图所示。

主要代码如下：

```
public void onClickBtn1(View v) {        //响应单击"调用百度地图的骑行导航功能"按钮
    //重庆中央公园北门的经度和纬度(106.590041,29.733082)
    //重庆中央公园南入口的经度和纬度(106.589102,29.711557)
    //此实例需要安装百度地图 App,模拟器无法测试
    LatLng myPoint1 =
            new LatLng(Double.valueOf(myEditPoint1A.getText().toString()),
                    Double.valueOf(myEditPoint1B.getText().toString()));
    LatLng myPoint2 =
```

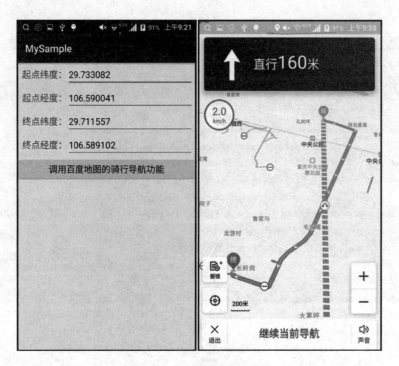

图 291-1

```
                new LatLng(Double.valueOf(myEditPoint2A.getText().toString()),
                        Double.valueOf(myEditPoint2B.getText().toString()));
        NaviParaOption myNaviParaOption = new NaviParaOption()
                .startPoint(myPoint1).endPoint(myPoint2)
                .startName("我的起点").endName("我的终点");
        try {
            //调用百度地图的骑行导航
            BaiduMapNavigation.openBaiduMapBikeNavi(myNaviParaOption, this);
        } catch (Exception e) {
            Toast.makeText(MainActivity.this,
                    e.getMessage().toString(), Toast.LENGTH_SHORT).show();
        }
    }
```

上面这段代码在 MyCode\MySampleX54\app\src\main\java\com\bin\luo\mysample\MainActivity.java 文件中。此外，此实例需要引入百度地图 SDK 库文件，并修改 AndroidManifest.xml 和 build.gradle 等文件，相关操作请参考实例 266。此实例的完整项目在 MyCode\MySampleX54 文件夹中。

292　使用百度 SDK 调用百度地图的 Web 导航

此实例主要通过使用百度地图 SDK 的 BaiduMapNavigation，实现根据起点和终点对驾车线路进行 Web 导航。当实例运行之后，在"起点："输入框中输入驾车起点，如"重庆中央公园"，在"终点："输入框中输入驾车终点，如"长寿湖风景区"，如图 292-1 的左图所示；然后单击"启动百度地图的 Web 导航"按钮，将通过浏览器启动百度地图的驾车导航功能，如图 292-1 的右图所示。

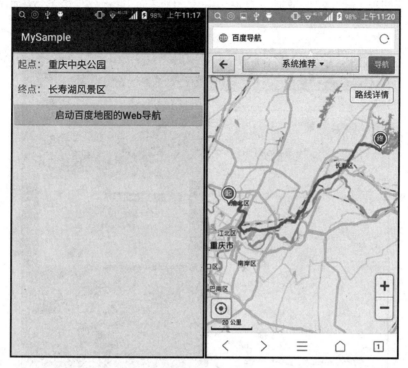

图 292-1

主要代码如下：

```
//响应单击"启动百度地图的 Web 导航"按钮
public void onClickBtn1(View v) {
    try {
        NaviParaOption myNaviParaOption = new NaviParaOption()
                    .startName(myEditStart.getText().toString())
                    .endName(myEditEnd.getText().toString());
        BaiduMapNavigation.openWebBaiduMapNavi(myNaviParaOption, this);
    } catch (Exception e) {
        Toast.makeText(MainActivity.this,
              e.getMessage().toString(), Toast.LENGTH_SHORT).show();
    }
}
```

上面这段代码在 MyCode\MySampleX62\app\src\main\java\com\bin\luo\mysamplc\ MainActivity. java 文件中。此外，此实例需要引入百度地图 SDK 库文件，并修改 AndroidManifest. xml 和 build. gradle 等文件，相关操作请参考实例 266。此实例的完整项目在 MyCode\MySampleX62 文件夹中。

293 使用百度 SDK 实现 POI 检索并分享相关地址

此实例主要通过使用百度地图 SDK 的 LocationShareURLOption，实现在指定的城市中搜索指定的兴趣点，并将兴趣点的地址分享给好友。当实例运行之后，在"城市："输入框中输入城市名称，如"渝北"，在"兴趣点（POI）："输入框中输入感兴趣的事物，如"火锅"，然后单击"搜索 POI 并分享兴趣点地址（URL）"按钮，将在百度地图中显示 POI 搜索结果，如图 293-1 的左图所示。任意单击悬浮的红色兴趣点图标,将该兴趣点的地址通过第三方分享组件(如 QQ)分享给好友,如图 293-1 的右图所示。

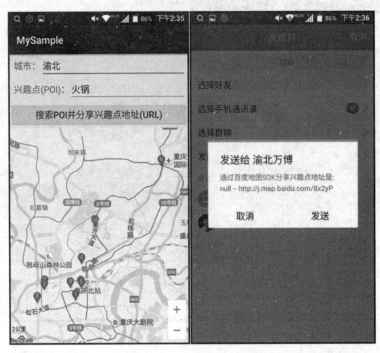

图 293-1

主要代码如下:

```
public void onClickBtn1(View view) {      //响应单击"搜索 POI 并分享兴趣点地址(URL)"按钮
    myCity = myEditCity.getText().toString();
    mySearchKey = myEditSearchKey.getText().toString();
    myPoiSearch.searchInCity((new PoiCitySearchOption()).city(myCity)
            .keyword(mySearchKey));
    Toast.makeText(this,
            "在" + myCity + "搜索 " + mySearchKey, Toast.LENGTH_SHORT).show();
}
@Override
public void onGetPoiResult(PoiResult result) {
    if (result == null || result.error != SearchResult.ERRORNO.NO_ERROR) {
        Toast.makeText(MainActivity.this,
                        "抱歉,未找到结果", Toast.LENGTH_LONG).show();
        return;
    }
    myBaiduMap.clear();
    PoiShareOverlay poiOverlay = new PoiShareOverlay(myBaiduMap);
    myBaiduMap.setOnMarkerClickListener(poiOverlay);
    poiOverlay.setData(result);
    poiOverlay.addToMap();
    poiOverlay.zoomToSpan();
}
@Override
```

```
public void onGetPoiDetailResult(PoiDetailResult poiDetailResult){ }
@Override
public void onGetPoiIndoorResult(PoiIndoorResult poiIndoorResult){ }
private class PoiShareOverlay extends PoiOverlay {
    public PoiShareOverlay(BaiduMap baiduMap){ super(baiduMap); }
    @Override
    public boolean onPoiClick(int i) {                //响应单击兴趣点
        //监听分享URL
        ShareUrlSearch myShareUrlSearch = ShareUrlSearch.newInstance();
        OnGetShareUrlResultListener myListener = new OnGetShareUrlResultListener(){
            @Override
            public void onGetPoiDetailShareUrlResult(ShareUrlResult result){ }
            @Override
            public void onGetRouteShareUrlResult(ShareUrlResult arg0){ }
            @Override
            public void onGetLocationShareUrlResult(ShareUrlResult arg0){
                Intent myIntent = new Intent(Intent.ACTION_SEND);
                myIntent.putExtra(Intent.EXTRA_TEXT, "通过百度地图SDK分享兴趣点地址是："
                        + myAddress + " -- " + arg0.getUrl());
                myIntent.setType("text/plain");
                startActivity(Intent.createChooser(myIntent, "百度地图兴趣点地址分享"));
            } };
        myShareUrlSearch.setOnGetShareUrlResultListener(myListener);
        PoiInfo myPoiInfo = getPoiResult().getAllPoi().get(i);
        LatLng myPoint = new LatLng(myPoiInfo.location.latitude,
                myPoiInfo.location.longitude);
        myShareUrlSearch.requestLocationShareUrl(new LocationShareURLOption()
                .location(myPoint).name("分享位置").snippet("123"));  //请求分享URL
        return true;
    }}
```

上面这段代码在 MyCode\MySampleX64\app\src\main\java\com\bin\luo\mysample\MainActivity.java 文件中。此外，此实例需要引入百度地图 SDK 库文件，并修改 AndroidManifest.xml 和 build.gradle 等文件，相关操作请参考实例 266。此实例的完整项目在 MyCode\MySampleX64 文件夹中。

294　使用百度 SDK 实现将公交线路分享给好友

此实例主要通过使用百度地图 SDK 的 ShareUrlSearch，实现将指定起点和终点的公交线路分享给好友。当实例运行之后，在"起点纬度："输入框中输入重庆中央公园的纬度，如"29.722571"，在"起点经度："输入框中输入重庆中央公园的经度，如"106.590326"，在"终点纬度："输入框中输入重庆园博园的纬度，如"29.686633"，在"终点经度："输入框中输入重庆园博园的经度，如"106.559096"，如图 294-1 的左图所示，然后单击"显示公交线路"按钮，将在百度地图中显示指定起点和终点之间的公交线路，如图 294-1 的右图所示。单击"分享公交线路"按钮，将启动第三方分享工具列表，在列表中选择 QQ 好友，然后单击"发送"按钮，如图 294-2 的左图所示，将把指定起点和终点之间的公交线路（地址短串）分享到 QQ 好友，如图 294-2 的右图所示。

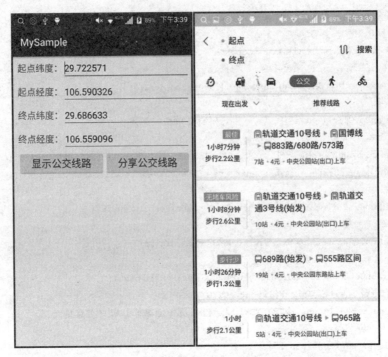

图 294-1

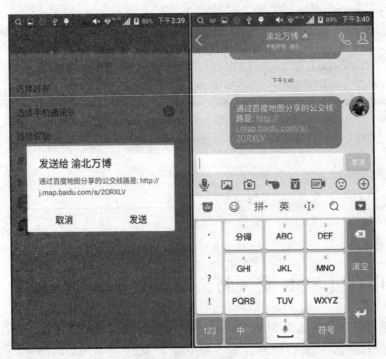

图 294-2

主要代码如下:

```
public void onClickBtn1(View v) {           //响应单击"显示公交线路"按钮
    //重庆中央公园的经纬度(106.590326,29.722571)
    //重庆园博园的经纬度(106.559096,29.686633)
    try {
```

```java
        Intent myIntent = Intent.getIntent("intent://map/direction?origin="
                + myEditPoint1A.getText().toString() + "," + myEditPoint1B.getText().toString()
                + "&destination=" + myEditPoint2A.getText().toString() + ","
                + myEditPoint2B.getText().toString()
                + "&mode=transit#Intent;scheme=bdapp;package=com.baidu.BaiduMap;end");
        //启动百度地图显示公交线路
        startActivity(myIntent);
    } catch (Exception e) {
        Toast.makeText(MainActivity.this,
                    e.getMessage().toString(), Toast.LENGTH_SHORT).show();
    }
}
public void onClickBtn2(View v) {                    //响应单击"分享公交线路"按钮
    //创建分享检索实例
    ShareUrlSearch myShareUrlSearch = ShareUrlSearch.newInstance();
    //创建、设置分享检索监听者
    myShareUrlSearch.setOnGetShareUrlResultListener(
        new OnGetShareUrlResultListener() {
            @Override
            public void onGetPoiDetailShareUrlResult(ShareUrlResult shareUrlResult){}
            @Override
            public void onGetLocationShareUrlResult(ShareUrlResult shareUrlResult) {}
            @Override
            //使用 Intent 执行分享动作
            public void onGetRouteShareUrlResult(ShareUrlResult shareUrlResult) {
                Intent myIntent = new Intent(Intent.ACTION_SEND);
                myIntent.putExtra(Intent.EXTRA_TEXT, "通过百度地图分享的公交线路是: "
                        + shareUrlResult.getUrl());
                myIntent.setType("text/plain");
                startActivity(Intent.createChooser(myIntent, "百度地图公交线路分享"));
            }});
    myStartNode = PlanNode.withLocation(new LatLng(
            Double.parseDouble(myEditPoint1A.getText().toString()),
            Double.parseDouble(myEditPoint1B.getText().toString())));
    myEndNode = PlanNode.withLocation(new LatLng(
            Double.parseDouble(myEditPoint2A.getText().toString()),
            Double.parseDouble(myEditPoint2B.getText().toString())));
    //设置分享模式为公交线路
    RouteShareURLOption.RouteShareMode myRouteShareMode =
            RouteShareURLOption.RouteShareMode.BUS_ROUTE_SHARE_MODE;
    myShareUrlSearch.requestRouteShareUrl(new RouteShareURLOption()
            .from(myStartNode).to(myEndNode).routMode(myRouteShareMode));
    //销毁分享检索实例;
    myShareUrlSearch.destroy();
}
```

上面这段代码在 MyCode\MySampleX68\app\src\main\java\com\bin\luo\mysample\MainActivity.java 文件中。此外，此实例需要引入百度地图 SDK 库文件，并修改 AndroidManifest.xml 和 build.gradle 等文件，相关操作请参考实例 266。此实例的完整项目在 MyCode\MySampleX68 文件夹中。

295 使用百度 SDK 实现将骑行线路分享给好友

此实例主要通过使用百度地图 SDK 的 ShareUrlSearch，实现将指定起点和终点的骑行（自行车）线路分享给好友。当实例运行之后，在"起点纬度："输入框中输入重庆中央公园的纬度，如

"29.722571",在"起点经度:"输入框中输入重庆中央公园的经度,如"106.590326",在"终点纬度:"输入框中输入重庆园博园的纬度,如"29.686633",在"终点经度:"输入框中输入重庆园博园的经度,如"106.559096",如图295-1的左图所示,然后单击"显示骑行线路"按钮,将在百度地图中显示指定起点和终点之间的骑行线路,如图295-1的右图所示。单击"分享骑行线路"按钮,则将启动第三方分享工具列表,在列表中选择QQ好友,然后单击"发送"按钮,如图295-2的左图所示,将把指定起点和终点之间的骑行线路(地址短串)分享到QQ好友,如图295-2的右图所示。

图　　295-1

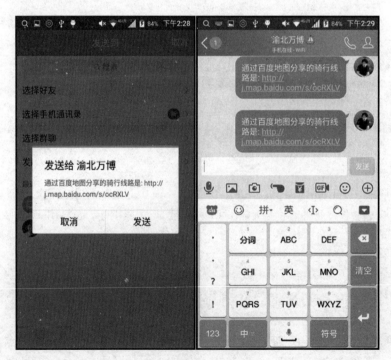

图　　295-2

主要代码如下：

```java
public void onClickBtn1(View v) {           //响应单击"显示骑行线路"按钮
  //重庆中央公园的经纬度(106.590326,29.722571)
  //重庆园博园的经纬度(106.559096,29.686633)
  try {
    Intent myIntent = Intent.getIntent("intent://map/direction?origin = "
      + myEditPoint1A.getText().toString() + "," + myEditPoint1B.getText().toString()
      + "&destination = " + myEditPoint2A.getText().toString() + ","
      + myEditPoint2B.getText().toString()
      + "&mode = riding#Intent;scheme = bdapp;package = com.baidu.BaiduMap;end");
    startActivity(myIntent);                //启动百度地图显示骑行线路
  } catch (Exception e) {
    Toast.makeText(MainActivity.this,
                   e.getMessage().toString(),Toast.LENGTH_SHORT).show();
  }
}
public void onClickBtn2(View v) {           //响应单击"分享骑行线路"按钮
  //创建分享检索实例
  ShareUrlSearch myShareUrlSearch = ShareUrlSearch.newInstance();
  //创建、设置分享检索监听者
  myShareUrlSearch.setOnGetShareUrlResultListener(
    new OnGetShareUrlResultListener() {
    @Override
    public void onGetPoiDetailShareUrlResult(ShareUrlResult shareUrlResult){}
    @Override
    public void onGetLocationShareUrlResult(ShareUrlResult shareUrlResult){}
    @Override
    //使用 Intent 执行分享动作
    public void onGetRouteShareUrlResult(ShareUrlResult shareUrlResult) {
      Intent myIntent = new Intent(Intent.ACTION_SEND);
      myIntent.putExtra(Intent.EXTRA_TEXT, "通过百度地图分享的骑行线路是："
              + shareUrlResult.getUrl());
      myIntent.setType("text/plain");
      startActivity(Intent.createChooser(myIntent, "百度地图骑行线路分享"));
    } });
  myStartNode = PlanNode.withLocation(new LatLng(
        Double.parseDouble(myEditPoint1A.getText().toString()),
        Double.parseDouble(myEditPoint1B.getText().toString())));
  myEndNode = PlanNode.withLocation(new LatLng(
        Double.parseDouble(myEditPoint2A.getText().toString()),
        Double.parseDouble(myEditPoint2B.getText().toString())));
  //设置分享模式为骑行线路
  RouteShareURLOption.RouteShareMode myRouteShareMode =
        RouteShareURLOption.RouteShareMode.CYCLE_ROUTE_SHARE_MODE;
  myShareUrlSearch.requestRouteShareUrl(new RouteShareURLOption()
        .from(myStartNode).to(myEndNode).routMode(myRouteShareMode));
  myShareUrlSearch.destroy();               //销毁分享检索实例
}
```

上面这段代码在 MyCode\MySampleX67\app\src\main\java\com\bin\luo\mysample\MainActivity.java 文件中。此外，此实例需要引入百度地图 SDK 库文件，并修改 AndroidManifest.xml 和 build.gradle 等文件，相关操作请参考实例 266。此实例的完整项目在 MyCode\MySampleX67 文件夹中。

296 使用百度 SDK 将当前地图分享给 QQ 好友

此实例主要通过使用百度地图 SDK,实现将指定位置的地图分享给 QQ、微信等好友。当实例运行之后,在"纬度:"输入框中输入"29.6876",在"经度:"输入框中输入"106.5979",然后单击按钮"定位",将在百度地图中显示以指定位置(29.6876,106.5979)为中心的地图,效果如图 296-1 的左图所示。单击"分享"按钮,将显示手机支持分享功能的应用列表,在应用列表中选择 QQ,然后选择接收分享的好友,该地图的链接将显示在好友对话中,效果如图 296-1 的右图所示。单击 QQ 对话中的地图链接,将在浏览器中显示分享的地图。

图 296-1

主要代码如下:

```java
public void onClickBtn1(View v) {           //响应单击"定位"按钮
    double myLatitude = Double.parseDouble(myEditLatitude.getText().toString());
    double myLongitude = Double.parseDouble(myEditLongitude.getText().toString());
    myLatLng = new LatLng(myLatitude, myLongitude);
    MapStatus myMapStatus =
            new MapStatus.Builder().target(myLatLng).zoom(16).build();
    //在地图上显示指定位置
    myBaiduMap.setMapStatus(MapStatusUpdateFactory.newMapStatus(myMapStatus));
}
public void onClickBtn2(View v) {           //响应单击"分享"按钮
    ShareUrlSearch myShareUrlSearch = ShareUrlSearch.newInstance();
    myShareUrlSearch.setOnGetShareUrlResultListener(
            new OnGetShareUrlResultListener(){
    @Override
    public void onGetPoiDetailShareUrlResult(ShareUrlResult shareUrlResult){
    @Override
```

```
public void onGetLocationShareUrlResult(ShareUrlResult result) {
    Intent myShareIntent = new Intent();
    //设置 action 属性字符串值
    myShareIntent.setAction(Intent.ACTION_SEND);
    //通过 Intent 发送位置信息
    myShareIntent.putExtra(Intent.EXTRA_TEXT, result.getUrl());
    myShareIntent.setType("text/plain");
    startActivity(myShareIntent);        //分享位置
}
@Override
public void onGetRouteShareUrlResult(ShareUrlResult shareUrlResult){ }
});
//请求将当前位置转换为分享 URL 字符串,并在回调函数中进行分享操作
myShareUrlSearch.requestLocationShareUrl(new LocationShareURLOption().
        location(myLatLng).name("我的分享位置").snippet("我的分享位置"));
}
```

上面这段代码在 MyCode \ MySample920 \ app \ src \ main \ java \ com \ bin \ luo \ mysample \ MainActivity.java 文件中。此外,此实例需要引入百度地图 SDK 库文件,并修改 build.gradle 和 AndroidManifest.xml 等文件,相关操作请参考实例 266。此实例的完整项目在 MyCode \ MySample920 文件夹中。

297 使用百度 SDK 实现在输入框滑出建议列表

此实例主要通过使用百度地图 SDK 实现在输入框输入部分文本时,自动在下拉列表中列出与此部分文本匹配的地址选项建议。当实例运行之后,在"地址:"输入框中输入部分文本,如"北京",将在滑出的下拉列表中显示与北京相关的地址选项建议,单击其中的选项,如"北京西站",该选项将自动填充到输入框中。然后单击"开始搜索"按钮,将在百度地图中显示该地址位置(即北京西站),效果分别如图 297-1 的左图和右图所示。

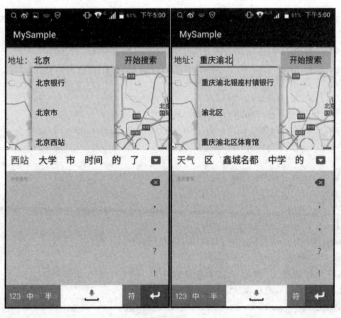

图 297-1

主要代码如下：

```java
public class MainActivity extends Activity {
    BaiduMap myBaiduMap;
    AutoCompleteTextView myEditText;
    SuggestionSearch mySuggestionSearch;
    @Override
    protected void onCreate(Bundle savedInstanceState) {
        super.onCreate(savedInstanceState);
        SDKInitializer.initialize(getApplicationContext());
        setContentView(R.layout.activity_main);
        MapView myMapView = (MapView) findViewById(R.id.myMapView);
        myBaiduMap = myMapView.getMap();
        mySuggestionSearch = SuggestionSearch.newInstance();
        mySuggestionSearch.setOnGetSuggestionResultListener(
                        new OnGetSuggestionResultListener(){
            @Override
            //获取建议数据
            public void onGetSuggestionResult(SuggestionResult result){
                ArrayList<String> mySuggestionList = new ArrayList();
                for(SuggestionResult.SuggestionInfo myInfo:result.getAllSuggestions()){
                    //将数据逐条添加至列表中
                    if(myInfo.key!= null) mySuggestionList.add(myInfo.key);
                }
                ArrayAdapter<String> myAdapter = new ArrayAdapter<String>(MainActivity.this,
                        android.R.layout.simple_dropdown_item_1line,mySuggestionList);
                myEditText.setAdapter(myAdapter);           //将适配器与输入框绑定
                myAdapter.notifyDataSetChanged();           //更新数据内容
        }});
        myBaiduMap = myMapView.getMap();
        myEditText = (AutoCompleteTextView)findViewById(R.id.myAutoCompleteTextView);
        myEditText.setThreshold(1);                 //设置输入多少(1)个字符后开始匹配数据
        myEditText.addTextChangedListener(new TextWatcher(){
            @Override
            public void beforeTextChanged(CharSequence s,int start,int count,int after){}
            @Override
            public void onTextChanged(CharSequence s,int start,int before,int count){
                if(count>=2){
                    //开始请求在线建议,并返回数据结果
                    mySuggestionSearch.requestSuggestion(
                            (new SuggestionSearchOption()).keyword(s.toString()).
                            city(myEditText.getText().toString().substring(0,2)));
                }}
            @Override
            public void afterTextChanged(Editable s){}
        });}
    public void onClickBtn1(View v) {               //响应单击"开始搜索"按钮
        GeoCoder myGeoCoder = GeoCoder.newInstance();
        myGeoCoder.setOnGetGeoCodeResultListener(new OnGetGeoCoderResultListener(){
            @Override
            //将搜索地址解析为经纬度值,并显示在百度地图上
            public void onGetGeoCodeResult(GeoCodeResult result){
                MapStatus myMapStatus = new MapStatus.Builder().target(
```

```
                                result.getLocation()).zoom(15).build();
            myBaiduMap.setMapStatus(MapStatusUpdateFactory.newMapStatus(myMapStatus));
        }
        @Override
        public void onGetReverseGeoCodeResult(ReverseGeoCodeResult result){}
    });
    //对输入地址进行解析操作
    myGeoCoder.geocode(new GeoCodeOption().city(myEditText.getText().toString().
            substring(0,2)).address(myEditText.getText().toString()));
} }
```

上面这段代码在 MyCode \ MySample923 \ app \ src \ main \ java \ com \ bin \ luo \ mysample \ MainActivity.java 文件中。此外，此实例需要引入百度地图 SDK 库文件，并修改 build.gradle 和 AndroidManifest.xml 等文件，相关操作请参考实例 266。此实例的完整项目在 MyCode \ MySample923 文件夹中。

298　使用百度 SDK 实现隐藏或显示地图比例尺

此实例主要通过使用百度地图 SDK 的 showScaleControl()方法，从而实现隐藏或显示百度地图左下角的比例尺。当实例运行之后，单击"显示比例尺"按钮，将在百度地图的左下角显示比例尺"1 公里"，单击右下角的"＋"或"－"按钮放大或缩小百度地图，比例尺值将会同步变化，如图 298-1 的左图所示。单击"隐藏比例尺"按钮，则在百度地图左下角的比例尺消失，如图 298-1 的右图所示。

图　298-1

主要代码如下：

```
public void onClickBtn1(View v) {          //响应单击"显示比例尺"按钮
    myMapView.showScaleControl(true);
```

```
}
public void onClickBtn2(View v) {    //响应单击"隐藏比例尺"按钮
    myMapView.showScaleControl(false);
}
```

上面这段代码在 MyCode\MySampleX73\app\src\main\java\com\bin\luo\mysample\MainActivity.java 文件中。此外,此实例需要引入百度地图 SDK 库文件,并修改 AndroidManifest.xml 和 build.gradle 等文件,相关操作请参考实例 266。此实例的完整项目在 MyCode\MySampleX73 文件夹中。

299 使用百度 SDK 实现隐藏或显示地图缩放按钮

此实例主要通过使用百度地图 SDK 的 showZoomControls() 方法,实现隐藏或显示百度地图右下角的放大和缩小按钮。当实例运行之后,单击"显示缩放按钮"按钮,将在百度地图的右下角显示放大和缩小按钮,如图 299-1 的左图所示。单击"隐藏缩放按钮"按钮,将隐藏百度地图右下角的放大和缩小按钮,如图 299-1 的右图所示。

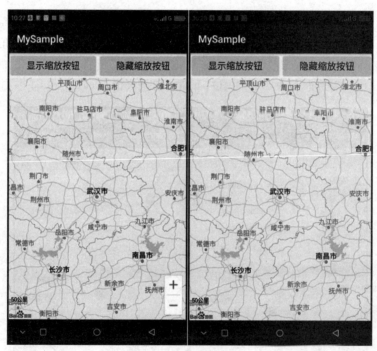

图 299-1

主要代码如下:

```
public void onClickBtn1(View v) {          //响应单击"显示缩放按钮"按钮
    myMapView.showZoomControls(true);
}
public void onClickBtn2(View v) {          //响应单击"隐藏缩放按钮"按钮
    myMapView.showZoomControls(false);
}
```

上面这段代码在 MyCode\MySampleX69\app\src\main\java\com\bin\luo\mysample\

MainActivity.java 文件中。此外,此实例需要引入百度地图 SDK 库文件,并修改 AndroidManifest. xml 和 build.gradle 等文件,相关操作请参考实例 266。此实例的完整项目在 MyCode \ MySampleX69 文件夹中。

300 使用百度 SDK 实现自定义地图缩放按钮的位置

此实例主要通过使用百度地图 SDK 的 setZoomControlsPosition()方法,实现自定义百度地图缩放按钮的显示位置。当实例运行之后,单击"在左上角显示百度地图的缩放按钮"按钮,将在百度地图的左上角显示缩放按钮(+/-),如图 300-1 的左图所示。单击"在右上角显示百度地图的缩放按钮"按钮,将在百度地图的右上角显示缩放按钮(+/-),如图 300-1 的右图所示。

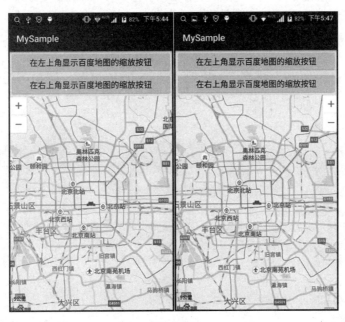

图 300-1

主要代码如下:

```
//响应单击"在左上角显示百度地图的缩放按钮"按钮
public void onClickBtn1(View v) {
    myMapView.setZoomControlsPosition(new Point(0,0));
}
//响应单击"在右上角显示百度地图的缩放按钮"按钮
public void onClickBtn2(View v) {
    DisplaymyDisplay = getWindowManager().getDefaultDisplay();
    int myWidth = myDisplay.getWidth();          //获取屏幕宽度
    myMapView.setZoomControlsPosition(new Point(myWidth - 100,0));
}
```

上面这段代码在 MyCode \ MySampleX75 \ app \ src \ main \ java \ com \ bin \ luo \ mysample \ MainActivity.java 文件中。此外,此实例需要引入百度地图 SDK 库文件,并修改 AndroidManifest. xml 和 build.gradle 等文件,相关操作请参考实例 266。此实例的完整项目在 MyCode \ MySampleX75 文件夹中。